Handbook of Materials Science

Handbook of Materials Science

Editor: Ricky Peyret

NY RESEARCH P R E S S

New York

Published by NY Research Press
118-35 Queens Blvd., Suite 400,
Forest Hills, NY 11375, USA
www.nyresearchpress.com

Handbook of Materials Science
Edited by Ricky Peyret

© 2018 NY Research Press

International Standard Book Number: 978-1-63238-579-6 (Hardback)

Cataloging-in-Publication Data

Handbook of materials science / edited by Ricky Peyret.
    p. cm.
Includes bibliographical references and index.
ISBN 978-1-63238-579-6
1. Materials science. I. Peyret, Ricky.
TA403 .H36 2018
620.11--dc23

Contents

Preface

Materials science studies the synthesis, structure and performance of materials. Research in this field focuses on the manufacture of materials that are developed out of metals, polymers, ceramics and their various composites. Applications of engineered materials are in varied sectors such as electronics and photonics, fuel cells, medical devices, etc. This book includes some of the vital pieces of work being conducted across the world, on various topics related to materials science. The extensive contents of this book provide the readers with a thorough understanding of the subject.

The information contained in this book is the result of intensive hard work done by researchers in this field. All due efforts have been made to make this book serve as a complete guiding source for students and researchers. The topics in this book have been comprehensively explained to help readers understand the growing trends in the field.

I would like to thank the entire group of writers who made sincere efforts in this book and my family who supported me in my efforts of working on this book. I take this opportunity to thank all those who have been a guiding force throughout my life.

Editor

Synthesis of Bi- and Trifunctional Cyclic Carbonates Based on Trimethylolpropane and Their Application to Networked Polyhydroxyurethanes

Hiroyuki Matsukizono[1] & Takeshi Endo[1]

[1] Molecular Engineering Institute, Kinki University, Japan

Correspondence: Takeshi Endo, Professor of Molecular Engineering Institute, Kinki University, 11-6 Kayanomori, Iizuka, Fukuoka, 820-8555, Japan. E-mail: tendo@moleng.fuk.kindai.ac.jp

Abstract

The reaction of trimethylolpropane (TMP) and diphenyl carbonate gives three types of TMP-based six-membered cyclic carbonates (TMPCs) via phosgene-free route. TMPC having one hydroxyl group (TMPC-OH) reacted with terephthaloyl chloride or trimesoyl chloride to give bifunctional (Ph-TMPC$_2$) or trifunctional cyclic carbonate monomers (Ph-TMPC$_3$), respectively. The ring-opening polyaddition of Ph-TMPC$_2$ and conventional diamines efficiently proceeded without the cleavage of ester bonds to afford linear polyhydroxyurethanes (PHUs) with well-controlled molecular weights and polydispersities via isocyanate-free route. Moreover, the polyaddition of Ph-TMPC$_2$ and diamine at TMPC$_2$/diamine feed ratio of 1.1 afforded PHUs having cyclic carbonate terminals, the hydroxyl side chains of which were easily reacted with acetic anhydride to give acetylated PHUs with cyclic carbonate terminals. On the other hand, the polyaddition at Ph-TMPC$_2$/diamine feed ratio of 0.91 gave PHUs with amine terminals. The copolymerization of Ph-TMPC$_2$, Ph-TMPC$_3$ and diamine gives PHUs comprising covalently-bridged networked structures. After DMF solutions of Ph-TMPC$_2$, Ph-TMPC$_3$ and diamines were simply kept at 60 °C overnight, networked PHU films with well transparency were successfully fabricated. The PHU films prepared at different diamines showed similar transparency and thermal stability, while the mechanical properties were significantly affected by the methylene spacers of diamines.

Keywords: trimethylolpropane, diphenyl carbonate, networked polyhydroxyurethane, polyhydroxyurethane film

1. Introduction

Poly(hydroxyurethane)s (PHUs) have been attracting great interest as the promising alternatives of widely-used polyurethanes because PHUs can be synthesized by ring-opening polyaddition of bifunctional cyclic carbonates and diamines without using toxic isocyanates (Nohra, 2013; Kathalewar, 2013). In particular, five-membered cyclic carbonates (5-CCs) can be derived from epoxides and CO$_2$ under mild conditions (Fleischer, 2013; Sheng, 2015) and therefore, numerous 5-CCs-derived PHUs have been synthesized using bisphenol A (Ochiai, 2014; Lambeth, 2013; Ochiai, 2007; Ochiai, 2005; Kihara, 1996) and bio-based polyols (Kathalewar, 2015; Maisonneuve, 2014; Annunziata, 2014; Guillaume, 2013; Guillaume, 2011) as starting materials. PHUs can be functionalized by the use of functional diamines or by chemical modifications via their hydroxyl side chains. Nowadays it has been reported that silicone-conjugated PHUs show lower glass transition temperatures and well water repelling properties (Ochiai, 2014) and networked PHU gels are obtainable by the reaction with metal alkoxides (Kihara, 1996) or by the free-radical polymerization of methacryl groups modified in their side chains (Ochiai, 2007). In addition, urethane linkages of PHUs can be degraded at alkaline conditions and further shows biodegradability, thus, PHUs are one of promising candidates for versatile materials based on biocompatibility and biodegradability. Meanwhile, PHUs from six-membered cyclic carbonates (6-CCs) are hardly reported except for a few case (He, 2011; Maisonneuve, 2014; Tomita, 2001). The ring-opening reaction of 6-CCs with amines requires the lower energy compared to 5-CCs because of the larger ring strain energy of 6-CCs (Tomita, 2001). This implies that PHUs can be synthesized at milder conditions. Besides, β-substituted 6-CCs generate only primary alcohols (hydroxymethyl groups) after the ring-opening reaction with amines, therefore, simple structured PHUs are obtainable. However, 6-CCs are generally synthesized using toxic phosgene or its derivatives (He, 2011; Maisonneuve, 2014; Tomita, 2001), which hinders the practical usage of 6-CCs and their application to PHU materials.

The development of synthetic route to 6-CCs without toxic phosgene and its derivatives is one of important issues in polycarbonate and polyurethane chemistry. It has been reported that the synthesis of mono- and bifunctional 6-CCs using diphenyl carbonate (DPC) and their application to networked polycarbonate and PHU films via their copolymerization (Matsukizono, 2015, 2016). Without phosgene and its derivatives, DPC can be synthesized from ethylene carbonates through the reaction with methanol and phenol (Nishihara, 2010; Okuyama, 2003) or from CO and phenol (Kanega, 2013; Murayama, 2012). Furthermore, Matsukizono and Endo (2016) have recently reported that the reaction of trimethylolpropane (TMP) with an excess amount of DPC gives three types of different structured TMP-based cyclic carbonates (TMPCs) shown in Scheme 1: TMPC bearing a hydroxyl group (TMPC-OH), TMPC bearing a phenoxycarbonylated hydroxyl group (TMPC-OCOPh) and TMPC dimers bridged by an acyclic carbonate bond (TMPC-dimer). In addition, they have reacted TMPC-dimer with conventional diamine to form poly(carbonate-hydroxyurethane)s and characterized their hydrolytic properties in basic aqueous media. Meanwhile, TMPC-OH contains one hydroxyl group, which can be reacted with multifunctional acyl chloride to synthesize multifunctional 6-CCs bridged by ester linkages. This approach lets us expect that networked PHU materials such as gels and films can be easily fabricated by the simple reaction of multifunctional 6-CCs with diamines.

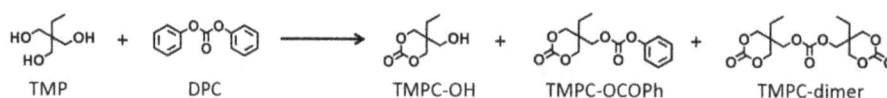

Scheme 1. Three types of different structured cyclic carbonates synthesized by the reaction of TMP and DPC

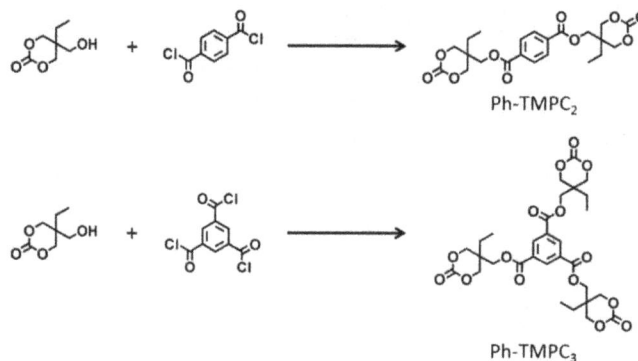

Scheme 2. Synthesis of bifunctional 6-CCs (Ph-TMPC$_2$) or trifunctional 6-CCs (Ph-TMPC$_3$) from TMPC-OH.

Here in, we describe the facile fabrication route to networked PHU material. At first, we synthesized bi- and trifunctional 6-CCs from TMPC-OH (Ph-TMPC$_2$ and Ph-TMPC$_3$) (See Scheme 2). Next, we performed the ring-opening polyaddition of Ph-TMPC$_2$ and conventional diamines to yield linear PHUs (Scheme 3a). Finally, we reacted Ph-TMPC$_2$, Ph-TMPC$_3$ and diamines at different feed ratios to build networked PHU films and characterized their transparency, thermal stability and mechanical properties (Scheme 3b). Monomers and polymers were analyzed by ^{1}H and ^{13}C nuclear magnetic resonance (NMR), Fourier translation infra-red (FT-IR) spectroscopy, size exclusion chromatography (SEC), thermogravimetric analysis (TGA), UV/Vis.-near infra-red (NIR) absorption spectroscopy and tensile tests.

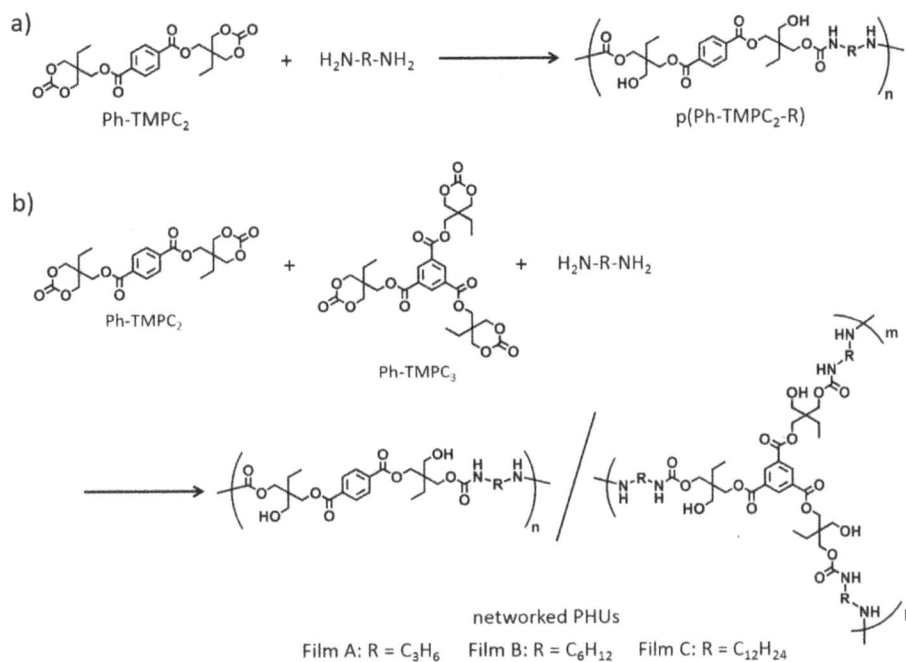

Scheme 3. a) Synthesis of linear PHUs (p(Ph-TMPC$_2$-R)) by ring-opening polyaddition of Ph-TMPC$_2$ with conventional diamines (R = C$_3$H$_6$, C$_6$H$_{12}$, C$_{12}$H$_{24}$). b) Fabrication of networked PHUs by the copolymerization of Ph-TMPC$_2$, Ph-TMPC$_3$ and diamines

2. Method

2.1 Reagents

TMP, terephthaloyl chloride, dehydrated pyridine, trimethylamine and acetic anhydride were purchased from Wako Pure Chemical Co., Ltd. DPC, trimesoyl chloride, 1,3-diaminopropane (C$_3$), 1,6-diaminehexane (C$_6$) and 1,12-diaminododecane (C$_{12}$) were obtained from Tokyo Chemical Industry Co., Ltd. Other reagents and solvents were commercially obtained and used without any purification.

2.2 Synthesis of TMPC-OH

TMPC-OH was synthesized by the same procedure reported previously (Matsukizono, 2016). TMP 13.4 g (100 mmol) was added slowly to DPC 85.7 g (400 mmol) melted by heating at 140 °C. After the mixture was stirred at 140 °C for 1-2 days, the mixture was cooled to ambient temperature and then dissolved in EtOAc (120 mL) and hexane (40 mL). TMPC-OH was isolated by silica gel column chromatograph (eluent: n-hexane/EtOAc volume ratio of 1/3). The crude product was used for the following reactions without further purification. Yield: 56.6%. ^{1}H NMR (400 MHz, CDCl$_3$, δ): 4.26 (dxd, 4H, J = 70, 11 Hz, - CH$_2$-OCOO- of cyclic carbonate), 3.67 (d, 2H, J = 4.8 Hz, HO-CH$_2$-), 2.77 (t, 1H, J = 5.4 Hz, HO-), 1.51 (q, 2H, J = 7.6 Hz, CH$_3$-CH$_2$-), 0.93 (t, 3H, J = 7.6 Hz, CH$_3$-). ^{13}C NMR (400 MHz, DMSO-d$_6$, δ): 148.6 (C=O), 72.8 (-CH$_2$-OCO-), 59.8 (HO-CH$_2$-), 35.8 (C-(CH$_2$O)$_3$), 22.5 (CH$_3$-CH$_2$-), 7.6 (CH$_3$). IR (ATR): ν = 3408 (m; ν(O-H)), 2968-2883 (w; ν(CH$_2$)), 1714 (s; ν(C=O)), 1472-1409 (m; δ(CH$_2$)), 1180-1110 cm^{-1} (s; ν(C-O) of carbonate and alcohol).

2.3 Synthesis of Ph-TMPC$_2$

TMPC-OH 3.37 g (21.0 mmol) and terephthaloyl chloride 2.04 g (10.0 mmol) were individually dissolved in dehydrated THF (15 mL). To the TMPC-OH solution at 0 °C, triethylamine 2.13 g (21.0 mmol) and the terephthaloyl chloride solution were slowly added and the mixture were stirred at ambient temperature for 14 h. Resulting precipitates were collected by suction filtration and washed with THF. After dried, the precipitates were dissolved in CH$_2$Cl$_2$ (100 mL) and then washed three times with distilled water (100 mL). The organic layer were dried over with anhydrous Na$_2$SO$_4$ and then concentrated. After drying under reduced pressure, Ph-TMPC$_2$ was obtained as a white solid. Yield: 3.09 g (68.5%). ^{1}H NMR (400 MHz, CDCl$_3$, δ): 8.10 (s, 4.0H, phenyl), 4.43-4.30 (m, 12.0H, -CH$_2$-OC(=O)-), 1.64 (q, 4.0H, J = 7.6 Hz, CH$_3$-CH$_2$-), 1.02 (t, 6.0H, J = 7.8 Hz, CH$_3$-). ^{13}C NMR (400 MHz, CDCl$_3$, δ): 165.1 (C=O of ester), 148.1 (C=O of carbonate), 133.5 (C1 of phenyl), 129.9 (C2 of phenyl), 72.7

(CH_2-O of carbonate), 64.1 (CH_2-O of ester), 35.0 (C-(CH_2O)$_3$), 23.9 (CH_3-CH_2), 7.5 (CH_3). IR (ATR): ν = 2966-2885 (w; $\nu(CH_2)$), 1756 (s; ν(C=O of ester)), 1721 cm^{-1} (s; ν(C=O of carbonate))

2.4 Synthesis of Ph-TMPC$_3$

TMPC-OH 5.29 g (33.0 mmol) and dehydrated pyridine 2.61 g (33.0 mmol) were dissolved in dehydrated THF (50 mL) and the solution was cooled to 0 °C in ice bath. To the solution was added slowly a dehydrated THF (30 mL) containing trimesoyl chloride 2.67 g (10.1 mmol) and the mixture was stirred at ambient temperature for 21 h. Resulting precipitates were collected by suction filtration and washed with THF. After dried, the precipitates were dissolved in CH_2Cl_2 (100 mL) and washed three times with distilled water (100 mL). After dried over with anhydrous Na_2SO_4, the organic layer was concentrated. The residues were purified by reprecipitation from acetone/methanol to give Ph-TMPC$_3$ as a white solid. Yield: 3.73 g (58.6%). ^{1}H NMR (400 MHz, CDCl$_3$, δ): 8.81 (s, 3.0H, phenyl), 4.52-4.31 (m, 18.0H, -CH_2-OC(=O)-), 1.64 (q, 6.0H, J = 7.6 Hz, CH$_3$-CH_2-), 1.02 (t, 9.0H, J = 7.6 Hz, CH$_3$-). ^{13}C NMR (400 MHz, CDCl$_3$, δ): 164.3 (C=O of ester), 148.2 (C=O of carbonate), 135.2 (C1 of phenyl), 130.8 (C2 of phenyl), 72.9 (CH_2-O of carbonate), 65.3 (CH_2-O of ester), 34.9 (C-(CH_2O)$_3$), 24.0 (CH_3-CH_2), 7.5 (CH_3). IR (ATR): ν = 2973-2882 (w; $\nu(CH_2)$), 1728 cm^{-1} (s; ν(C=O of carbonate and ester)).

2.5 Synthesis of linear PHUs

2.5.1 Synthesis of p(Ph-TMPC$_2$-C$_3$)

To DMF (1.5 mL) containing Ph-TMPC$_2$ 212 mg (0.47 mmol) was added a DMF (1.5 mL) of C$_3$ 32 mg (0.43 mmol) at ambient temperature. The mixture was stirred at ambient temperature for 24 h and the fraction was analyzed by ^{1}H NMR spectroscopy and SEC measurements. After that, the mixture was added in distilled water (100 mL) and then the resulting precipitates were washed thoroughly with distilled water. The precipitates were collected by dissolving in acetone and drying under reduced pressure to obtain p(Ph-TMPC$_2$-C$_3$) having TMPC terminals as colorless liquids. Yield: 109 mg (44.7%). ^{1}H NMR (400 MHz, CDCl$_3$, δ): 8.11-8.05 (m, 4.3H, phenyl), 7.05 (br, 1.9H, NH), 4.68 (br, 2.0H, OH), 4.42-4.31 (m, 1.4H, -CH_2-OC(=O)- of terminal cyclic carbonates), 4.12 (s, 4.0H, -CH_2-OC(=O)-Ph), 3.92 (s, 3.8H, -CH_2-OC(=O)-NH), 3.37 (s, 3.9H, -CH_2-OH), 3.03-2.87 (m, 4.4H, NH-CH_2-CH$_2$-), 1.54-1.37 (m, 6.4H, CH$_3$-CH_2- and NH–CH$_2$-CH_2-), 0.87-0.81 (m, 6.6H, CH$_3$-). IR (ATR): ν = 3360 (w; ν(OH)), 2963-2883 (w; $\nu(CH_2)$), 1713 (s; ν(C=O of carbonate and ester)), 1268 cm^{-1} (s; ν(C-O)).

2.5.2 Synthesis of p(Ph-TMPC$_2$-C$_6$)

p(Ph-TMPC$_2$-C$_6$) was synthesized by the same procedure of p(Ph-TMPC$_2$-C$_3$) and obtained as white solids. Yield: 91.8%. ^{1}H NMR (400 MHz, DMSO-d$_6$, δ): 8.09-8.05 (m, 4.5H, phenyl), 7.04 (br, 1.8H, NH), 4.73-4.68 (br, 1.9H, OH), 4.42-4.31 (m, 1.5H, -CH_2-OC(=O)- of terminal cyclic carbonates), 4.12 (s, 4.0H, -CH_2-OC(=O)-Ph), 3.92 (s, 3.7H, -CH_2-OC(=O)-NH), 3.37 (s, 3.9H, -CH_2-OH), 3.00-2.85 (m, 4.0H, NH-CH_2-CH$_2$-), 1.51-1.11 (m, 12.4H, CH$_3$-CH_2- and NH–CH$_2$-(CH_2)$_4$-), 0.87-0.81 (m, 6.6H, CH$_3$-). IR (ATR): ν = 3339 (w; ν(OH)), 2962-2881 (w; $\nu(CH_2)$), 1700 (s; ν(C=O of carbonate and ester)), 1246 cm^{-1} (s; ν(C-O)).

2.5.3 Synthesis of p(Ph-TMPC$_2$-C$_{12}$)

p(Ph-TMPC$_2$-C$_{12}$) was synthesized by the same procedure of p(Ph-TMPC$_2$-C$_3$) and obtained as white solids. Yield: 69.3%. ^{1}H NMR (400 MHz, DMSO-d$_6$, δ): 8.09-8.05 (m, 4.4H, phenyl), 7.03 (br, 1.8H, NH), 4.73-4.67 (br, 1.9H, OH), 4.42-4.31 (m, 1.5H, -CH_2-OC(=O)- of terminal cyclic carbonates), 4.12 (s, 4.0H, -CH_2-OC(=O)-Ph), 3.92 (s, 3.8H, -CH_2-OC(=O)-NH), 3.37 (s, 3.9H, -CH_2-OH), 3.01-2.85 (m, 4.0H, NH-CH_2-CH$_2$-), 1.51-1.27 (m, 8.6H, CH$_3$-CH_2- and NH–CH$_2$-CH_2-), 1.13 (s, 16.0H, NH-(CH$_2$)$_2$-(CH_2)$_8$-), 0.87-0.80 (m, 6.6H, CH$_3$-). IR (ATR): ν = 3345 (w; ν(OH)), 2964-2881 (w; $\nu(CH_2)$), 1698 (s; ν(C=O of carbonate and ester)), 1243 cm^{-1} (s; ν(C-O)).

2.5.4 Synthesis of p(Ph-TMPC$_2$-C$_3$-OAc)

p(Ph-TMPC$_2$-C$_3$) was synthesized by the reaction of Ph-TMPC$_2$ 202 mg (0.45 mmol, 1.1 equiv.) and C$_3$ 30 mg (0.41 mmol, 1.0 equiv.) in DMF (2.0 mL). Without purification, p(Ph-TMPC$_2$-C$_3$-OAc) was synthesized. To the DMF solution of p(Ph-TMPC$_2$-C$_3$), acetic anhydride 315 mg (3.09 mmol) and pyridine 245 mg (3.10 mmol) were added at ambient temperature and the mixture was stirred at 40 °C for 41 h. After DMF was removed, the yellow residues were dissolved in CH_2Cl_2 (100 mL) and then washed three times with distilled water (100 mL). After drying with anhydrous Na_2SO_4, CH_2Cl_2 was evaporated. After the yellow residues were dried under reduced pressure, p(Ph-TMPC$_2$-C$_3$-OAc) was obtained as pale yellow solids. Yield: 253 mg (95.0%). ^{1}H NMR (400 MHz, CDCl$_3$, δ): 8.08 (m, 4.4H, phenyl), 5.23 (br, 2.1H, NH), 4.44-4.41 (m, 1.2H, -CH_2-OC(=O)- of terminal cyclic carbonates), 4.32-4.11 (m, 11.6H, -CH_2-OC(=O)), 3.27-3.12 (s, 4.0H , NH-CH_2-CH$_2$-), 2.06 (s, 5.8H, CH$_3$-(C=O)), 1.67-1.57 (m, 6.4H, CH$_3$-CH_2- and NH–CH$_2$-CH_2-), 1.04-0.93 (m, 6.6H, CH_3-CH$_2$-). IR (ATR): ν = 3367 (w; ν(OH of residual hydroxyl group or adsorbed water)), 2965-2880 (w; $\nu(CH_2)$), 1720 (s; ν(C=O of carbonate and ester)), 1241 cm^{-1} (s; ν(C-O)).

2.5.5 Synthesis of p(Ph-TMPC$_2$-C$_3$)'

A DMF solution (1.0 mL) of C$_3$ 163 mg (2.2 mmol, 1.1 equiv.) was added to a DMF solution (2.0 mL) of Ph-TMPC$_2$ 901 mg (2.0 mmol) at 70 °C. At the temperature, the mixture was stirred for 17 h and then cooled to ambient temperature. After concentrated, the mixture was added to distilled water (200 mL) and the precipitates were washed with distilled water. After the precipitates were dissolved in acetone, the solution was concentrated and then dried under reduced pressure to give p(Ph-TMPC$_2$-C$_3$)' as white solids. Yield: 95.2%. ^{1}H NMR (400 MHz, CDCl$_3$, δ): 8.05 (s, 4.1H, phenyl), 7.05-6.72 (m, 2.0H, NH), 4.68 (br, 1.9H, OH), 4.12 (s, 4.0H, -CH$_2$-OC(=O)-Ph), 3.92 (s, 4.0H, -CH$_2$-OC(=O)-NH), 3.37 (s, 4.0H, -CH$_2$-OH), 3.01-2.85 (m, 4.3H, NH-CH$_2$-CH$_2$-), 1.55-1.37 (m, 6.2H, CH$_3$-CH$_2$- and NH–CH$_2$-CH$_2$-), 0.81 (t, 6.1H, J = 7.2 Hz, CH$_3$-). IR (ATR): ν = 3364 (w; ν(OH and NH$_2$)), 2964-2881 (w; ν(CH$_2$)), 1698 (s; ν(C=O of carbonate and ester)), 1243 cm^{-1} (s; ν(C-O)).

2.6 Fabrication of Networked PHU Films

Ph-TMPC$_2$ and Ph-TMPC$_3$ were dissolved in DMF (1 mL) at 70 °C to prepare the Ph-TMPC$_2$/Ph-TMPC$_3$ solution at different molar ratios (5-20 mol% of Ph-TMPC$_3$ with respect to the total carbonate monomers (Ph-TMPC$_2$ + Ph-TMPC$_3$), [TMPC] = 0.5 M). To the solutions, C$_3$ dissolved in DMF (0.5 mL, [C$_3$] = 0.25 M) was added and then the mixed solutions (TMPC/NH$_2$ = 1.0) were stirred at 70 °C for within 5 min. After that, the solutions were poured in glass petri-dishes with 6.0 cm in diameter and allowed to stand at 60 °C overnight. The resulting films were carefully removed and immersed in distilled water for purification. After drying, the networked PHU films (Film A) were obtained. Similarly, Film B and Film C were prepared using 20 mol% Ph-TMPC$_3$ and C$_6$ or C$_{12}$, respectively.

2.7 Measurements

^{1}H and ^{13}C NMR spectra were recorded with a JEOL ECS-400 NMR spectrometer operating at 400 MHz using a tetramethylsilane (TMS) as an internal reference. FT-IR spectroscopy was conducted with a Thermo Fisher Scientific Nicolet iS10 equipped with an ATR instrument. SEC were performed with a Tosoh HLC-8320GPC using DMF as eluents operating at a flow rate of 0.5 mL min^{-1}. Number averaged molecular weight (M_n), Weight averaged molecular weight (M_w) and polydispersity (M_w/M_n) were determined from SCE traces using polystyrene standards. PHU films were fabricated by gradual drying of monomer solutions under ambient atmosphere using a Sanso Vacuum Drying Oven SVD10P (width, 20 cm; depth, 25 cm; height, 20 cm). TGA was carried out with a TG/DTA 6200 (Seiko Instrument Inc.) at a heating rate of 10 °C min^{-1} under nitrogen gas at a flow rate of 200 mL min^{-1}. UV/Vis.-NIR spectra were recorded on a Jasco V570 UV/VIS/NIR spectrophotometer equipped with Pbs power supply instruments. Stress-strain curves were measured with a Shimazu EZ Test EZ-LX with an operation rate of 50 mm min^{-1}. Young's modulus (E), stress at yield point (σ_{yield}), stress at break point (σ_{break}) and strain at break point (ε_{break}) of the films were analyzed with a TRAPEZIUM X software. The tensile tests were performed 1-3 times and then mechanical parameters were averaged.

3. Results and Discussion

3.1 Synthesis of bi- and Trifunctional 6-CCs Based on TMPC

The reaction of TMP and DPC gives mainly three types of different structured TMPCs (Matsukizono, 2016). These TMPCs can be to some extent individually produced by the modulation of synthetic conditions. Longer reaction time leads to the formation of TMPC-OCOPh and TMPC-dimer, while shorter reaction periods afford dominantly TMPC-OH. In this case, TMP was reacted with 4 equiv. of DPC at 140 °C for 1-2 days and consequently, TMPC-OH was obtained in 56.6 % yield after purification by column chromatography. The ^{1}H and ^{13}C NMR spectra of TMPC-OH are shown in Figure S1 in the appendix section.

The bifunctional TMP-based cyclic carbonate (Ph-TMPC$_2$) was synthesized by the simple reaction of TMPC-OH, terephthaloyl chloride and triethylamine in anhydrous THF. While the reaction proceeded, Ph-TMPC$_2$ and triethylamine hydrochloride precipitated. After washing with distilled water followed by methanol, white solids were isolated in 68.5% yield. The ^{1}H NMR spectrum of the solids is shown in Figure 1. The signal at 3.6 ppm assigned to methylene protons adjacent to a hydroxyl group completely disappears, while the singlet signal ascribed to methylene protons neighboring to an ester bond newly appears at 4.4 ppm. The singlet signal at 8.1 ppm originates from benzene protons, the integral ratio of which is ca. 4.0 and equals to the theoretical value. In addition, the ^{13}C NMR spectrum of the solids gives a signal at 165 ppm originating from ester carbons and signals at 133 and 130 ppm assigned to benzene carbons (Figure S2). FT-IR spectrum of the solids exhibits no absorption at around 3500 cm^{-1} based on hydroxyl groups, while the characteristic absorption of ester bonds appears at 1756 cm^{-1} (Figure S3). These results clearly indicate that Ph-TMPC$_2$ is obtained as white solids. Similarly, Ph-TMPC$_3$ was synthesized from TMPC-OH and trimesoyl chloride and isolated as white solids in 58.6% yield. The ^{1}H NMR spectrum of Ph-TMPC$_3$ exhibits the signals assigned to the benzene protons at 8.81 ppm and methylene protons adjacent to ester bonds at 4.49 ppm (Figure 2). In FT-IR spectrum of Ph-TMPC$_3$, the streching vibration of hydroxyl groups disappears and ^{13}C NMR spectrum of Ph-TMPC$_3$ shows the signals assigned to carbonyl carbons of ester bonds and benzene carbons (Figure S3 and S4), indicating the formation and purification of Ph-TMPC$_3$.

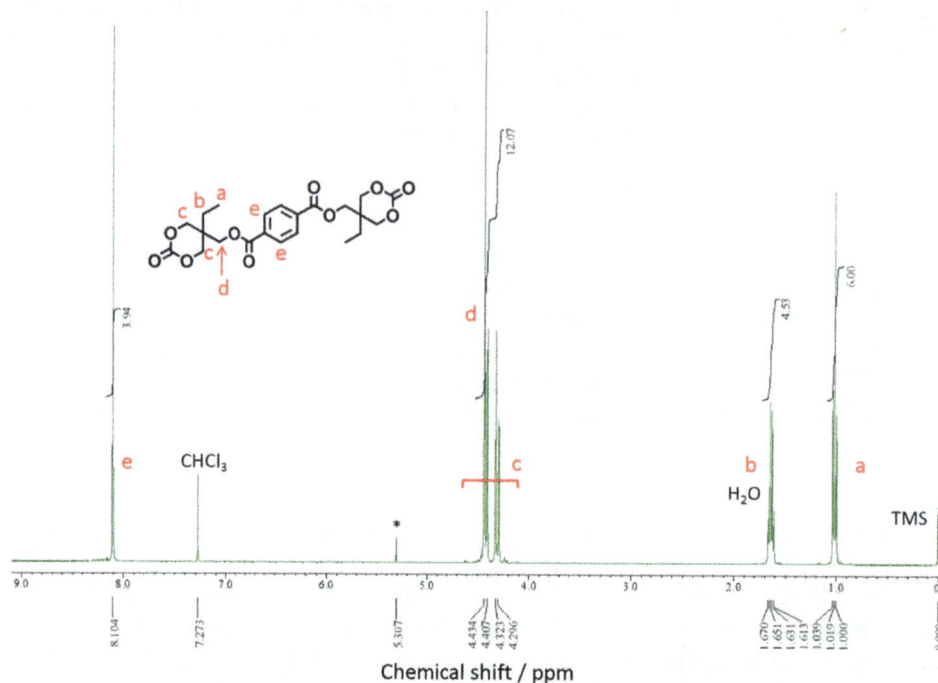

Figure 1. ^{1}H NMR spectrum of Ph-TMPC$_2$ in CDCl$_3$ containing 0.03 v/v% of TMS. An asterisk denotes residual CH$_2$Cl$_2$ used for purification

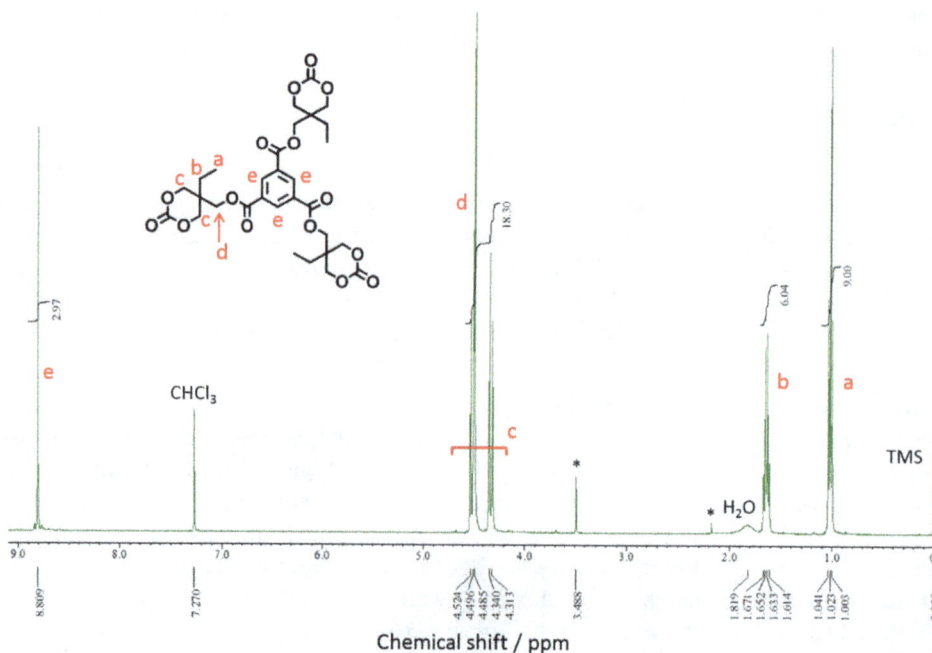

Figure 2. ^{1}H NMR spectrum of Ph-TMPC$_3$ in CDCl$_3$ containing 0.03 v/v% of TMS. Asterisks mean residual MeOH and acetone used for purification

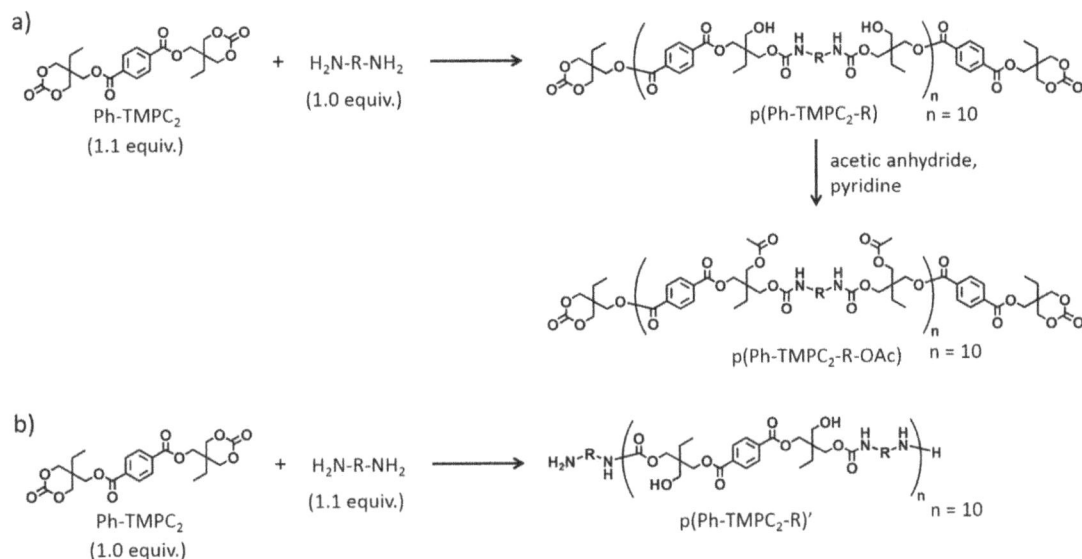

Scheme 4. Synthesis of PHUs with different terminal structures from Ph-TMPC$_2$ and diamine (R = C$_3$H$_6$, C$_6$H$_{12}$ or C$_{12}$H$_{24}$) at different Ph-TMPC$_2$/diamine feed ratios. a) PHUs with cyclic carbonate terminals (p(Ph-TMPC$_2$-R)) by the polyaddition of 1.1 equiv. of Ph-TMPC$_2$ and 1.0 equiv. of diamine and acetylated PHUs with same terminal structures (p(Ph-TMPC$_2$-R-OAc)) after acetylation. b) PHUs with amine terminals (p(Ph-TMPC$_2$-R)') by the polyaddition of 1.0 equiv. of Ph-TMPC$_2$ and 1.1 equiv. of diamine

3.2 Synthesis and Characterization of PHUs Synthesized from Ph-TMPC$_2$ and Conventional diamines

Before the fabrication of networked PHUs, we investigated whether the ring-opening polyaddition of Ph-TMPC$_2$ and diamines proceeds to form linear PHUs with well-controlled structures or not. The reaction of Ph-TMPC$_2$ with an equivalent amount of diamine, in which Ph-TMPC$_2$/diamine feed ratio is 1.0, leads to the formation of long-chained PHUs, however, the chain ends of the PHUs are unclear. On the other hand, slight different Ph-TMPC$_2$/diamine feed ratios provide shorter-chained PHUs with cyclic carbonate or amine terminals (Scheme 4). In this paper, we preferred the synthesis of shorter-chained PHUs with well-controlled terminal structures to the formation of longer-chained PHUs.

At first, we carried out the polyaddition of 1.1 equiv. of Ph-TMPC$_2$ with 1.0 equiv. of C$_3$ in DMF at 70 °C for 1 day. The reaction mixture was analyzed by ^{1}H NMR spectroscopy and SEC measurements. The conversion of Ph-TMPC$_2$ was determined by the change in the integral ratio of signals at 4.4-4.3 ppm assigned to methylene protons adjacent to ester and carbonate bonds. In actual, the integral ratio is 1.62, which includes the integral ratio of a slightly excess amount (0.1 equiv.) of Ph-TMPC$_2$ as well as unreacted ones, and then the conversion of Ph-TMPC$_2$ is calculated to be 96.5%. After the reaction mixture was added in distilled water, the resulting precipitate was analyzed by ^{1}H NMR, FT-IR spectroscopy and SEC measurement. Figure 3 exhibits ^{1}H NMR spectrum of the precipitate. The signals at 4.4-4.3 ppm decrease and three signals at 4.1, 3.9 and 3.4 ppm newly appear. These signals are ascribed to methylene protons adjacent to ester, urethane and hydroxyl groups, respectively. In addition, the signals based on NH and OH bonds appear at 7.0-6.7 and 4.7 ppm, respectively. The conversion after purification is calculated to be 98.0%, which is slightly higher than that before purification. This is due to the purification treatment. On the other hand, M_n and M_w/M_n values of the precipitate calculated by its SEC trace are 4,400 g mol^{-1} and 2.11, respectively. These values are close to their theoretical ones (M_n = 4,100 g mol^{-1}, M_w/M_n = 2.00). These results apparently indicate that the reaction of Ph-TMPC$_2$ with C$_3$ proceeds efficiently to form PHUs with cyclic carbonate terminals without the cleavage of ester bonds. When the synthesis of PHUs from Ph-TMPC$_2$ and C$_3$ was performed at lower temperature, the PHUs with similar M_n and M_w/M_n values were obtained (data are not shown).

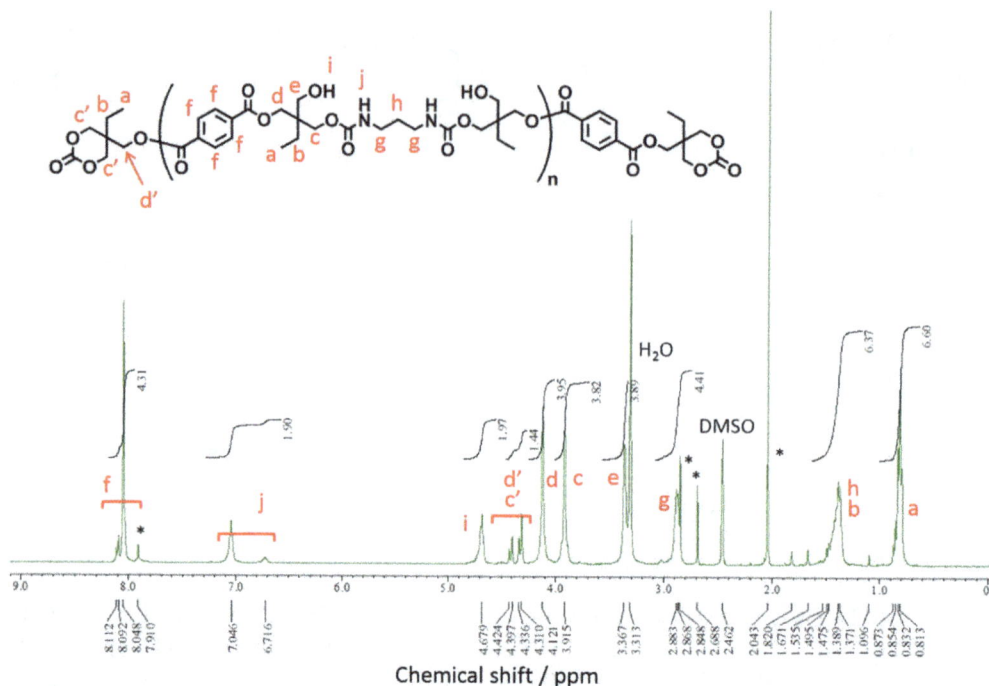

Figure 3. ^{1}H NMR spectrum of p(Ph-TMPC$_2$-C$_3$) in DMSO-d$_6$. Asterisks mean residual DMF and acetone

Table 1. Characterization of PHUs synthesized from Ph-TMPC$_2$ and diamines with different methylene spacers

Entry	PHUs	r [a]	Conv. / % [b]	$M_{n,theor}$ / g mol^{-1} [c]	M_n / g mol^{-1} [d]	M_w/M_n [d]
1	p(pH-TMPC$_2$-C$_3$)	1.1	98.0	4,100	4,400	2.11
2	p(pH-TMPC$_2$-C$_6$)	1.1	99.0	4,300	3,600	2.07
3	p(pH-TMPC$_2$-C$_{12}$)	1.1	97.6	4,700	5,000	2.13
4	p(pH-TMPC$_2$-C$_3$-OAc)	1.1	> 99.9	7,000	3,400	1.92
5	p(pH-TMPC$_2$-C$_3$)'	0.91	> 99.9	5,300	5,300	2.19

[a] Ph-TMPC$_2$/diamine feed ratio. [b] Reactivity of Ph-TMPC$_2$ calculated by ^{1}H NMR spectroscopy. [c] Determined using the conversion values. [d] Determined by SEC traces using polystyrene standards.

Similarly, using C$_6$ or C$_{12}$ instead of C$_3$, the synthesis of PHUs was conducted. The PHUs obtained were characterized by ^{1}H NMR, FT-IR spectroscopy and SEC measurements and the results are summarized in Entry 1-3 in Table 1. The ^{1}H NMR spectrum of these PHUs is similar to that of p(Ph-TMPC$_2$-C$_3$) (Figure S5 and S6). Conversion values are close to 100% and M_n and M_w/M_n values are well corresponded to those of theoretical values.

PHUs can be modified with various functional groups via their hydroxyl side chains. After PHUs are reacted with acetic anhydride, acetylated PHUs can be obtained. Actually, we performed the acetylation of p(Ph-TMPC$_2$-C$_3$) and acetylated PHU (p(Ph-TMPC$_2$-C$_3$-OAc)) was spectroscopically characterized. The ^{1}H NMR spectrum shown in Figure S7 reveals that the acetylation was adequately performed. On the other hand, M_n values determined by its SEC trace is lower than that theoretically calculated by its ^{1}H NMR spectrum (Entry 4 in Table 1), which would be caused by the potential difference of hydrodynamic radius from standard polystyrenes.

As shown in Scheme 4b, it is expected that the polyaddition of Ph-TMPC$_2$ and a slightly excess amount of diamine gives PHUs with amine terminals. We reacted 1.0 equiv. of Ph-TMPC$_2$ with 1.1 equiv. of C$_3$. The ^{1}H NMR spectrum of the resulting PHUs is shown in Figure S8. Differing from the spectrum of p(Ph-TMPC$_2$-C$_3$) (Figure 3), the spectrum exhibits no signals at ca. 4.4 ppm ascribed to cyclic carbonate terminals. The signal at 3.0-2.5 ppm derived from amine terminals is unclear because of the overlapping with other signals, however, its M_n and M_w/M_n

values are close to theoretical values (Entry 5 in Table 1). These results apparently suggest that PHUs with different terminal structures are individually synthesized depending on the changes in Ph-TMPC$_2$/diamine feed ratios.

3.3 Fabrication of Networked PHU Films

To examine the formative ability of networked structures, we at first reacted Ph-TMPC$_3$ with C$_3$ in DMF at 70 °C. The reaction mixture (TMPC/NH$_2$ = 1.0) immediately changed to viscous solution and finally formed gels within 1 h. This behavior apparently indicates the formation of networked structures. Next, the DMF solutions of Ph-TMPC$_2$/Ph-TMPC$_3$/C$_3$, in which mole ratios of Ph-TMPC$_3$ were 5-20 mol% with respect to total carbonate monomer and the TMPC/NH$_2$ ratio was 1.0, were kept in glass petri-dishes at 70 °C overnight. During this treatment, DMF was slowly evaporated and concomitantly the reaction of Ph-TMPC$_2$/Ph-TMPC$_3$ with C$_3$ proceeded to form networked PHU films (Film A) in one-step. These films were quite stable and able to detach from the dishes. These films gave stretching vibrations assigned to carbonyl groups of urethane bonds at ca. 1650 cm^{-1} (data are not shown), indicating the formation of hydroxyurethane linkages. The photographs of these films are shown in Figure 4. All films are well transparent and flexible. Similarly, Film B and Film C, which were prepared using 20 mol% Ph-TMPC$_3$ with C$_6$ or C$_{12}$, respectively, are well transparent and flexible (Figure 5e,f). UV/Vis-NIR absorption spectra of Film A-C are shown in Figure 5. These films show more than 80% transmittance in 440-1600 nm. The Film A prepared at different amounts of Ph-TMPC$_3$ also show same absorption spectra (Figure S9).

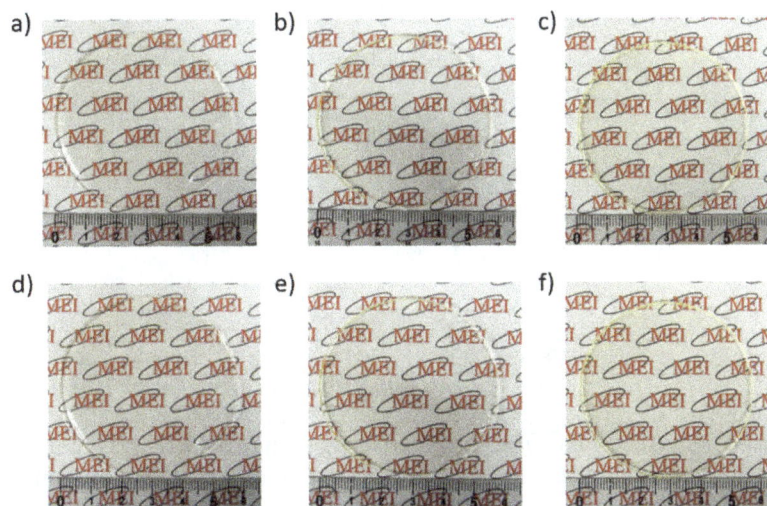

Figure 4. a-d) Photographs of Film A prepared from Ph-TMPC$_2$/Ph-TMPC$_3$/C$_3$. The mole ratios of Ph-TMPC$_3$ are a) 5, b) 10, c) 15 or d) 20 mol% with respect to total carbonate monomers (Ph-TMPC$_2$ + Ph-TMPC$_3$). e) Film B prepared using 20 mol% Ph-TMPC$_3$ and C$_6$. f) Film C prepared using 20 mol% Ph-TMPC$_3$ and C$_{12}$

TGA profiles of these films are shown in Figure 6._For Film A, a slight weight loss occurred at 100-200, which would be caused by the removal of adsorbed water molecules. After that, weights abruptly decreased at 250-300 °C and then the film was completely decomposed at 300-520 °C. Since urethane bonds of polyurethane are decomposed at 300-500 °C (Nayak, 1997), the weight loss at 250-300 °C would be caused by the decomposition of diamine moieties. On the other hand, Film B and Film C gave gradual weight loss at 300-500 °C and showed the slightly higher thermal stability compared to Film A. This could be due to the stronger Van der Waals interactions among longer methylene spacers of diamines. TGA profiles of Film A prepared at different amounts of Ph-TMPC$_3$ are also shown in Figure S10. Without depending on the amounts of Ph-TMPC$_3$, the same thermal decomposition behavior was observed. Since even networked PHUs prepared using 100 mol% Ph-TMPC$_3$ gave the similar TGA profile (data are not shown), the density and amount of cross-linked structures do not seem to affect the thermal properties.

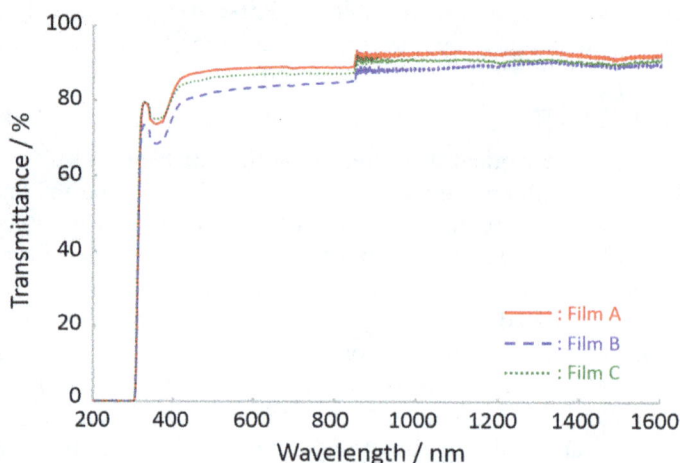

Figure 5 UV/Vis.-NIR absorption spectra of networked PHU films prepared from 20 mol% Ph-TMPC$_3$ and different diamines. Red solid line: Film A. Blue dashed line: Film B. Green dotted line: Film C

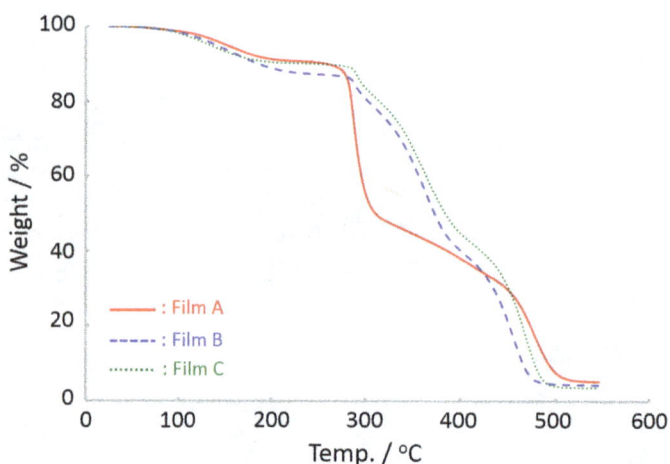

Figure 6 TGA profiles of networked PHU films prepared using 20 mol% Ph-TMPC$_3$ and different diamines. Red solid line: Film A. Blue dashed line: Film B. Green dotted line: Film C

These networked PHU films were investigated for their mechanical properties by tensile test. Representative stress-strain curves of these films are shown in Figure 7 and averaged mechanical parameters obtained from the curves are summarized in Table 2. Although these films showed similar transparency and thermal properties, their mechanical properties were quite different. Film A gave the yield point at 42.4 MPa and then broken at an elongation of 50.1%. Film A prepared using different amounts of Ph-TMPC$_3$ showed similar curves (data are not shown). On the other hands, Film B and Film C elongated gradually without showing yield points and finally broken at longer than ca. 290% elongation points. From these results, it is apparent that mechanical properties of networked PHU films can be regulated by tuning the structures of diamines.

Table 2. Averaged mechanical parameters determined from Stress-strain curves of networked PHU films prepared using 20 mol% Ph-TMPC$_3$ and different diamines.

Entry	Diamine	E / MPa	σ_{yild} / MPa	σ_{break} / MPa	ε_{break} / %
Film A	C$_3$	7.1	42.4	24.3	50.1
Film B	C$_6$	0.38	-	8.6	296
Film C	C$_{12}$	0.70	-	11.0	402

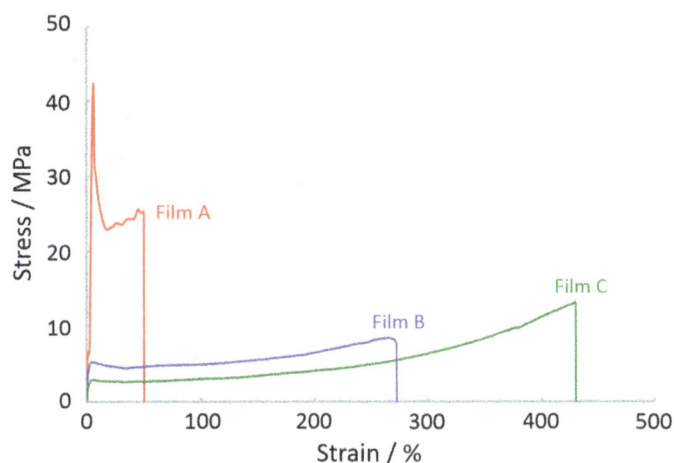

Figure 7 Representative stress-strain curves of networked PHU films prepared using 20 mol% Ph-TMPC$_3$ and different diamines. Red line: Film A. Blue line: Film B. Green line: Film C

4. Conclusions

In this paper, we have described the followings: i) the synthesis of bi- and trifunctional TMP-based cyclic carbonates (Ph-TMPC$_2$ or Ph-TMPC$_3$) from TMPC-OH and the corresponding acyl chlorides, ii) the synthesis of linear PHUs by the ring-opening polyaddition of Ph-TMPC$_2$ and conventional diamines, iii) the fabrication of networked PHU materials by the copolymerization of Ph-TMPC$_2$, Ph-TMPC$_3$ and diamines. The reaction of TMP and 4 equiv. of DPC at 140 °C for 1-2 days gives mainly TMPC-OH and the reaction of TMPC-OH and terephthaloyl chloride or trimesoyl chloride affords Ph-TMPC$_2$ or Ph-TMPC$_3$, respectively. Ring-opening polyaddition of Ph-TMPC$_2$ and conventional diamines effectively proceeded without the cleavage of ester bonds to give PHUs with well-controlled molecular weights and polydispersities. By changing the Ph-TMPC$_2$/diamine feed ratios, we could obtain PHUs with different terminal structures. The copolymerization of Ph-TMPC$_2$, small amounts of Ph-TMPC$_3$ and diamines forms effectively covalently-crosslinked structures and then networked PHU films are readily fabricated. The resulting films showed similar transparency and thermal stability, while the mechanical properties were significantly affected by the methylene spacers of diamines.

In our approach, PHUs are derived from non-expensive reagents via non-phosgene derivatives and non-isocyanate routes in whole processes. Networked PHUs are simply obtainable by copolymerization of bi- and trifunctional cyclic carbonates and diamines in one step without further chemical modification of hydroxyl side chines. Our approach will be one of the fundamental for designing PHU-based networked films.

Acknowledgments

This work was financially supported by JSR Co., Ltd.

References

Annunziata, L., Diallo, A. K., Fouquay, S., Michaud, G., Simon, F., Brusson, J.-M., Carpentier, J.-F., & Guillaume, S. M. (2014). α,ω-Di(glycerol carbonate) telechelic polyesters and polyolefins as precursors to polyhydroxyurethanes: an isocyanate-free approach. *Green Chem.*, *16*, 1947-1956. http://dx.doi.org/10.1039/C3GC41821A

Fleischer, M., Blattmann, H., & Mülhaupt, R. (2013). Glycerol-, pentaerythritol- and trimethylolpropane-based polyurethanes and their cellulose carbonate composites prepared via the non-isocyanate route with catalytic carbon dioxide fixation. *Green Chem.*, *15*, 934-942. http://dx.doi.org/10.1039/C3GC00078H

Guillaume, S. M. (2013). Recent advances in ring-opening polymerization strategies toward α,ω-hydroxyl telechelic polyesters and resulting copolymers. *Eur. Polym. J., 49*, 768-779. http://dx.doi.org/10.1016/j.eurpolymj.2012.10.011

He, Y., Keul, H., & Möller, M. (2011). Synthesis, characterization, and application of a bifunctional coupler containing a five- and a six-membered ring carbonate. *React. Funct. Polym., 71*, 175-186. http://dx.doi.org/10.1016/j.reactfunctpolym.2010.11.031

Helou, M., Carpentier, J.-F., & Guillaume, S. M. (2011). Poly(carbonate-urethane): an isocyanate-free procedure from α,ω-di(cyclic carbonate) telechelic poly(trimethylene carbonate)s. *Green Chem.*, *13*, 266-271. http://dx.doi.org/10.1039/C0GC00686F

Kanega, R., Hayashi, T., & Yamanaka, I. (2013). Pd(NHC) electrocatalysis for phosgene-free synthesis of diphenyl carbonate. *ACS Catal.*, *3*, 389-392. http://dx.doi.org/10.1021/cs300725j

Kathalewar, M., Joshi, P., Sabnis, A., & Malshe, V. (2013). Non-isocyanate polyurethanes: from chemistry to applications. *RSC Adv.*, *3*, 4110-4129. http://dx.doi.org/10.1039/C2RA21938G

Kathalewar, M., & Sabnis, A. (2015). Preparation of novel CNSL-based urethane polyol via nonisocyanate route: Curing with melamine-formaldehyde resin and structure-property relationship. *J. Appl. Polym. Sci.*, *132*, 41391. http://dx.doi.org/10.1002/APP.41391

Kihara, N., Kushida, Y., & Endo, T. (1996). Optically active poly(hydroxyurethane)s derived from cyclic carbonate and L-lysine derivatives. *J. Polym. Sci. Part A: Polym. Chem.*, *34*, 2173-2179. http://dx.doi.org/10.1002/(SICI)1099-0518(199608)34:11<2173::AID-POLA10>3.0.CO;2-C.

Lambeth, R. H., & Henderson, T. J. (2013). Organocatalytic synthesis of (poly)hydroxyurethanes from cyclic carbonates and amines. *Polymer*, *54*, 5568-5573. http://dx.doi.org/10.1016/j.polymer.2013.08.053

Maisonneuve, L., More, A. S., Foltran, S., Alfos, C., Robert, F., Landais, Y., Tassaing, T., Grau, E., & Cramail, H. (2014). Novel green fatty acid-based bis-cyclic carbonates for the synthesis of isocyanate-free poly(hydroxyurethane)s. *RSC Adv.*, *4*, 25795-25803. http://dx.doi.org/10.1039/C4RA03675A

Maisonneuve, L., Wirotius, A.-L., Alfos, C., Grau, E., & Cramail, H. (2014). Fatty acid-based (bis) 6-membered cyclic carbonates as efficient isocyanate free poly(hydroxyurethane) precursors. *Polym. Chem.*, *5*, 6142-6147. http://dx.doi.org/10.1039/C4PY00922C

Matsukizono, H., & Endo, T. (2015). Mono- and bifunctional six-membered cyclic carbonates synthesized by diphenyl carbonate toward networked polycarbonate films. *J. Appl. Polym. Sci.*, *132*, 41956. http://dx.doi.org/10.1002/app.41956

Matsukizono, H., & Endo, T. (2016). Ring-opening polymerization of six-membered cyclic carbonates initiated by ethanolamine derivatives and their application to protonated- or quaternary ammonium salt-functionalized polycarbonate films. *J. Polym. Sci. Part A: Polym. Chem.*, *54*, 487-497. http://dx.doi.org/10.1002/pola.27922

Matsukizono, H., & Endo, T. (2015). Synthesis of polyhydroxyurethanes from di(trimethylolpropane) and their application to quaternary ammonium chloride-functionalized films. *RSC Adv.*, *5*, 71360-71369. http://dx.doi.org/10.1039/C5RA09885H

Matsukizono, H., & Endo, T. (2016). Synthesis and hydrolytic properties of water-soluble poly(carbonate-hydroxyurethane)s from trimethylolpropane. *Polym. Chem.*, *7*, 958-969. http://dx.doi.org/10.1039/C5PY01733E

Murayama, T., Hayashi, T., Kanega, R., & Yamanaka, I. (2012). Phosgene-free method for diphenyl carbonate synthesis at the Pd^0/Ketjenblack anode. *J. Phys. Chem. C*, *116*, 10607-10616. http://dx.doi.org/10.1021/jp300809s

Nayak, P., Mishra, D. K., Parida, D., Sahoo, K. C., Nanda, M., Lenka, S., & Nayak, P. L. (1997). Polymers from renewable resources. IX. Interpenetrating polymer networks based on castor oil polyurethane poly(hydroxyethyl methacrylate): Synthesis, chemical, thermal, and mechanical properties. *J. Appl. Polym. Sci.*, *63*, 671-679. http://dx.doi.org/10.1002/(SICI)1097-4628(19970131)63:5<671::AID-APP15>3.0.CO;2-X

Nishihara, R. (2010). The novel process for diphenyl carbonate and polycarbonate production. *Catal. Surv. Asia*, *14*, 140-145. http://dx.doi.org/10.1007/s10563-010-9095-3

Nohra, B., Candy, L., Blanco, J.-F., Guerin, C., Raoul, Y., & Mouloungui, Z. (2013). From petrochemical polyurethanes to biobased polyurethanes. *Macromolecules*, *46*, 3771-3792. http://dx.doi.org/10.1021/ma400197c

Ochiai, B., Kojima, H., & Endo, T. (2014). Synthesis and properties of polyhydroxyurethane bearing silicone backbone. *J. Polym. Sci. Part A: Polym. Chem.*, *52*, 1113-1118. http://dx.doi.org/10.1002/pola.27091

Ochiai, B., Sato, S., & Endo, T. (2007). Crosslinkable polyurethane bearing a methacrylate structure in the side chain. *J. Polym. Sci. Part A: Polym. Chem.*, *45*, 3400-3407. http://dx.doi.org/10.1002/pola.22092

Ochiai, B., Sato, S., & Endo, T. (2007). Synthesis and properties of polyurethanes bearing urethane moieties in the side chain. *J. Polym. Sci. Part A: Polym. Chem.*, *45*, 3408-3414. http://dx.doi.org/10.1002/pola.22093

Ochiai, B., Satoh, Y., & Endo, T. (2005). Nucleophilic polyaddition in water based in chemo-selective reaction of cyclic carbonate with amine. *Green Chem.*, *7*, 765-767. http://dx.doi.org/10.1039/B511019J

Ochiai, B., Inoue, S., & Endo, T. (2005). One-pot non-isocyanate synthesis of polyurethanes from bisepoxide, carbon dioxide, and diamine. *J. Polym. Sci. Part A: Polym. Chem.*, *43*, 6613-6618. http://dx.doi.org/10.1002/pola.21103

Ochiai, B., Nakayama, J., Mashiko, M., Kaneko, Y., Nagasawa, T., & Endo, T. (2005). Synthesis and crosslinking reaction of poly(hydroxyurethane) bearing a secondary amine structure in the main chain. *J. Polym. Sci. Part A: Polym. Chem.*, *43*, 5899-5905. http://dx.doi.org/10.1002/pola.21078

Okuyama, K., Sugiyama, J., Nagahata, R., Asai, M., Ueda, M., & Takeuchi, K. (2003). An environmentally benign process for aromatic polycarbonate synthesis by efficient oxidative carbonylation catalyzed by Pd-carbene complexes. *Green Chem.*, *5*, 563-566. http://dx.doi.org/10.1039/B304878K

Sheng, X., Ren, G., Qin, Y., Chen, X., Wang, X., & Wang, F. (2015). Quantitative synthesis of bis(cyclic carbonate)s by iron catalyst for non-isocyanate polyurethane synthesis. *Green Chem.*, *17*, 373-379. http://dx.doi.org/10.1039/C4GC01294A

Tomita, H., Sanda, F., & Endo, T. (2001). Polyaddition behavior of bis(five- and six-membered cyclic carbonate)s with diamine. *J. Polym. Sci. Part A: Polym. Chem.*, *39*, 860-867. http://dx.doi.org/10.1002/1099-0518(20010315)39:6<860::AID-POLA1059>3.0.CO;2-2

Tomita, H., Sanda, F., Endo, T. (2001). Reactivity comparison of five- and six-membered cyclic carbonates with amines: Basic evaluation for synthesis of poly(hydroxyurethane). *J. Polym. Sci. Part A: Polym. Chem.*, *39*, 162-168. http://dx.doi.org/10.1002/1099-0518(20010101)39:1<162::AID-POLA180>3.0.CO;2-O

Appendix

a)

b)

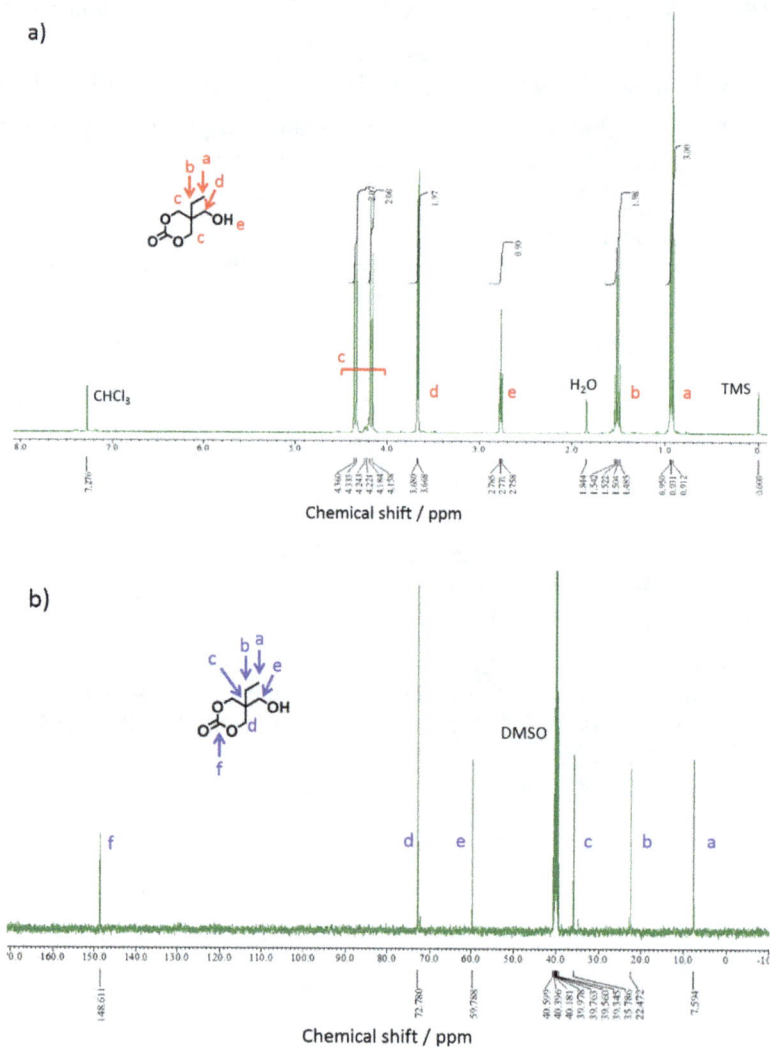

Figure S1. NMR spectra of TMPC-OH after purification by column chromatography. a) ^{1}H NMR spectrum in CDCl$_3$ with 0.03 v/v% of TMS. b) ^{13}C NMR spectrum in DMSO-d$_6$

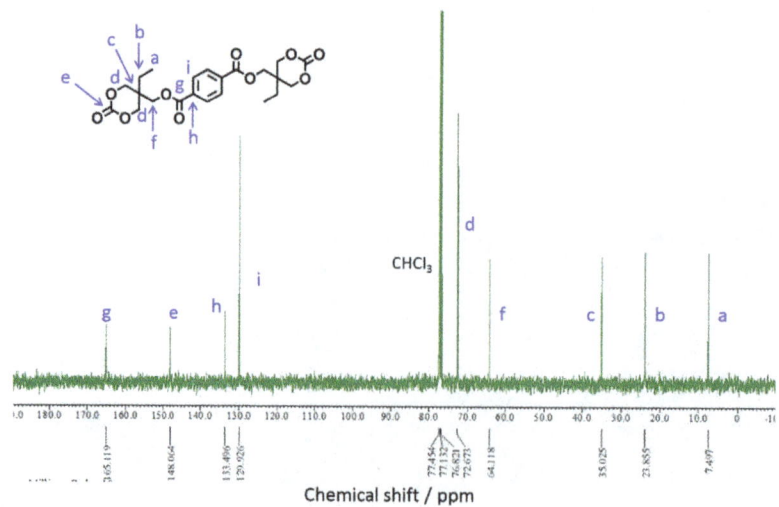

Figure S2. ^{13}C NMR spectrum of Ph-TMPC$_2$ in CDCl$_3$ containing 0.03 v/v% of TMS.

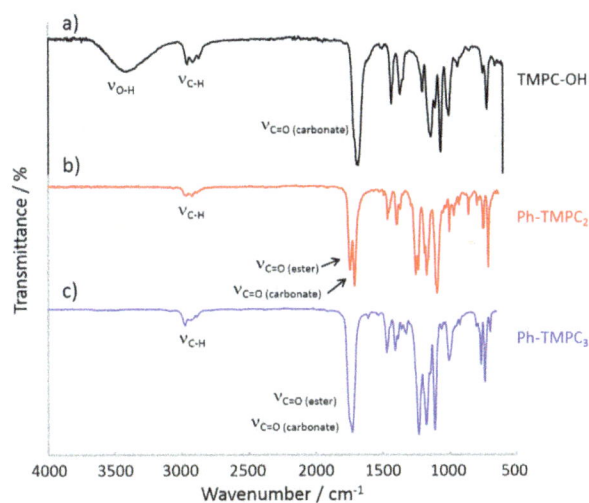

Figure S3. FT-IR spectra of a) TMPC-OH, b) Ph-TMPC$_2$ and c) Ph-TMPC$_3$.

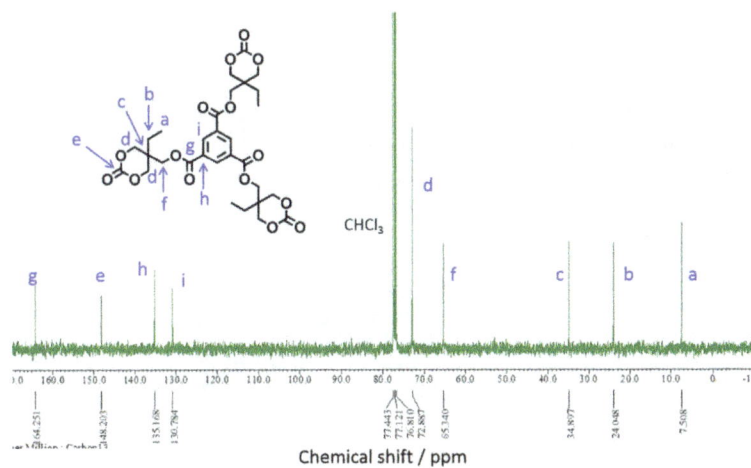

Figure S4. ^{13}C NMR spectrum of Ph-TMPC$_3$ in CDCl$_3$ containing 0.03 v/v% of TMS.

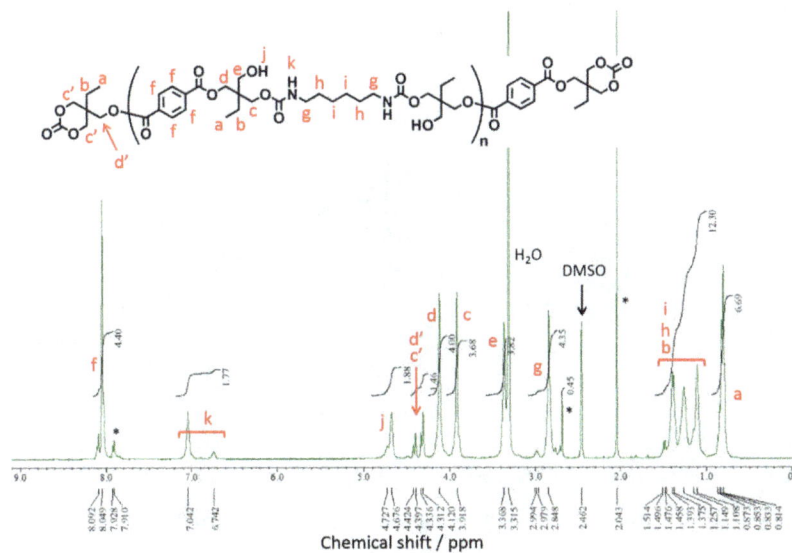

Figure S5. ^{1}H NMR spectrum of p(Ph-TMPC$_2$-C$_6$) in DMSO-d$_6$. Asterisks mean residual solvents (DMF and acetone)

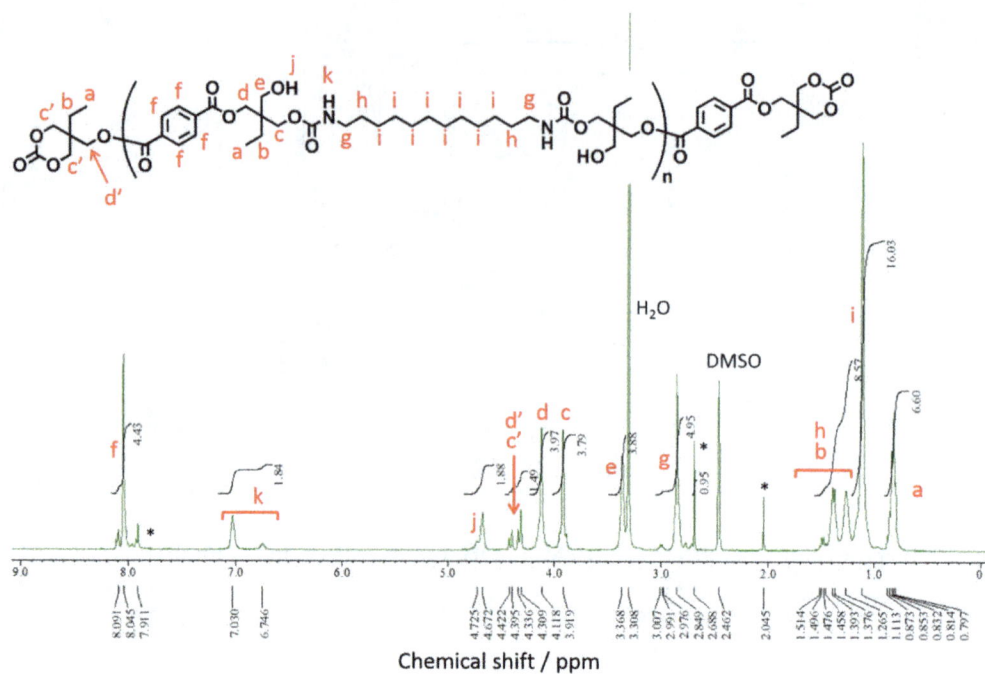

Figure S6. ^{1}H NMR spectrum of p(Ph-TMPC$_2$-C$_{12}$) in DMSO-d$_6$. Asterisks mean residual solvents (DMF and acetone).

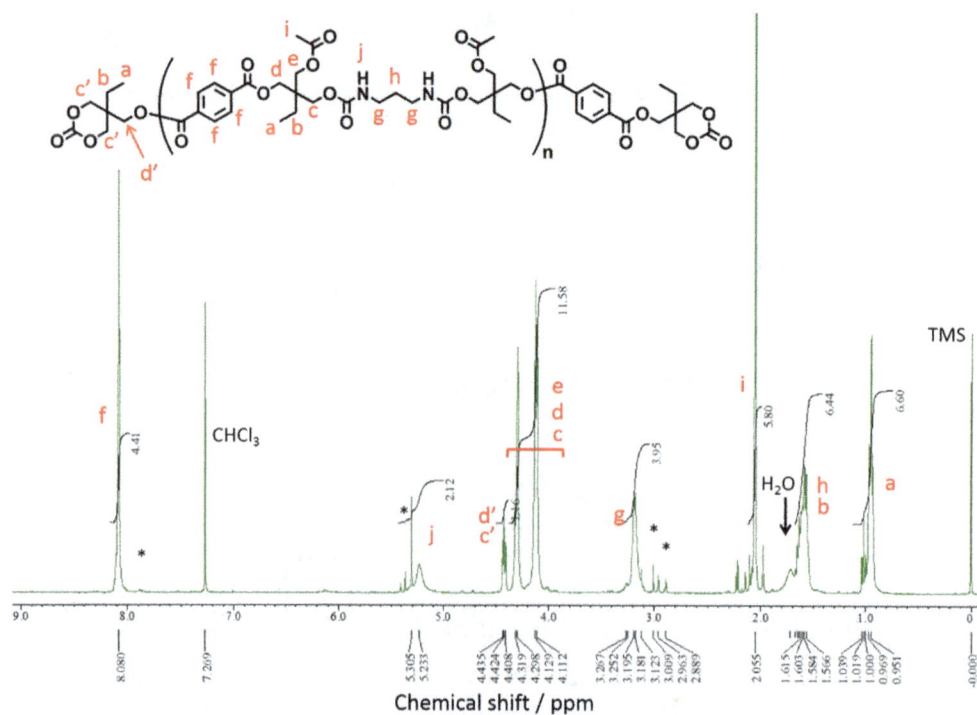

Figure S7. ^{1}H NMR spectrum of p(Ph-TMPC$_2$-C$_3$-OAc) in CDCl$_3$ containing 0.03 v/v% of TMS. Asterisks denote residual solvents (CH$_2$Cl$_2$ and DMF).

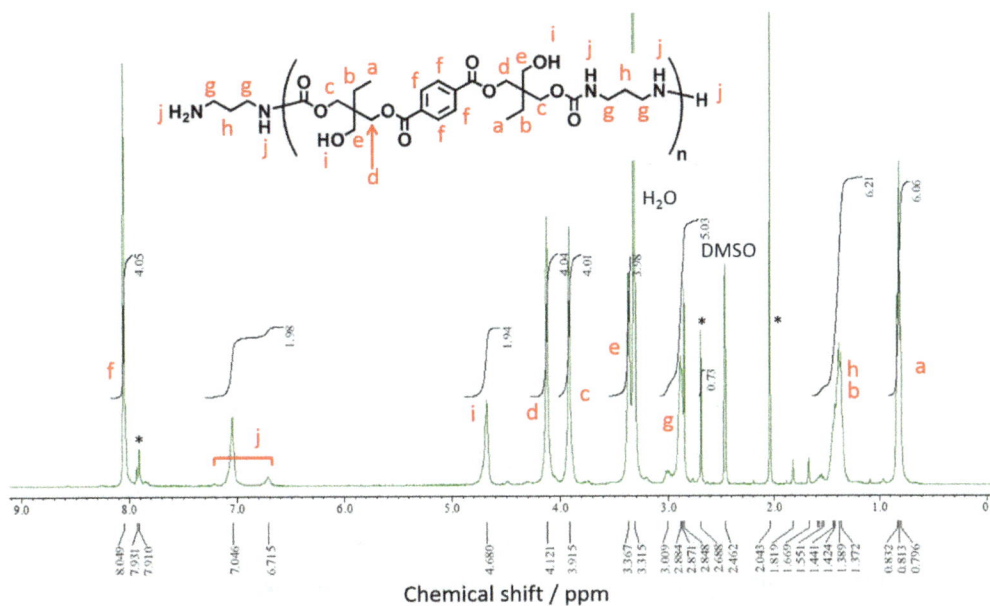

Figure S8. ^{1}H NMR spectrum of p(Ph-TMPC$_2$-C$_3$)' in DMSO-d$_6$. Asterisks denote residual solvents (DMF and acetone)

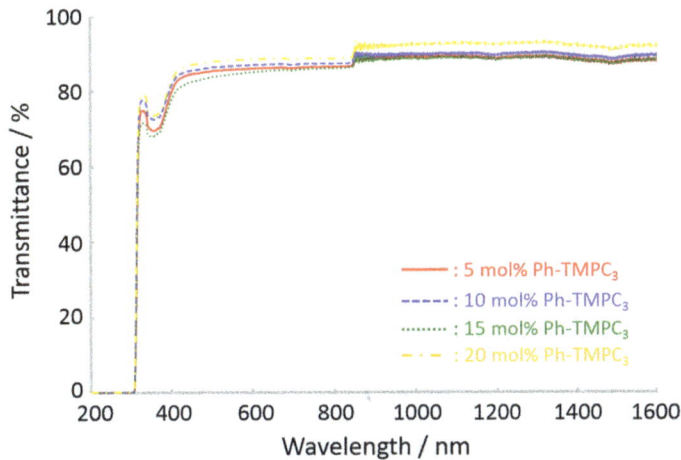

Figure S9. UV/Vis.-NIR absorption spectra of Film A prepared at different Ph-TMPC$_2$/Ph-TMPC$_3$/C$_3$ feed ratios. The mole ratios of Ph-TMPC$_3$ are 5 (red solid line), 10 (blue dashed line), 15 (green dotted line) or 20 mol% (orange dashed and dotted line) with respect to total carbonate monomers (Ph-TMPC$_2$ + Ph-TMPC$_3$)

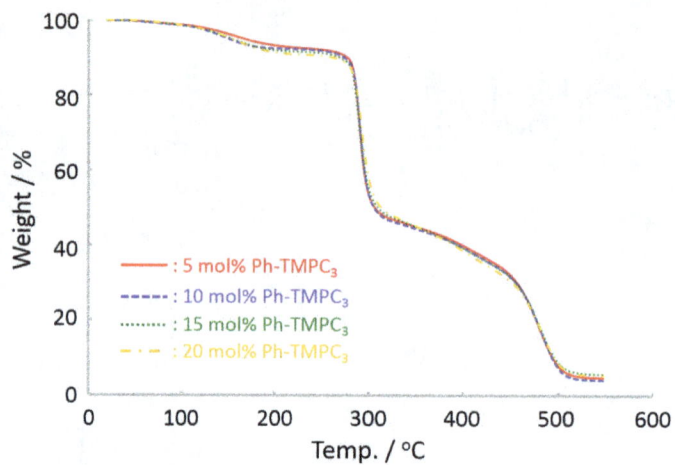

Figure S10. TGA profiles of Film A prepared at different Ph-TMPC$_2$/Ph-TMPC$_3$/C$_3$ feed ratios. The mole ratios of Ph-TMPC$_3$ are 5 (red solid line), 10 (blue dashed line), 15 (green dotted line) or 20 mol% (orange dashed and dotted line) with respect to total carbonate monomers (Ph-TMPC$_2$ + Ph-TMPC$_3$).

2

Ni-Ba$_{0.5}$Sr$_{0.5}$Ce$_{0.6}$Zr$_{0.2}$Gd$_{0.1}$Y$_{0.1}$O$_{3-\delta}$ Anode Composites for Proton Conducting Solid Oxide Fuel Cells (H-SOFCs)

Kalpana Singh[1], Ashok Kumar Baral[1] & Venkataraman Thangadurai[1]

[1] Department of Chemistry, University of Calgary, 2500 University Drive NW, Calgary, Alberta, T2N 1N4, Canada

Correspondence: Venkataraman Thangadurai, Department of Chemistry, University of Calgary, 2500 University Drive NW, Calgary, Alberta, T2N 1N4, Canada. E-mail: vthangad@ucalgary.ca

Abstract

In this paper, we report the electrochemical properties of Ni-Ba$_{0.5}$Sr$_{0.5}$Ce$_{0.6}$Zr$_{0.2}$Gd$_{0.1}$Y$_{0.1}$O$_{3-\delta}$ (BSCZGY) anode composites in 3% H$_2$O-H$_2$ for proton conducting solid oxide fuel cells (H-SOFCs). Ni-BSCZGY composites with volume ratio of 30:70, 40:60, and 50:50 were synthesised through mechanical mixing and combustion methods. In combustion method, auto-ignition step led to brown coloured ash, which was calcined at 1000 °C for 5 h to form NiO-BSCZGY powder. Screen-printing, co-firing and reduction process were used to prepare the symmetrical cell: Ni-BSCZGY/BSCZGY/Ni-BSCZGY. Ni50-BSCZGY anode exhibited the lowest polarisation resistance (Rp) of 0.8 Ω.cm^2 and 1.9 Ω.cm^2 at 710 °C under 3% H$_2$O-H$_2$, for both mechanically mixed and combustion methods, respectively.

Keywords: composite anodes, combustion synthesis, proton conducting solid oxide fuel cells, symmetrical cell, polarisation resistance, hydrogen oxidation reaction (HOR).

1. Introduction

Acceptor-doped ACeO$_3$ (A = Sr, Ba) perovskites show appreciable proton (H$^+$) conductivity under humid atmospheres in the temperature range of 500-700 °C (Iwahara, Esaka, Uchida, & Maeda, 1981; Iwahara, Uchida, Ono, & Ogaki, 1988). Ceramic proton conductors are being considered as alternative electrolyte materials to replace conventional oxide ion conducting yttrium-stabilised zirconia (YSZ) electrolyte in solid oxide fuel cells (SOFCs), which exhibits high conductivity only at very high temperatures (800-1000 °C) (Fergus, 2006). However, poor chemical stability of acceptor-doped BaCeO$_3$ under water and CO$_2$ containing atmospheres mars their widespread use as an electrolyte in SOFCs (Scholten & Schoonman, 1993; Bhide, & Virkar, 1999; Wu & Liu, 1997; Sneha & Thangadurai, 2007; Trobec & Thangadurai, 2008; Gill, Kannan, Maffei, & Thangadurai, 2013; Kan, Lussier, Bieringer, & Thangadurai, 2014). Recently, Ba$_{0.5}$Sr$_{0.5}$Ce$_{0.6}$Zr$_{0.2}$Gd$_{0.1}$Y$_{0.1}$O$_{3-\delta}$ (BSCZGY) was reported to exhibit good chemical stability under water vapor and CO$_2$ containing atmospheres (Kannan, Singh, Gill, Fürstenhaupt, & Thangadurai, 2013). At 700 °C for H$_2$-air cell, BSCZGY electrolyte with an un-optimised Pt electrodes exhibited an OCV and power density of 1.15 V and 18 mWcm^{-2}, respectively (Kannan et al., 2013). Thus, appropriate anode and cathode materials are needed for its success as an electrolyte membrane for proton conducting SOFCs (H-SOFCs). Thus, aim of this work is to study the phase and microstructure of Ni-BSCZGY anodes, and to investigate the electrochemical processes occurring at the Ni-BSCZGY anodes under wet hydrogen through electrochemical ac impedance spectroscopy (EIS) (Singh, 2016).

By analogy with Ni-YSZ composite anode, most of the studies in H-SOFCs are also reported on Ni-proton conducting ceramic phase composites such as Ni-SrCe$_{0.9}$Yb$_{0.1}$O$_{3-\delta}$ (SCYb), Ni-BaCe$_{0.9}$Y$_{0.1}$O$_{3-\delta}$ (BCY10), and Ni-BaZr$_{0.8}$Y$_{0.2}$O$_{3-\delta}$ (BZY20) (Mather al., 2003; Essoumhi, Taillades, Taillades-Jacquin, Jones, & Rozière, 2008; Bi, Fabbri, Sun, & Traversa, 2011). Symmetrical cells studies on Ni33-SCYb (33 vol. % Ni), Ni35-BCY10 (35 vol. %) and Ni50-BZY20 (50 wt. % Ni) showed area specific polarization resistance (ASR) of about 10, 0.32 and 0.37 Ω.cm^2 at 600 °C under humid fuels (Mather et al., 2003; Essoumhi et al., 2008; Bi et al., 2011).

Mechanical mixing of the ceramic electrolyte with NiO powder and one-step nitrate-based combustion methods are being employed for preparing composite anodes for H-SOFCs (Mather et al., 2003; Essoumhi et al., 2008; Bi et al., 2011, Song et al., Moon, Lee, Dorris, & Balachandran, 2008; Mather, Figueiredo, Jurado, & Frade, 2003). Nitrate-based combustion method produce ultra-fine anode composite powders and possess a uniform distribution of NiO in the electrolyte matrix. This results in enhancement of the electrochemical performance towards fuel oxidation, due to the increase in triple-phase boundary length for electrochemical reaction (Bi et al., 2011; Narendar, Mather, Dias, & Fagg, 2013). On the other hand, mechanical mixing method results in nonhomogeneous distribution of NiO and electrolyte phases, and thus, resulted in lower electrochemical performance (Bi et al., 2011). Here, solid state and combustion methods have been employed to prepare anode composites in Ni to BSCZGY volume ratio of 30:70, 40:60, and 50:50. Anode composites before reduction are labelled here as NiO-BSCZGY and after reduction as Ni-BSCZGY.

2. Materials Preparation and Experimental Procedures

NiO-BSCZGY anode composites with volume ratio of 30:70, 40:60 and 50:50 were prepared through mechanical mixing and combustion methods. The compositions of oxides were chosen, so that after reduction of NiO to Ni metal, the volume fraction of Ni would be about 30, 40 and 50% in Ni-BSCZGY. For the mechanical mixing, BSCZGY powders were first synthesized through a solid-state reaction (Kannan et al., 2013; Singh, 2016). Then, desired amounts of BSCZGY powders were mixed with commercial NiO (Alfa Aeser, 99.9%), and milled for 6 h with isopropanol, dried and used for making symmetrical cells (Singh, 2016).

In the combustion method, NiO-BSCZGY composite anodes were prepared in one step through auto-ignition. Stoichiometric amounts of high purity $Ba(NO_3)_2.6H_2O$ (>99%), $Sr(NO_3)_2.6H_2O$ (99.0%), $Ce(NO_3)_3.6H_2O$ (99.5%), $ZrO(NO_3)_2.xH_2O$ (99.9%), $Y(NO_3)_3.6H_2O$ (99.9%), $Gd(NO_3)_3.6H_2O$ (99.9%), and $Ni(NO_3)_2.6H_2O$ (98.0%) from Alfa Aeser were dissolved in distilled water. Citric acid monohydrate (99.5%, Alfa Aeser) as a complexing agent was added to the mixed solution in the molar ratio of 1:1.5 (metal ion: citric acid). The temperature of the solution was maintained at 90 °C throughout the stirring process until most of the water content evaporated and the gel was formed. Finally, the gel burnt into a brown colored ash through the auto-ignition process. The brown colored ash was calcined at 1000 °C for 5 h to get fine NiO-BSCZGY composite powders. Powder X-ray diffraction (PXRD) (Bruker D8 Powder X-ray diffractometer, CuKα) was used to examine the chemical reactivity of NiO and BSCZGY powders (Singh, 2016).

BSCZGY discs for symmetrical cell measurements were prepared by pressing the as-prepared BSCZGY powders into pellets and calcining them at 1400 °C for 24 h. NiO-BSCZGY pastes were prepared by mechanically mixing the composite powders with appropriate amount of α-Terpineol and ethyl cellulose as solvent and binder, respectively. Symmetrical cell (NiO-BSCZGY/BSCZGY/NiO-BSCZGY) assemblies were fabricated by screen printing technique. The BSCZGY pellet was first screen printed with NiO-BSCZGY composite pastes on both sides. Then, it was fired at 1200 °C for 3h in air at the heating and cooling rate of 2 °C/min. The sintered symmetrical cells were painted with Au paste (LP A88-11S, Heraeus Inc., Germany), as the current collector during the symmetrical cell measurements. The symmetrical cells were reduced at 900 °C for 3 h in 3% H_2O/H_2 at the flow rate of 50 SCCM.

The electrochemical performance of the anode symmetrical cells was studied through electrochemical impedance spectroscopy (0.1 Hz to 1MHz) by using VersaSTAT 3 instrument. OTF1200X furnace was used to hold the cell at desired temperature. The morphology of anode composite before and after the reduction was analysed through scanning electron microscopy (Philips XL 30).

3. Results and Discussion

3.1 Phase and Microstructure

Figure 1 shows the PXRD patterns of mechanically mixed composite powders in 30:70, 40:60, and 50:50 ratio (volume ratio after the reduction), and the mixtures heat treated at 1200 °C for 3 h. The effect of high temperature firing on the phase and lattice parameters of BSCZGY in NiO-BSCZGY composites was investigated by PXRD. PXRD patterns of both the as-prepared NiO-BSCZGY composites and their respective heat-treated composite samples prepared by mechanical mixing show only the diffraction peaks corresponding to BSCZGY and NiO phases (Figure 1 (a-c)). It has been demonstrated that small amount of Ni might substitute the Zr due to the similar ionic radius ($r_{Zr(IV)}$ = 0.72 Å, $r_{Ni(II)}$ = 0.69 Å) during mixing through milling and high temperature firing process (Bi et al., 2011; Shannon, 1976; Babilo & Haile, 2005; Tong, Clark, Hoban, & O'Hayre, 2010). However, PXRD patterns did not show a shift in 2θ peak positions, and the lattice constant for BSCZGY phase almost remained constant ~ 4.298 Å for all the three anode compositions (Table 1). This indicates that the stoichiometry of BSCZGY seems to be retained after mixing with NiO and firing at 1200 °C in air. Even if Ni substitution does happen into the BSCZY lattice, it will be beneficial for promoting anode reactions (hydrogen oxidation reaction, HOR), as it may introduce some electronic conductivity into the BSCZGY phase (Babilo & Haile, 2005).

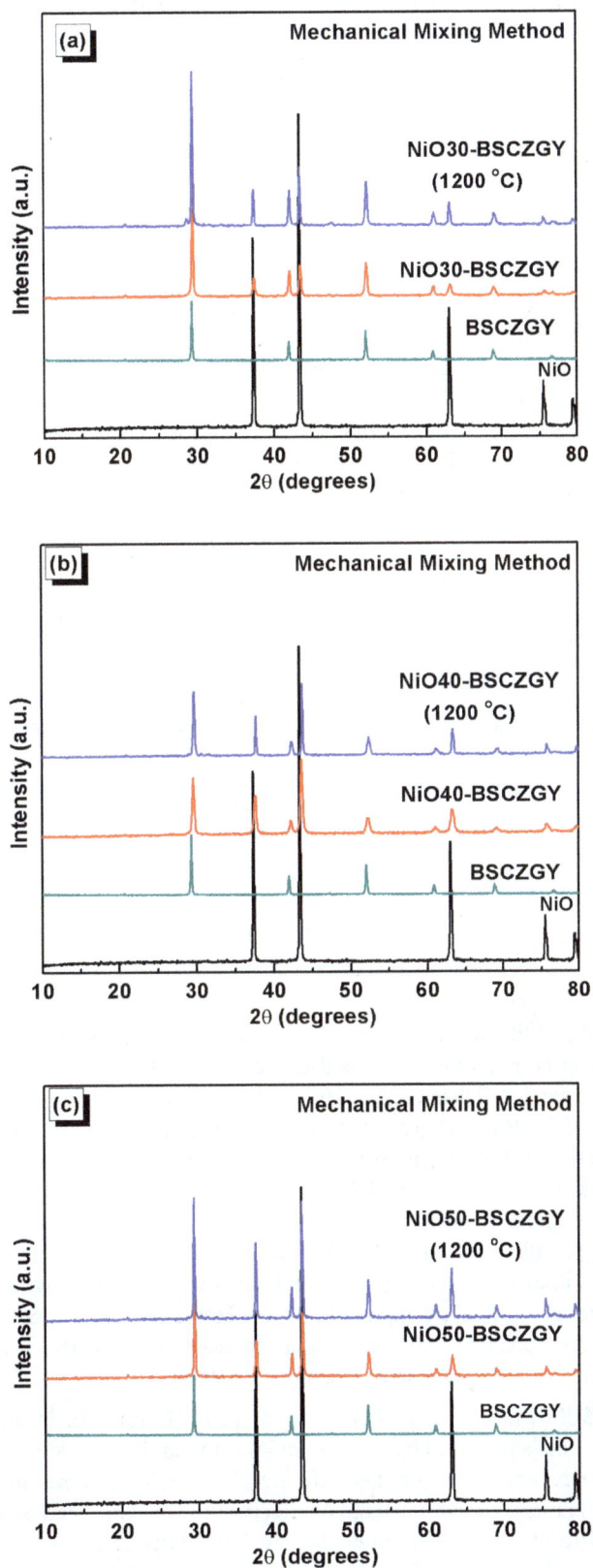

Figure 1. Powder X-ray diffraction (PXRD) patterns of NiO, BSCZGY, as-prepared mechanically mixed composite powders, and the mixtures sintered at 1200 °C for 3 h. (a) NiO30-BSCZGY, (b) NiO40-BSCZGY, and (c) NiO50-BSCZGY

Table 1. Variation of experimental lattice constants (Å) for BSCZGY in NiO-BSCZGY anodes prepared through mechanical mixing method.

Composition	Lattice Constant (Å) as-prepared	Lattice Constant (Å) after firing at 1200 °C
NiO30-BSCZGY	4.296 (2)	4.301 (8)
NiO40–BSCZGY	4.293 (2)	4.295 (2)
NiO50–BSCZGY	4.295 (7)	4.302 (8)
BSCZGY (as-prepared)	4.298 (9)	--

Figure 2 shows the PXRD patterns of NiO-BSCZGY (30:70, 40:60, and 50:50) composites prepared by combustion method and calcined at 1000 °C for 5 h, 1100 °C for 5 h, and 1200 °C for 3h. The PXRD patterns of NiO-BSCZGY powders prepared by combustion method show impurity phases of BaY_2NiO_5 and SrY_2O_4 at all sintering temperatures, along with the NiO and perovskite BSCZGY phases (Figure 2 (a-c)). In combustion method, the lattice constants for the BSCZGY phase in all compositions are found to be higher than the lattice parameter of stoichiometric as-prepared BSCZGY (4.298 Å) (Table 2). Presence of BaY_2NiO_5 impurity phase indicates the potential loss of Ba and Y, and hence the deviation from nominal stoichiometry. One should expect a reduction in lattice parameter of BSCZGY phase in anode composite due to the loss of large ionic sized Ba^{2+} (1.35 Å) and Y^{3+} (0.9 Å) to form BaY_2NiO_5 (Shannon, 1976). Instead, increase in lattice parameter of BSCZGY phase was observed for combustion-synthesised composites. The higher lattice constant could be attributed to the formation of SrY_2O_4, in addition to BaY_2NiO_5, and former compound has smaller Sr^{2+} ions (1.18 Å) compared to Ba^{2+} (1.35 Å) (Shannon, 1976).

Table 2. Variation of experimental lattice constants (Å) for BSCZGY in NiO-BSCZGY anodes prepared through combustion method.

Composition	Lattice Constant (Å) after firing at 1000 °C	Lattice Constant (Å) after firing at 1200 °C
NiO30–BSCZGY	4.333 (2)	4.320 (7)
NiO40–BSCZGY	4.340 (2)	4.309 (2)
NiO50–BSCZGY	4.330 (2)	4.328 (6)
BSCZGY (as-prepared)	4.298 (9)	--

Literature studies have shown that the deviation from stoichiometry, presence of impurities, and Ba^{2+} and Y^{3+} loss results in a decrease of proton conductivity in doped-$BaCeO_3$ and $BaZrO_3$ (Narendar et al., 2013; Magrez & Schober, 2004; Yamazaki, Hernandez-Schanchez, & Haile, 2010). Also, the presence of BaY_2NiO_5 impurity phase was found during synthesis of $Ni-BaZr_{1-x}Y_xO_{3-\delta}$ composite anodes prepared through nitrate-based combustion method, and when Ni was used as a sintering aid for doped-$BaZrO_3$ (Bi et al., 2011; Babilo & Haile, 2005; Magrez & Schober, 2004; Tong, Clark, Bernau, Sanders, & O'Hayre, 2010; Bi, Fabbri, Sun, & Traversa, 2011). Literature studies have suggested that the impurity phase BaY_2NiO_5 form at temperatures above 900 °C, and remains even at high temperatures of 1500 °C (Tong et al., 2010). Some researchers have shown that BaY_2NiO_5 in the $BaZr_{1-x}Y_xO_{3-\delta}$ phases led to improved densification and enhanced grain-growth along with enhanced total electrical conductivity (Mather et al., 2003; Tong et al., 2010). However, subsequent work by Tong et al. showed that BaY_2NiO_5 impurity hampers proton conductivity and increases the activation energy for proton conduction (Narendar et al., 2013; Tong et al., 2010; Ricote & Bonanos, 2010).

Figure 3 shows the typical SEM micrographs after firing at 1200 °C for 3 h, of anode/electrolyte layer of the NiO40-BSCZGY symmetrical cells prepared by mechanical mixing and combustion methods. The thicknesses of the sintered anode layers were about 48 µm and 26 µm for mechanical mixing and combustion method, respectively. Also, proper attachment of anode layers on electrolyte was observed by the absence of delamination and cracks between composite electrode/electrolyte layers. The purpose of preparing anodes by combustion method was to obtain a uniform distribution of NiO in the BSCZGY. From Figure 3 (b), it can be seen that a better microstructure has been obtained by combustion method.

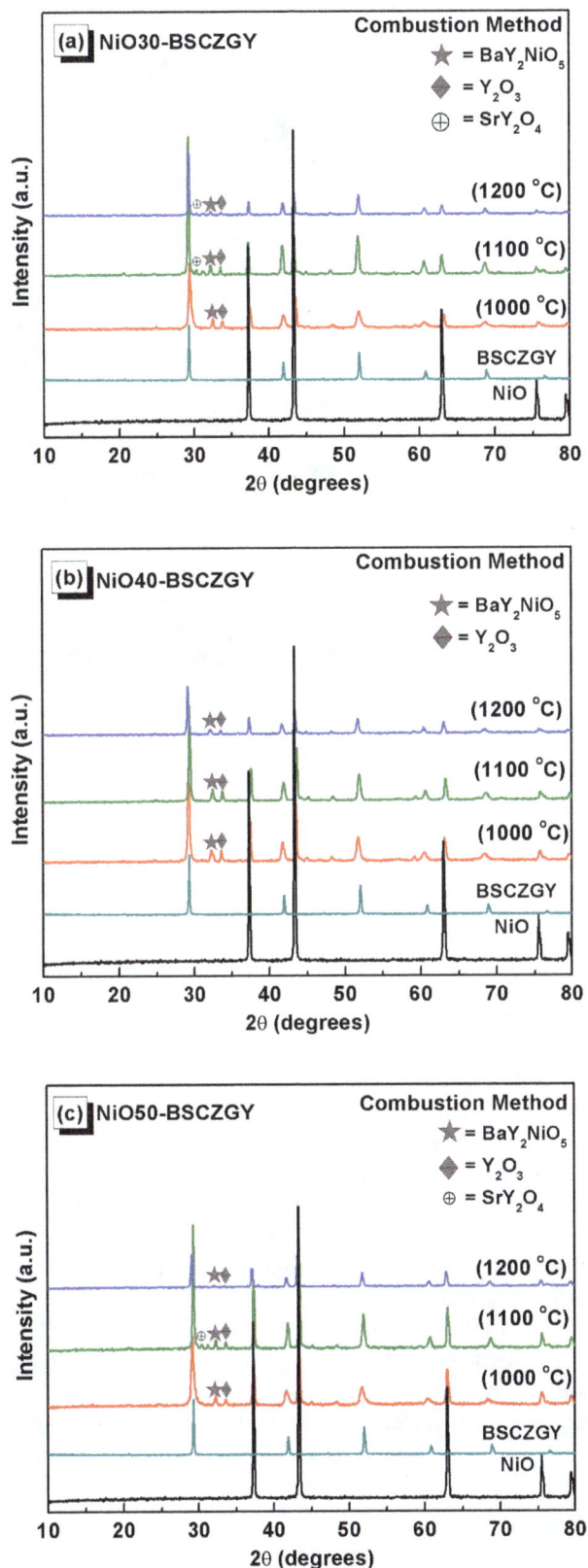

Figure 2. Powder X-ray diffraction (PXRD) patterns of NiO, BSCZGY, (a) NiO30-BSCZGY, (b) NiO40-BSCZGY, and (c) NiO50-BSCZGY prepared by combustion method and calcined at 1000 ºC (5 h), 1100 ºC (5 h) and 1200 ºC (3h)

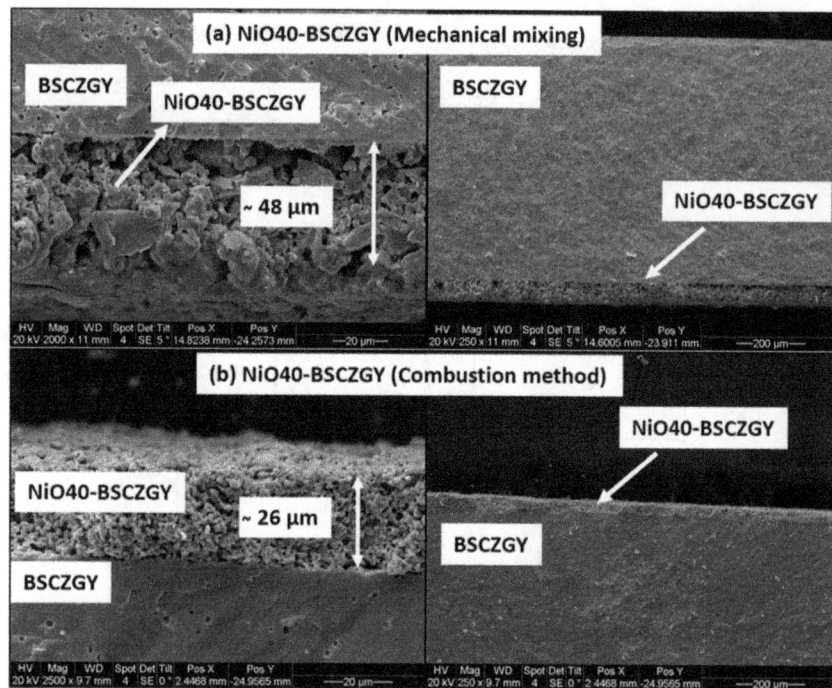

Figure 3. Typical SEM images of cross-section view of sintered NiO40-BSCZGY/BSCZGY anode interface prepared by (a) mechanical mixing and (b) combustion method before reduction. Different size meshes were used for screen-printing the mechanically mixed and combustion method synthesised anode pastes

3.2 Electrochemical Analysis

Figures 4 and 5 show the typical Nyquist plots of the Ni50-BSCZGY symmetrical cells measured in $H_2/3\%$ H_2O as a function of temperature for electrodes formed by the combustion and mechanical mixing methods, respectively. Electrochemical measurements were made in the direction of decreasing temperature after reduction of anode composites at 900 °C in H_2. At 500 °C, in the case of the symmetrical cells of Ni50-BSCZGY anode formed by the combustion method, three semicircles corresponding to the bulk contribution of BSCZGY electrolyte at high frequency, followed by two distinct electrode responses at low frequencies are visible (Figure 4 (a)). On increasing the temperature to 650 °C, the bulk response disappears from impedance spectra, and only two low-frequency electrode responses are visible (Figure 4 (b)). The increase in temperature resulted in a reduction of the diameter of both semicircles gradually. The total resistance of the cell can be extracted from the low frequency (LF) (0.1 Hz) intercept. Contributions from two arcs were separated by fitting the equivalent circuit, as outlined in the inset of Figure 4 (a-c). In the equivalent circuits at 710 °C, Rs is due to ohmic resistance, and R2 and R3 correspond to mid frequency (MF) and LF arcs, respectively due to the anode processes. The capacitance values in the range of 10^{-6} – 10^{-5} F for MF arc associated with the resistance R2, suggests charge transfer between anode and electrolyte interface. LF arc with the capacitance in the order of 10^{-2} –10^{-1} F related to the resistance R3 suggests surface electrochemical reactions at the anode (Bi et al., 2011; Narendar et al., 2013). The Nyquist plots of Ni30-BSCZGY and Ni40-BSCZGY anode composites showed similar Nyquist plots as Ni50- BSCZGY composite under the same conditions. The capacitance values obtained for the electrode process are in accord with the previous studies on Ni-BaZr$_{0.8}$Y$_{0.2}$O$_{3-\delta}$ and Ni–BaZr$_{0.85}$Y$_{0.15}$O$_{3-\delta}$ electrodes (Bi et al., 2011; Narendar et al., 2013). The ASR of LF arc could be due to dissociative adsorption of H_2 gas at anode with the following reaction steps (Bi et al., 2011; Narendar et al., 2013).

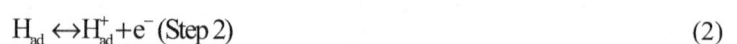

$$H_2(g) \rightarrow 2H_{ad} \text{ (Step 1)} \tag{1}$$

$$H_{ad} \leftrightarrow H_{ad}^+ + e^- \text{ (Step 2)} \tag{2}$$

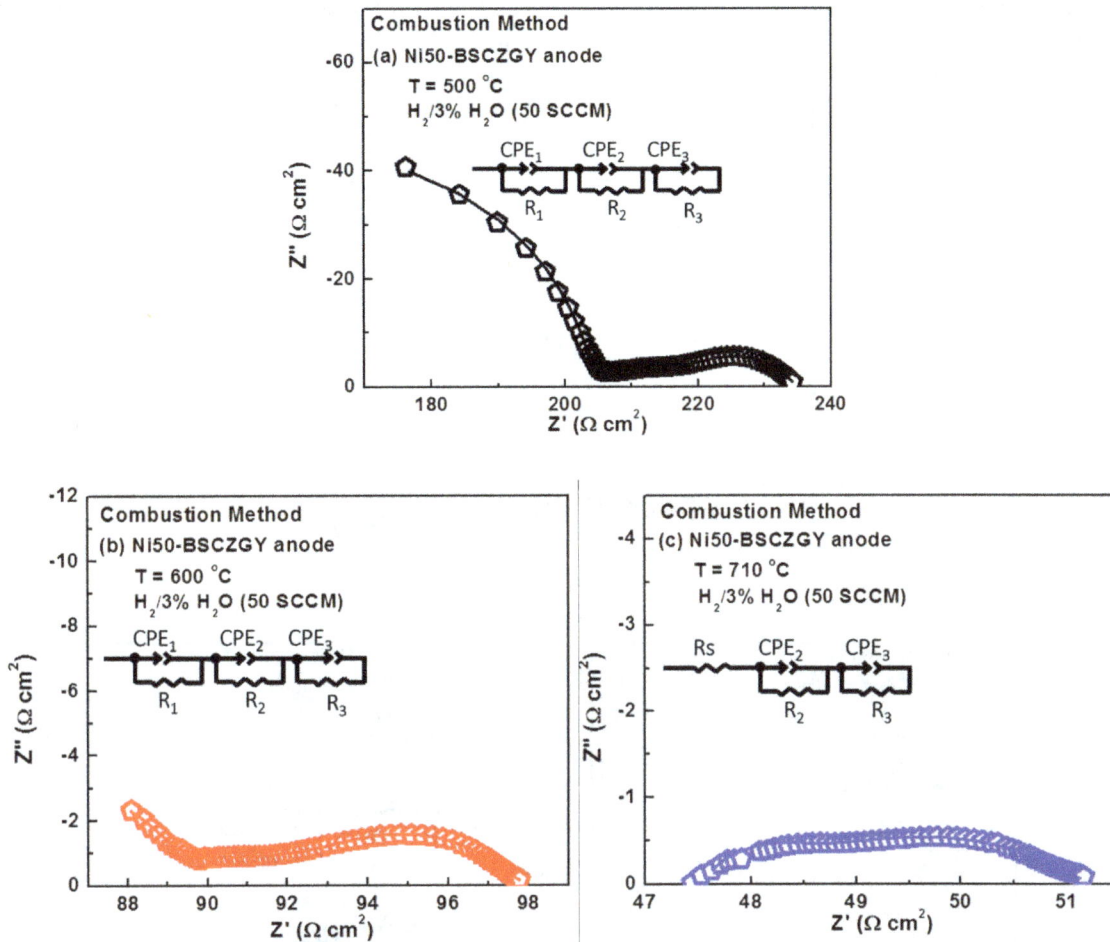

Figure 4. Typical Nyquist plots for the Ni50-BSCZGY/BSCZGY/Ni50-BSCZGY symmetrical cells (with anode prepared by the combustion method) in H₂/3% H₂O. (a) 500 °C, (b) 600 °C, and (c) 710 °C. Insets show the equivalent circuits used for fitting impedance data, and open symbols represent experimental data and solid lines represents fitted data. Porous Au was used as current collector during symmetrical cell measurements

For the symmetrical cells of Ni50-BSCZGY electrode prepared by the mechanical mixing method, only two responses due to electrode contributions are visible (Figure 5 (a-c)). Moreover, they also show the capacitance values in the range of $10^{-6}-10^{-5}$ F, and $10^{-2}-10^{-1}$ F for MF and LF arcs from 550 to 710 °C. Figure 6 (a-b) displays the variation of area specific resistance (ASR) as a function of temperature for Ni-BSCZGY anodes with different Ni contents. For anode composite prepared by the combustion method, ASR values of 1.9, 3.75, and 5.15 $\Omega.cm^2$ were obtained for Ni50-BSCZGY, Ni40-BSCZGY, and Ni30-BSCZGY, respectively at 710 °C (Figure 6 (a)). The ASR values for anode composites prepared by the mechanical mixing method are found to be lower than the combustion method at all temperatures. At 710 °C the ASR values are 0.8, 2.6 and 2.1 $\Omega.cm^2$ for Ni50-BSCZGY, Ni40-BSCZGY, and Ni30-BSCZGY anodes, respectively (Figure 6 (b)). From the Figure 6 (a-b), it can be seen that among the samples investigated, the lowest ASR value for HOR, and hence, the best electrochemical performance is shown by Ni50-BSCZGY anode for both synthetic methods at all temperature ranges. Also, the electrochemical performances of the NiO-BSCZGY anodes prepared by the mechanical mixing method are found to be better than those of the anodes prepared by the combustion method.

The presence of impurity phases and thinner anode layer might have contributed towards the lower electrochemical performance of investigated composites prepared by combustion method. Firstly, BaY₂NiO₅ impurity phase along with Ba and Y loss deviated the BSCZGY stoichiometry, which may have resulted in higher R2 resistance. The presence of impurity phase might have also increased the resistance of anode composite for proton conduction, especially in bulk and grain-boundary regions of BSCZGY phase, leading to a higher R2 and HOR ASR. Secondly, the thickness of the anode layer for Ni-BSCZGY anode composite prepared by the combustion method is smaller (~26 μm) than that of the anode layer prepared by the mechanical mixing method

(~48 μm). In some literature work, enhancement in electrochemical performance of porous composite anodes for HOR has been achieved by increasing the anode layer thickness (Barbucci et al., 2008; Tanner et al., Fung & Virkar, 1997). By increasing the thickness of the anode layer, better electrochemical performance may be achieved. Additionally, the formation of BaY_2NiO_5 impurity can be avoided by employing a nitrate-free combustion method as suggested by Narendar et al. (2013).

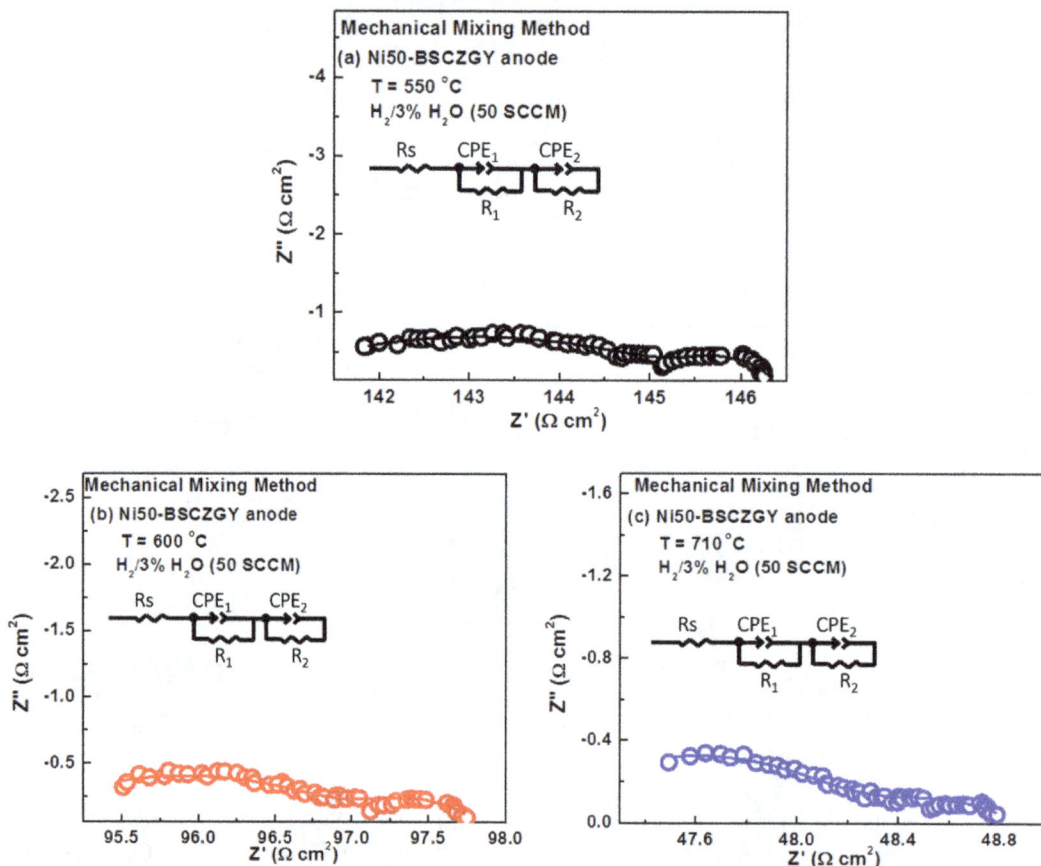

Figure 5. Typical Nyquist plots for the Ni50-BSCZGY/BSCZGY/Ni50-BSCZGY symmetrical cells (with anode prepared by the mechanical mixing method) in H_2/3% H_2O and equivalent circuits at (a) 500 °C, (b) 600 °C, and (c) 710 °C. Insets show the equivalent circuits used for fitting impedance data, and open symbols represent experimental data and solid lines represents fitted data

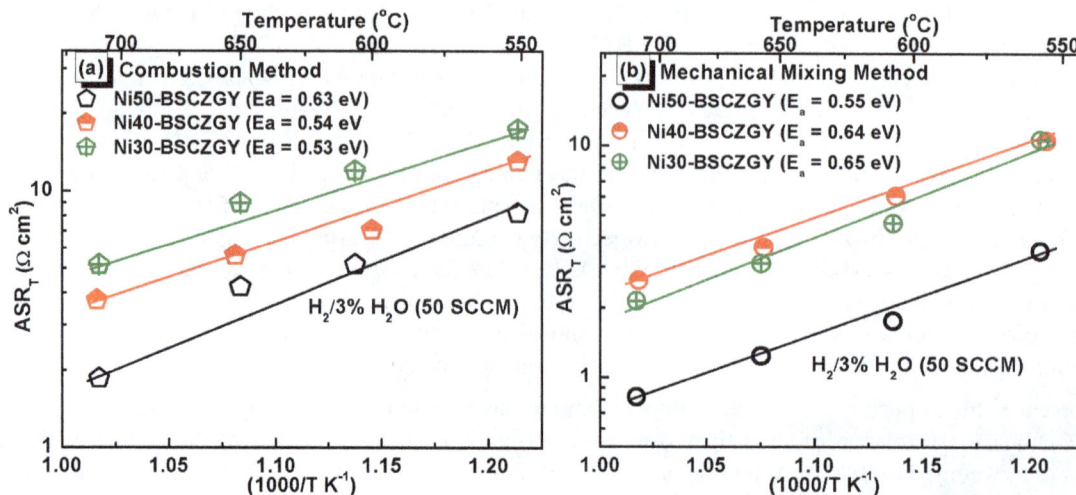

Figure 6. Total ASR values as a function of temperature for the Ni-BSCZGY/BSCZGY/Ni-BSCZGY symmetrical cells formed by (a) combustion method and (b) mechanical mixing method with different Ni contents

4. Conclusion

Two synthetic routes for the synthesis of Ni-BSCZGY anode composites with different Ni to BSCZGY ratio were employed, and their electrochemical performance for HOR ASR was compared. Conventional mechanical mixing and nitrate based combustion synthesis methods were used for preparing the anode powders. PXRD analysis of as-prepared and heat treated anode powders revealed that the Ni-BSCZGY anodes prepared through the mechanical mixing route were free from impurity phases under the studied conditions. Conversely, anode powders prepared by the combustion route possessed BaY_2NiO_5 and SrY_2O_4 impurity phases along with NiO and BSCZGY phases. Among the compositions studied, Ni50-BSCZGY showed the lowest HOR ASR values for both the mechanical mixing and combustion methods. However, the combustion route synthesized anode composites showed higher HOR ASR (1.9 Ω.cm^2) values than the mechanically mixed one (0.8 Ω.cm^2) at 710 °C under H_2/3% H_2O for Ni50-BSCZGY. The formation of impurity phases and a thinner anode layer might have limited the electrochemical performance of anodes prepared by the combustion method.

Acknowledgment

This research was supported through Discovery Grants from the Natural Science and Engineering Research Council of Canada (NSERC) (Grant Number: RGPIN-2016-03853).

References

Babilo, P., & Haile, S. M. (2005). Enhanced sintering of yttrium-doped barium zirconate by addition of ZnO. *J. Am. Ceram. Soc., 88*(9), 2362–2368. http://dx.doi.org/10.1111/j.1551-2916.2005.00449.x

Barbucci, A., Carpanese, M., Reverberi, A. P., Cerisola, G., Blanes, M., Cabot, P. L., Nicolella, C. (2008). Influence of electrode thickness on the performance of composite electrodes for SOFC. *J. Appl. Electrochem., 38*, 939–945. http://dx.doi.org/ 10.1007/s10800-008-9500-z

Bhide, S. V., & Virkar, A.V. (1999). Stability of $BaCeO_3$-based proton conductors in water-containing atmospheres. *J. Electrochem. Soc., 146*(6), 2038–2044. http://dx.doi.org/ doi: 10.1149/1.1391888

Bi, L., Fabbri, E., Sun, Z., & Traversa, E. (2011). $BaZr_{0.8}Y_{0.2}O_{3-\delta}$-NiO composite anodic powders for proton-conducting SOFCs prepared by a combustion method. *J. Electrochem. Soc., 158*(7), B797–B803. http://dx.doi.org/ doi: 10.1149/1.3591040

Bi, L., Fabbri, E., Sun, Z., & Traversa, E. (2011). Sinteractive anodic powders improve densification and electrochemical properties of $BaZr_{0.8}Y_{0.2}O_{3-\delta}$ electrolyte films for anode-supported solid oxide fuel cells. *Energy Environ. Sci., 4*(4), 1352–1357. http://dx.doi.org/ 10.1039/C0EE00387E

Essoumhi, A., Taillades, G., Taillades-Jacquin, M., Jones, D. J., & Rozière, J. (2008). Synthesis and characterization of Ni-cermet/proton conducting thin film electrolyte symmetrical assemblies. *Solid State Ionics, 179*, 2155–2159. http://dx.doi.org/10.1016/j.ssi.2008.07.025

Fergus, J. W. (2006). Electrolytes for solid oxide fuel cells. *J. Power Sources, 162*, 30–40. http://dx.doi.org/10. 1016/j.jpowsour.2006.06.062

Gill, S., Kannan, R., Maffei, N., & Thangadurai, V. (2013). Synthesis, structure, chemical and physical properties of perovskite-type $BaCe_{0.8-x}Zr_xGd_{0.15}Pr_{0.05}O_{3-d}$ (BCZGP) and $BaCe_{0.85-x}Zr_xSm_{0.15}O_{3-d}$ (BCZS) (0.01 < x < 0.3). *RSC Advances, 3*, 3599-3605. http://dx.doi.org/10.1039/C2RA22097K

Iwahara, F., Esaka, T., Uchida, H., & Maeda, N. (1981). Proton conduction in sintered oxides and its application to steam electrolysis for hydrogen production. *Solid State Ionics, 3/4*, 359–363. http://dx.doi.org/doi:10.1016/0167-2738(81)90113-2

Iwahara, H., Uchida, H., Ono, K., & Ogaki, K. (1988). Proton conduction in sintered oxides based on $BaCeO_3$. *J. Electrochem. Soc., 135*(2), 529–533. http://dx.doi.org/ 10.1149/1.2095649

Kannan, R., Singh, K., Gill, S., Fürstenhaupt, T., & Thangadurai, V. (2013). Chemically stable proton conducting doped $BaCeO_3$-no more fear to SOFC wastes. *Sci. Rep., 3*, 2138. http://dx.doi.org/ 10.1038/srep02138

Kan, W. H., Lussier, J., Bieringer, M., & Thangadurai, V. (2014). Studies on polymorphic sequence during the formation of the 1:1 ordered perovskite-type $BaCa_{0.335}M_{0.165}Nb_{0.5}O_{3-\delta}$ (M = Mn, Fe, Co) using in-situ and ex-situ powder x-ray diffraction. *Inorg. Chem., 53*(19), 10085–10093. http://dx.doi.org/10.1021/ic501270k

Magrez, A., & Schober, T. (2004). Preparation, sintering, and water incorporation of proton conducting $Ba_{0.99}Zr_{0.8}Y_{0.2}O_{3-\delta}$: comparison between three different synthesis techniques. *Solid State Ionics, 175*, 585–588. http://dx.doi.org/10.1016/j.ssi.2004.03.045

Mather, G. C., Figueiredo, F. M., Fagg, D. P., Norby, T., Jurado, J. R., & Frade, J. R. (2003). Synthesis and characterisation of Ni–SrCe$_{0.9}$Yb$_{0.1}$O$_{3-\delta}$ cermet anodes for protonic ceramic fuel cells. *Solid State Ionics, 158,* 333–342. http://dx.doi.org/10.1016/S0167-2738(02)00904-9

Mather, G. C., Figueiredo, F. M., Jurado, J. R., & Frade, J. R. (2003). Synthesis and characterisation of cermet anodes for SOFCs with a proton-conducting ceramic phase. *Solid State Ionics, 162-163,* 115–120. http://dx.doi.org/10.1016/S0167-2738(03)00250-9

Narendar, N., Mather, G. C., Dias, P. A. N., & Fagg, D. P. (2013). The importance of phase purity in Ni–BaZr$_{0.85}$Y$_{0.15}$O$_{3-\delta}$ cermet anodes–novel nitrate-free combustion route and electrochemical study. *RSC Advances, 3,* 859–869. http://dx.doi.org/ 10.1039/C2RA22301E

Ricote, S., & Bonanos, N. (2010). Enhanced sintering and conductivity study of cobalt or nickel doped solid solution of barium cerate and zirconate. *Solid State Ionics, 181,* 694–700. http://dx.doi.org/10.1016/j.ssi.2010.04.007

Scholten, M. J., & Schoonman, J. (1993). Synthesis of strontium and barium cerate and their reaction with carbon dioxide. *Solid State Ionics, 61,* 83–91. http://dx.doi.org/10.1016/0167-2738(93)90338-4

Shannon, R. D. (1976). Revised effective ionic radii and systematic studies of interatomic distances in halides and chalcogenides. *Acta. Cryst., A32,* 751–767. http://dx.doi.org/ 10.1107/S0567739476001551

Singh, K. (2016). *Tailoring Perovskite- and Fluorite-Type Oxides for Solid Oxide Fuel Cells (SOFCs)* (Doctoral dissertation, April 2016, University of Calgary, Calgary, Canada). http://hdl.handle.net/11023/2986

Sneha, B. R., & Thangadurai, V. (2007). Synthesis of nano-sized crystalline oxide ion conducting fluorite-type Y$_2$O$_3$-doped CeO$_2$ using perovskite-like BaCe$_{0.9}$Y$_{0.1}$O$_{2.95}$ (BCY) and study of CO$_2$ capture properties of BCY. *J. Solid State Chem., 180,* 2661-2669. http://dx.doi.org/10.1016/j.jssc.2007.07.016

Song, S. J., Moon, J. H., Lee, T. H., Dorris, S. E., & Balachandran, U. (2008). Thickness dependence of hydrogen permeability for Ni–BaCe$_{0.8}$Y$_{0.2}$O$_{3-\delta}$. *Solid State Ionics, 179,* 1854–1857. http://dx.doi.org/10.1016/j.ssi.2008.05.012

Tanner, C. W., Fung, K. Z., & Virkar, A. V. (1997). The effect of porous composite electrode structure on solid oxide fuel cell performance I. Theoretical analysis. *J. Electrochem. Soc., 144,* 21–30. http://dx.doi.org/ 10.1149/1.1837360

Tong, J. H., Clark, D., Hoban, M., & O'Hayre, R. (2010). Cost-effective solid-state reactive sintering method for high conductivity proton conducting yttrium-doped barium zirconium ceramics. *Solid State Ionics, 181,* 496–503. http://dx.doi.org/10.1016/j.ssi.2010.02.008

Tong, J., Clark, D., Bernau, L., Sanders, M., & O'Hayre, R. (2010). Solid-state reactive sintering mechanism for large-grained yttrium-doped barium zirconate proton conducting ceramics. *J. Mater. Chem., 20,* 6333–6341. http://dx.doi.org/ 10.1039/C0JM00381F

Trobec, F., & Thangadurai, V. (2008). Transformation of proton-conducting perovskite-type into fluorite-type fast oxide ion electrolytes using a CO$_2$ capture technique and their electrical properties. *Inorg.Chem., 47,* 8972-8984. http://dx.doi.org/ 10.1021/ic8010025

Wu, Z., & Liu, M. (1997). Stability of BaCe$_{0.8}$Gd$_{0.2}$O$_3$ in a H$_2$O-containing atmosphere at intermediate temperatures. *J. Electrochem. Soc., 144*(6), 2170–2175. http://dx.doi.org/ 10.1149/1.1837759

Yamazaki, Y., Hernandez-Schanchez, R., & Haile, S. M. (2010). Cation non-stoichiometry in yttrium-doped barium zirconate: phase behavior, microstructure, and proton conductivity. *J. Mater. Chem., 20,* 8158–8166. http://dx.doi.org/ 10.1039/C0JM02013C

3

Performance of Sulfuric Acid Leaching of Titanium from Titanium-Bearing Electric Furnace Slag

XiaoMing Qu[1], YuFeng Guo[1], FuQiang Zheng[1], Tao Jiang[1] & GuanZhou Qiu[1]

[1] School of Minerals Processing and Bioengineering, Central South University, Changsha 410083, Hunan, China.

Correspondence: YuFeng Guo, School of Minerals Processing and Bioengineering, Central South University, Changsha 410083, Hunan, China. E-mail: guoyufengcsu@163.com

FuQiang Zheng, School of Minerals Processing and Bioengineering, Central South University, Changsha 410083, Hunan, China. E-mail: zhengfuqiangcsu@163.com

Abstract

The sulfuric acid leaching of titanium from titanium-bearing electric furnace slag (TEFS) was investigated under different experimental conditions. In the sulfuric acid leaching process, the $M_xTi_{3-x}O_5(0 \leq x \leq 2)$ and diopside could react with sulfuric acid. The optimum conditions of sulfuric acid leaching process were particle size at < 0.045mm, sulfuric acid concentration at 90 wt.%, acid/slag mass ratio at 1.6:1, feeding temperature at 120 °C, reaction temperature at 220 °C, reaction time at 120minute, curing at 200°C for 120 minute. The [TiO_2] concentration of the water leaching was 150 g/L, and leaching temperature at 60℃ for 120 minute. Ti extraction could reach 84.29 %. F of titanium-bearing solution was 2.15, and the Ti^{3+}/TiO_2 of the titanium-bearing solution was 0.068. The TiO_2 content of the leaching residue was 18.32 wt.%. The main mineral phases of the leaching residue were calcium sulphate, spinel, diopside and little $M_xTi_{3-x}O_5$.

Keywords: Sulfuric acid leaching, Titanium-bearing electric furnace slag, Ti extraction

1. Introduction

Vanadium titanomagnetite ore is an important resource of iron, vanadium and titanium. The Panzhihua-Xichang region of China is rich in vanadium titanomagnetite resources with reserves of about 9.66 billion tons which contained 11.6% of vanadium and 35.17% of titanium all over the world (Zheng et al., 2016; U.S. Geological Survey, 2015). At present, the vanadium titanomagnetite ore is separated as ilmenite concentrate, vanadium titanomagnetite concentrate and other minerals in Panzhihua-Xichang region. The vanadium titanomagnetite concentrate containing about 52 % titanium and 89 % vanadium of the vanadium titanomagnetite ore is used to produce iron and extract vanadium by the conventional blast furnace process. However, the titanium remains in the titanium-bearing blast furnace slag (22-25 wt.% TiO_2) without effective utilization. The sodium salt roasting-direct reduction-electric furnace smelting, direct reduction-magnetic separation and some other methods has been studied to utilize the titanium resource in the vanadium titanomagnetite concentrate, but these methods have been in the stage of laboratory research (Peng & Hwang, 2015; Samanta, Mukherjee, & Dey, 2015; Chen et al., 2015; Taylor & Shuey, 2006). The direct reduction-electric furnace smelting process has been commercialized in South Africa and New Zealand (Jen, Dresler, & Reilly, 1995) due to large production scale and mature recovery method of vanadium in molten iron. The direct reduction-electric furnace smelting process is also more environmentally friendly than blast furnace process. The titanium-bearing electric furnace slag (TEFS) of the direct reduction-electric smelting process contains 40-60 wt.% TiO_2 which is much higher than the TiO_2 content of titanium-bearing blast furnace slag. Therefore, it is crucial to effectively extract titanium from TEFS for the development of direct reduction-electric furnace smelting and the utilization of the vanadium titanomagnetite concentrate.

Currently, few effort has been made to recover titanium from TEFS (Li et al., 2013; Zheng et al., 2016). Other studies focused on recover titanium from titanium-bearing blast furnace slag. Ma et al. (2000) separated titanium from treated slag by gravity separating or flotation. Hydrometallurgical methods include leaching of titanium slag with sulfuric acid (Farzaneh et al., 2013) or hydrochloric acid (Wang et al., 1994). Wang et al. (2012) recovered titanium from titanium-bearing blast furnace slag by ammonium sulfate melting method. Photocatalytic materials (Lü et al., 2013) and Ti-Si ferroalloy (Li et al., 1996) were prepared using blast furnace titaniferous slag. Wang et al. (2006) separated titanium components from titanium bearing blast furnace slag by the selective separation technique. Liu (2009) recovered titanium by high-temperature carbonization and low-temperature chlorination from modified titanium bearing blast furnace slag. Although some of processes have been proved to extract titanium from titanium slag, the application of these processes may result in high cost, poor recovery and environmental pollution problem. Among these methods, apart from the sulfuric acid leaching process which is a mature titanium dioxide process, other methods are all in the stage of laboratory research. Recovery of titanium by sulfuric acid can not only treat titanium slag in large-scale, but also promote the process of direct reduction-electric furnace smelting by utilizing the titanium resource in the vanadium titanomagnetite concentrate. Therefore, studying the performance of sulfuric acid leaching of titanium from TEFS is very necessary.

The performance of sulfuric acid leaching affected by a variety of parameters. In the present study, a list of conditions of sulfuric acid leaching of TEFS was investigated, with an emphasis on the effects of particle sizes, liquid/solid ratio, sulfuric acid concentration, feeding temperature, reaction temperature, Curing time, cooling rate, TiO2 concentration of water leaching, water leaching temperature time. The findings will provide a technical basis for titanium dioxide preparation by sulfuric acid leaching process of TEFS.

2. Experimental

2.1 Materials

The TEFS used in the present work was provided by Chongqing Steel (Sichuan Province, China). The chemical analysis of the sample is listed in Table 1. It can be seen that the TiO_2 content of the TEFS is 47.35 wt.%, and the major impurities are aluminum, silicon, magnesium, and calcium-bearing diopside and spinel. The mineral composition of the sample was investigated by X-ray diffractometer (D/max2550PC, Japan Rigaku Co., Ltd). The XRD pattern of the slag is showed in Figure 1. The major phases in the TEFS are $M_xTi_{3-x}O_5(0 \leqslant x \leqslant 2)$, diopside and spinel. All reagents used in the experiments were of at least analytical grade provided by Hunan Chashang Chemical Reagents Factory.

Table 1. Chemical composition of the original titanium slag

Component	TiO_2	TFe	V_2O_5	Al_2O_3	SiO_2	MgO	CaO
wt.%	47.35	3.16	0.49	12.10	16.08	9.88	9.97

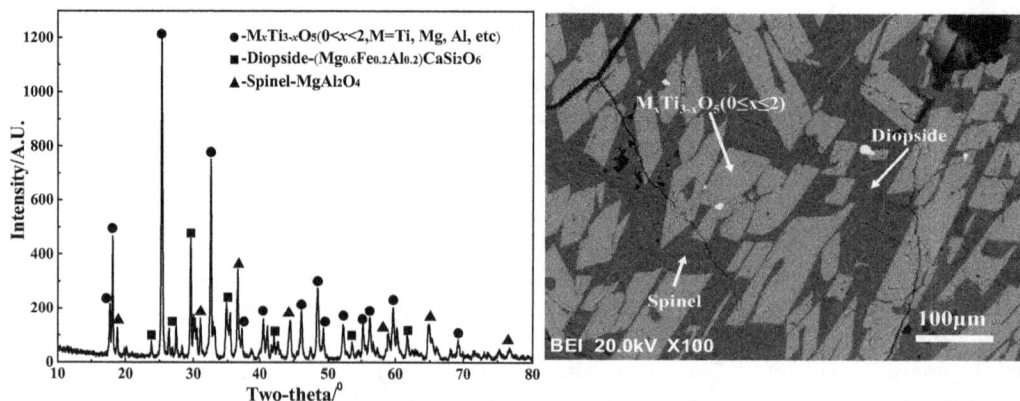

Figure 1. XRD pattern and micrograph of the original TEFS

2.2 Experimental Procedure

TEFS was crushed to a certain size and screened by tyler sieve. A certain amount of concentrated sulfuric acid were heated to the feed temperature in a beaker by an oil bath pan, then a certain quality of sample was added into the sulfuric acid followed by stirring at 20 rpm. Then the temperature of the leaching system was attempered to a certain reaction temperature in 2 minute. After the reaction finishing, the beaker was transferred to the muffle furnace at curing temperature for a certain time. Then the curing material was cooled to room temperature, and a certain quality of water was added to dissolve the curing material at a certain temperature for a required time. At last, the leaching solution was filtered to obtain the titanium-bearing solution and leaching residue. The leaching residue was dried in an oven at 105 °C for 4 h. The property of titanium-bearing solution and leaching residue were analyzed by EDTA titration.

The Ti extraction is defined as follows:

$$\eta_{Ti} = \left[1 - \left(\frac{m \times \omega_{Ti}}{m_0 \times \omega_0} \right) \right] \tag{1}$$

Where η_{Ti} is the Ti extraction, m is the quality of the leaching residue (g, dried solids), w_{Ti} is the content of Ti in the leaching residue (wt.%, dried solids), m_0 is the initial mass of slag sample (g, dried solids), w_0 is the content of Ti in the slag sample (wt.%, dried solids).

2.3 Experiment Principle

The main mineral of titanium slag is MxTi3-xO5(0≤x≤2), which reacts with sulfuric acid can be simply expressed as:

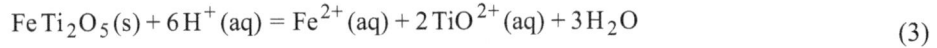

$$MgTi_2O_5(s) + 6H^+(aq) = Mg^{2+}(aq) + 2TiO^{2+}(aq) + 3H_2O \tag{2}$$

$$FeTi_2O_5(s) + 6H^+(aq) = Fe^{2+}(aq) + 2TiO^{2+}(aq) + 3H_2O \tag{3}$$

The main gangues in the slag are spinel and diopside. The main reactions could occur in the solution can be simply expressed as:

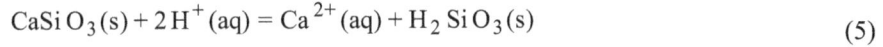

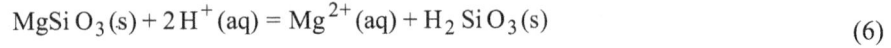

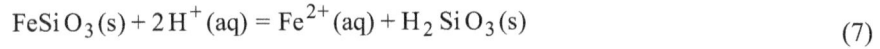

$$MgAl_2O_4(s) + 8H^+(aq) = Mg^{2+}(aq) + 3Al^{3+}(aq) + 4H_2O \tag{4}$$

$$CaSiO_3(s) + 2H^+(aq) = Ca^{2+}(aq) + H_2SiO_3(s) \tag{5}$$

$$MgSiO_3(s) + 2H^+(aq) = Mg^{2+}(aq) + H_2SiO_3(s) \tag{6}$$

$$FeSiO_3(s) + 2H^+(aq) = Fe^{2+}(aq) + H_2SiO_3(s) \tag{7}$$

$$Al_2SiO_5(s) + 6H^+(aq) = 2Al^{3+}(aq) + H_2SiO_3(s) + 2H_2O \tag{8}$$

3. Results and discussion

3.1 Effect of the Particle Size on Sulfuric Acid Leaching

Several leaching experiments were performed in 90 wt.% H_2SO_4 solution with different particle sizes, acid/slag mass ratio at 1.6:1, feeding temperature at 100°C, reaction temperature at 200°C, curing temperature at 200 °C, curing time for 120minute, and the water leaching conditions were leaching [TiO_2] concentration at 150 g/L, water leaching temperature at 60°C , water leaching time for 120 minute. The results are listed in Table 2.

Table 2. Effect of the particle size on sulfuric acid leaching

Particle size/mm	Ti extraction/%	Ti^{3+}/TiO_2	F
>0.074	80.11	0.075	2.46
0.045-0.074	80.76	0.076	2.44
<0.045	81.93	0.077	2.40

The results in Table 2 show that the Ti extraction and Ti^{3+}/TiO_2 in the leaching solution are increased with decreasing of slag particle size. F decreases with decreasing of particle size. More slag was decomposed by H_2SO_4, less effective acid was remained in the leaching solution which lead to the decreasing of F.

3.2 Effect of the Acid/Solid Ratio on Sulfuric Acid Leaching

The effect of liquid/solid ratio was investigated in 90 wt.% H_2SO_4 solution with different particle sizes < 0.045 mm, feeding temperature at 100 °C, reaction temperature at 200 °C, curing temperature at 200 °C, curing time for 120 minute, and the water leaching conditions were leaching [TiO_2] concentration at 150 g/L, water leaching temperature at 60 °C , water leaching time for 120 minute. T Results are presented in Figure 2.

Ti extraction and F increased with the increasing of acid/slag ratio, because more sulfuric acid could improve the acidolysis of TEFS, at the same time, more effective acid were remained in the leaching solution. When the acid/slag ratio increased from 1.6 to 1.7, the F increased sharply from 2.4 to 2.8. Higher F would be a big obstacle for Ti^{4+} hydrolyzing. Thus the optimum acid/slag ratio was selected as 1.6:1 and applied in the subsequent experiments.

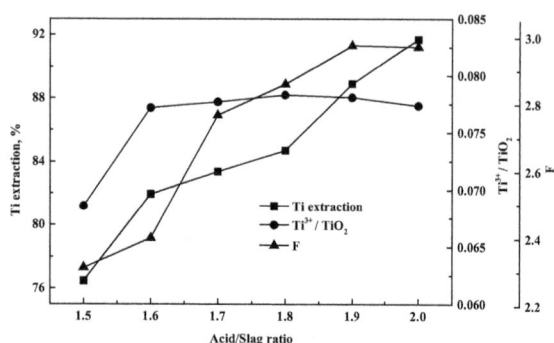

Figure 2. Effect of the Liquid/Solid ratio on sulfuric acid leaching

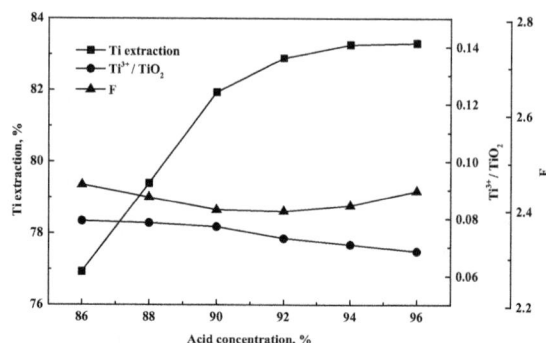

Figure 3. Effect of the Acid concentration ratio on sulfuric acid leaching

3.3 Effect of the sulfuric Acid Concentration on Sulfuric Acid Leaching

Figure 3 shows the results of experiments which carried out with acid/slag mass ratio at 1.6:1, feeding temperature at 100 °C, reaction temperature at 200 °C, curing temperature at 200 °C, curing time for 120 minute, and the water leaching conditions were leaching [TiO_2] concentration at 150 g/L, water leaching temperature at 60 °C, water leaching time for 120 min and the H_2SO_4 concentration varied from 86% to 96% with particle size <0.045mm.

The results in Figure 3 showed that increasing H_2SO_4 concentration had an effectively effect on the Ti extraction. When the H_2SO_4 concentration was at 86 wt.%, the Ti extraction was 76.9%. However, 83.3 % of Ti extraction was obtain when the H_2SO_4 concentration was at 94 wt.%. On the other hand, the Ti^{3+}/TiO_2 decreased with the increasing of H_2SO_4 concentration. The F was slightly increased when the H_2SO_4 concentration was higher than 92 wt.%. The optimum sulfuric acid concentration was 94%.

3.4 Effect of the Feeding Temperature on Sulfuric Acid Leaching

A series of experiments was carried out in 90 wt.% H_2SO_4 solution with varied feeding temperature from 25 °C to 140°C, and the other conditions was particle size < 0.045mm, acid/slag mass ratio of at 1.6:1, reaction temperature at 200 °C, curing temperature at 200 °C, curing time for 120 minute. And the water leaching conditions were leaching [TiO_2] concentration at 150 g/L, water leaching temperature at 60 °C, water leaching time for 120 minute.

The results in Figure 4 showed that with the increasing of feeding temperature from 20°C to 120°C, Ti extraction increased to 83.64%, Ti^{3+}/TiO_2 dropped to 0.068, F dropped to 2.40. While the feed temperature raised to 140°C, Ti extraction dropped quickly, Ti^{3+}/TiO_2 and F reached the peak values. The optimum Feed temperature was 120°C.

3.5 Effect of the Reaction Temperature on Sulfuric Acid Leaching

The effect of Reaction temperature was investigated in 94 wt.% H_2SO_4 solution with particle size < 0.045mm, acid/slag mass ratio of at 1.6:1, curing at 200°C for 120 min and the water leaching conditions were leaching [TiO_2] concentration at 150 g/L, water leaching at 60 °C for 120 minute. Reaction temperature varied from 160 °C to 250 °C. Results are presented in Figure 5.

The results in Figure 5 showed that increasing reaction temperature has effectively effects on the Ti extraction. When the reaction temperature increased to 220°C, the Ti extraction raised to 84.29%, the Ti^{3+}/TiO_2 dropped to

0.067. When the reaction temperature reached 250°C, the Ti extraction dropped and the Ti^{3+}/TiO_2 raised. The F decreased with the increasing of reaction temperature. The optimum reaction temperature was 220°C.

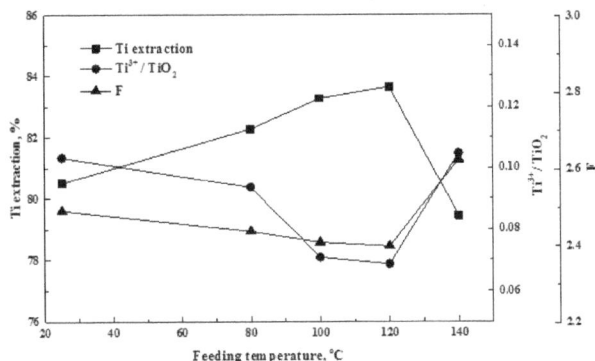

Figure 4. Effect of the feed temperature on sulfuric acid leaching

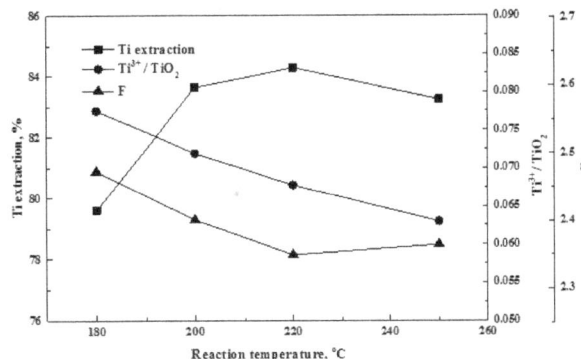

Figure 5. Effect of the reaction temperature on sulfuric acid leaching

3.6 Effect of the Curing Time on Sulfuric Acid Leaching

The effect of curing time was investigated in 90 wt.% H_2SO_4 solution with particle size < 0.045mm, acid/slag mass ratio of at 1.6:1, feeding temperature at 100 °C, reaction temperature at 200 °C, curing temperature at 200 °C, curing time varied from 0 to 180 minute, and the water leaching conditions were leaching [TiO_2] concentration at 150 g/L, water leaching temperature at 60 °C, water leaching time for 120 minute.

The Figure 6 showed that with the increased of the curing time, the Ti extraction increased, the Ti^{3+}/TiO_2 and F decreased. Before the time of 120min, the Ti extraction, Ti^{3+}/TiO_2 and F varied quickly. After 120 minute, the Ti extraction increased and the Ti^{3+}/TiO_2 varied not obviously. The optimum curing time was 120 minute.

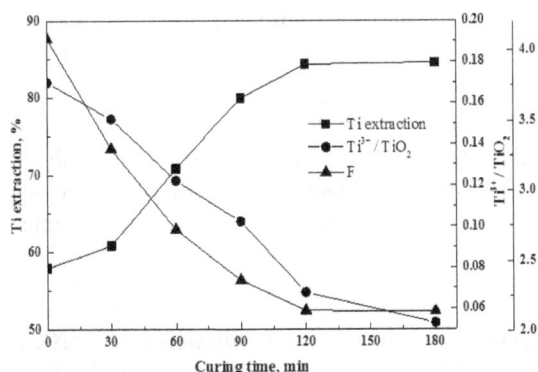

Figure 6. Effect of the curing time on sulfuric acid leaching

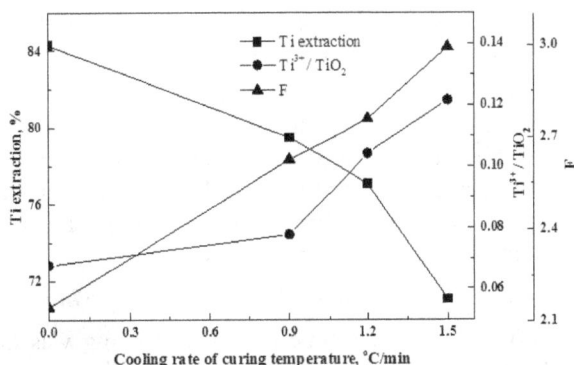

Figure 7. Effect of the cooling rate of curing temperature on sulfuric acid leaching

3.7 Effect of the Cooling Rate of Curing Temperature on Sulfuric Acid Leaching

For the present experiments, the initial curing temperature were 200°C with the cooling rate ranged from 0 to 1.5 °C/minute. Particle size < 0.045mm in 90 wt.% H_2SO_4 solution, acid/slag mass ratio of at 1.6:1, feeding temperature at 100 °C, reaction temperature at 200 °C, and the water leaching conditions were leaching [TiO_2] concentration at 150 g/L, water leaching temperature at 60 °C, water leaching time for 120 minute. Figure 7 shows the influences of cooling rate of curing temperature on sulfuric acid leaching. With the increasing of cooling rate, the Ti extraction decreased, the Ti^{3+}/TiO_2 and F increased. To ensuring and increasing leaching performance, the cooling rate should keep as low as possible.

3.8 Effect of the Leaching Concentration on Water Leaching

The effect of leaching concentration on water leaching was investigated in 90 wt.% H_2SO_4 solution with particle size < 0.045mm, acid/slag mass ratio of at 1.6:1, feeding temperature at 100 °C, reaction temperature at 200 °C, curing temperature at 200 °C, curing time for 120 minute, but the water leaching conditions of leaching $[TiO_2]$ concentration varied from120 g/L to 150 g/L, water leaching temperature at 60 °C , water leaching time for 120 minute.

The results in Figure 5 showed that leaching $[TiO_2]$ concentration had obvious improvement on Ti^{3+}/TiO_2 and F. The leaching $[TiO_2]$ concentration from 150 g/L to 170 g/L, the Ti extraction decreased and the Ti^{3+}/TiO_2 increased sharply. High leaching concentration was good for the concentration process. High leaching $[TiO_2]$ concentration also made the solid phase products dissolved and the Ti extraction decreased.

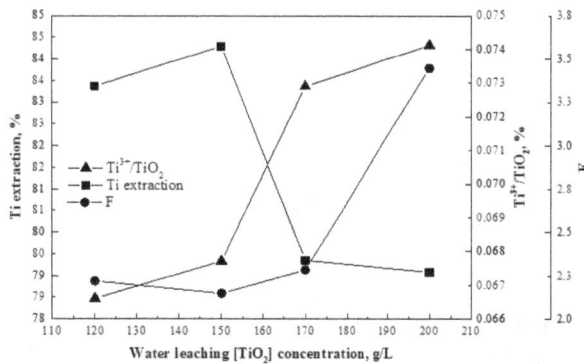

Figure 8. Effect of the leaching $[TiO_2]$ concentration on water leaching

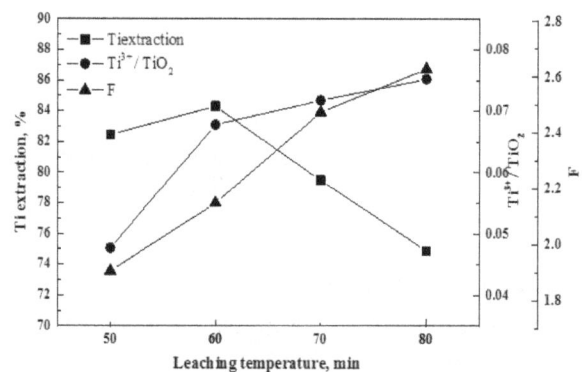

Figure 9. Effect of the leaching temperature on water leaching

3.9 Effect of the Leaching Temperature on Water Leaching

The effect of leaching temperature on water leaching was investigated in 90 wt.% H_2SO_4 solution with particle size < 0.045mm, acid/slag mass ratio of at 1.6:1, feeding temperature at 100 °C, reaction temperature at 200 °C, curing temperature at 200 °C, curing time for 120 min and the water leaching conditions were leaching $[TiO_2]$ concentration at 150 g/L, water leaching temperature ranged from 50 °C to 80 °C, water leaching time for 120 minute.

The results in Figure 9 showed that Ti^{3+}/TiO_2 and F were increased with the increasing of water leaching temperature. When the water leaching temperature was above 60 °C, Ti extraction decreased sharply. In the leaching process, most of the salt dissolved in water releasing quantity of heatwhich leaded to the temperature increased. Excessively high water leaching temperatures leaded to titanium hydrolysis ahead of time. The optimum leaching temperature was 60 °C.

3.10 Effect of the Leaching Time on Water Leaching

The effect of leaching time on water leaching was investigated in 90 wt.% H_2SO_4 solution with particle size <0.045mm, acid/slag mass ratio of at 1.6:1, feeding temperature at 100°C, reaction temperature at 200°C, curing temperature at 200°C, curing time for 120 minute and the water leaching conditions were leaching $[TiO_2]$ concentration at 150 g/L, water leaching temperature at 60°C, water leaching time for 60 minute, 120 minute, 180 minute respectively.

The results in Figure 10 showed that Ti^{3+}/TiO_2 and Ti extraction were increased with the rise in temperature, the F decreased with the rise in temperature. The optimum leaching temperature is 2 hours.

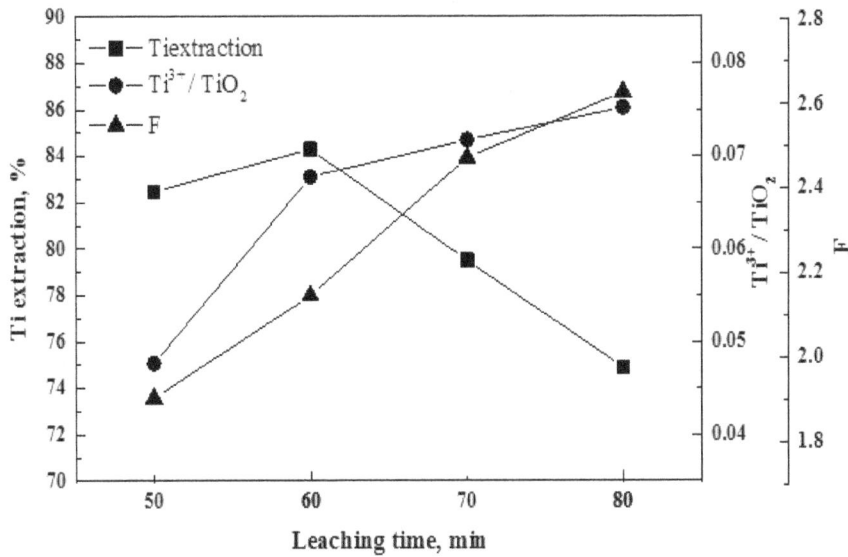

Figure 10. Effect of the leaching time on water leaching

3. 11 The Titanium Solution and Leaching Residue

The optimum conditions of sulfuric acid leaching were particle size at < 0.045 mm, sulfuric acid concentration at 90 wt.%, acid/slag mass ratio at 1.6:1, feeding temperature at 120 °C, reaction temperature at 220 °C, reaction time of 120 minute, curing temperature at 200 °C for 120 minute. The water leaching [TiO_2] concentration were 180 g/L, leaching temperature at 60 °C for 120 minute. Ti extraction could reach 84.29%, F of lixivium was 2.15, Ti^{3+}/TiO_2 was 0.068, Ti^{3+} contend was higher than the required quantity, further processing was required to reduce the Ti^{3+} contend. The chemical analysis of the residue and titanium solution are listed in Table 3 and Table 4 respectively. The XRD pattern of the leaching residue is showed in Figure 11.

The main mineral phases of the leaching residue were calcium sulphate, spinel, diopside and little $M_xTi_{3-x}O_5$. 18.32 wt.% TiO_2 were remained in the leaching residue. The reasons were as followed. First, sulfuric acid reacted with calcium-bearing silicate and formed $CaSO_4$ which was not soluble. $CaSO_4$ covered on the slag particles and hindered the reaction. Second, $M_xTi_{3-x}O_5$ were so stable that it could not be completely dissolved by sulfuric acid.

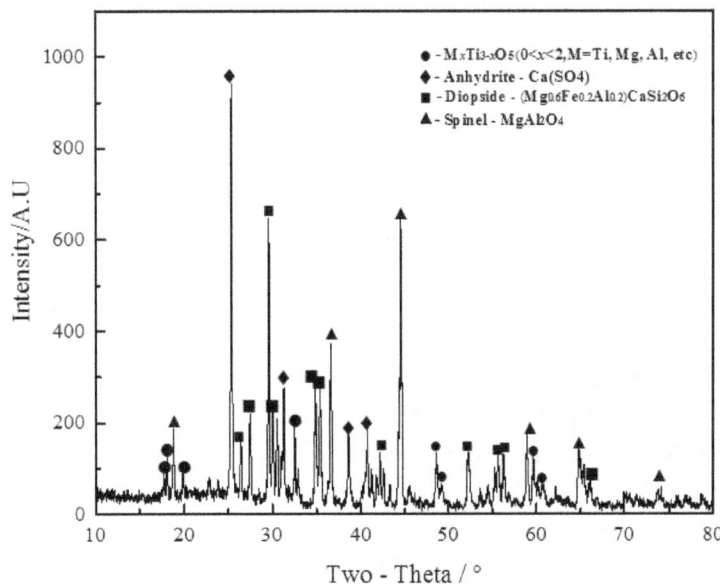

Figure 11. XRD pattern of the sulfuric leaching residue

Table 3. Chemical composition of the leaching residue

Component	TiO$_2$	MgO	Fe$_2$O$_3$	Al$_2$O$_3$	SiO$_2$	CaO
wt.%	18.321	12.448	2.252	14.601	26.579	13.333

Table 4. Chemical composition of titanium solution

Component	TiO$_2$	MgO	Fe2O$_3$	Al$_2$O$_3$	SiO$_2$	CaO
Extraction %	84.44%	52.86%	43.06%	58.04%	5.38%	2.22%
Content g/L	156.210	21.938	2.676	31.734	2.376	0.477

4. Conclusion

The major phases in the TEFS are $M_xTi_{3-x}O_5(0 \leqslant x \leqslant 2)$, diopside and spinel. In the sulfuric acid leaching process, the $M_xTi_{3-x}O_5(0 \leqslant x \leqslant 2)$ and diopside could react with sulfuric acid. The optimum condition of sulfuric acid leaching were particle size at < 0.045mm, 90 wt.% sulfuric acid concentration, acid/slag mass ratio at 1.6:1, feeding temperature at 120 °C, reaction temperature at 220 °C, reaction time at 120minute, curing at 200 ℃ for 120 minute. The water leaching [TiO$_2$] concentration were 150 g/L, water leaching at 60℃ for 120 minute. Ti extraction could reach 84.29%, F of lixivium was 2.15, Ti^{3+}/TiO_2 was 0.068. The main mineral phases of the leaching residue were calcium sulphate, spinel, diopside and little $M_xTi_{3-x}O_5$During the leaching process, CaSO4 were generated and covered on the surface of slag particles.

Acknowledgments

The authors are grateful to the Program for New Century Excellent Talents in University from Chi-nese Ministry of Education (NCET-10-0834) and the Visiting Scholar Program from China Scholarship Council ([2013]3018) for supporting this research.

References

Chen, D., Zhao, H., Hu, G., Qi, T., Yu, H., Zhang, G., ... & Wang, W. (2015). An extraction process to recover vanadium from low-grade vanadium-bearing titanomagnetite. *Journal of hazardous materials, 294*, 35-40. https://doi.org/10.1016/j.jhazmat.2015.03.054

Jena, B. C., Dresler, W., & Reilly, I. G. (1995). Extraction of titanium, vanadium and iron from titanomagnetite deposits at pipestone lake, Manitoba, Canada. *Minerals engineering, 8*(1), 159-168. https://doi.org/10.1016/0892-6875(94)00110-X

Li, Y., Yue, Y., Que, Z. Q., Zhang, M., & Guo, M. (2013). Preparation and visible-light photocatalytic property of nanostructured fe-doped tio2 from titanium containing electric furnace molten slag. *International Journal of Minerals Metallurgy and Materials, 20*(10), 1012-1020. https://doi.org/10.1007/s12613-013-0828-y

Li, Z., Xu, C., Li, Z., & Zhou, Y. (1996). The study on smelting ti-si ferroalloy by dc electrothermal process using pisc blast furnace titaniferous slag. *Journal of Chongqing University.*

Liu, X. H. (2009). *Study on High-temperature Carbonization and Low-temperature Chlorination on Modified Titanium Bearing Blast Furnace Slag.* Northeastern University, Shenyang (in Chinese).

Lü, H., Li, N., Wu, X., Li, L., Gao, Z., & Shen, X. (2013). A novel conversion of ti-bearing blast-furnace slag into water splitting photocatalyst with visible-light-response. *Metallurgical and Materials Transactions B, 44*(6), 1317-1320.

Ma, J., Sui, Z., & Chen, B. (2000). *Separating titanium from treated slag by gravity separation/flotation. Nonferrous Metals-Beijing-, 52*(2), 26-31.

Peng, Z., & Hwang, J. Y. (2015). Microwave-assisted metallurgy. *International Materials Reviews, 60*(1), 30-63. https://doi.org/10.1179/1743280414Y.0000000042

Samanta, S., Mukherjee, S., & Dey, R. (2015). Upgrading metals via direct reduction from poly-metallic titaniferous magnetite ore. *JOM, 67*(2), 467-476. https://doi.org/10.1007/s11837-014-1203-9

Taylor, P. R., Shuey, S. A., Vidal, E. E., & Gomez, J. C. (2006). Extactive metallurgy of vanadium-containing titaniferous magnetite ores: A review. *Minerals and Metallurgical Processing, 23*(2), 80-86.

U.S. Geological Survey. (2015). *Mineral commodity summaries 2015.* Center for Integrated Data Analytics Wisconsin Science Center. https://doi.org/10.3133/70140094

Valighazvini, F., Rashchi, F., & Nekouei, R. K. (2013). Recovery of titanium from blast furnace slag. *Industrial and Engineering Chemistry Research, 52*(4), 1723-1730. https://doi.org/10.1021/ie301837m

Wang, D. K., Lei, M. L., Li, B. E., & Jiang, Z. Q. (1994). *A process for comprehensive utilization of blast furnace slag with dilute hydrochloric acid.* Chinese Patent CN94108086.2.

Wang, M. Y., Zhang, L. N., Zhang, L., Sui, Z. T., & Tu, G. F. (2006). Selective enrichment of tio2 and precipitation behavior of perovskite phase in titania bearing slag. *Transactions of Nonferrous Metals Society of China, 16*(2), 421-425. https://doi.org/10.1016/S1003-6326(06)60072-1

Wang, S., Zhang, Y., Xue, X., & Yang, H. (2012). Recovery of titanium from titanium-bearing blast furnace slag by ammonium sulfate melting method. *Ciesc Journal, 63*(3), 991-995. https://doi.org/10.3969/j.issn.0438-1157.2012.03.045

Zheng, F., Chen, F., Guo, Y., Jiang, T., Travyanov, A. Y., & Qiu, G. (2016). Kinetics of hydrochloric acid leaching of titanium from titanium-bearing electric furnace slag. *Jom the Journal of the Minerals Metals and Materials Society, 30*(30), 15-68. https://doi.org/10.1007/s11837-015-1808-7

4

Comparison of Dislocation Density Tensor Fields Derived from Discrete Dislocation Dynamics and Crystal Plasticity Simulations of Torsion

Reese E. Jones[1], Jonathan A. Zimmerman[1] & Giacomo Po[2]

[1] Sandia National Laboratories, Livermore, CA 94550, USA

[2] University of California, Los Angeles, CA 90095, USA

Correspondence: Reese E. Jones, Sandia National Laboratories, Livermore, CA 94550, USA. E-mail: rjones@sandia.gov

Abstract

The importance of accurate simulation of the plastic deformation of ductile metals to the design of structures and components is well-known. Many techniques exist that address the length scales relevant to deformation processes, including dislocation dynamics (DD), which models the interaction and evolution of discrete dislocation line segments, and crystal plasticity (CP), which incorporates the crystalline nature and restricted motion of dislocations into a higher scale continuous field framework. While these two methods are conceptually related, there have been only nominal efforts focused on the system-level material response that use DD-generated information to enhance the fidelity of plasticity models. To ascertain to what degree the predictions of CP are consistent with those of DD, we compare their global and microstructural response in a number of deformation modes. After using nominally homogeneous compression and shear deformation dislocation dynamics simulations to calibrate crystal plasticity flow rule parameters, we compare not only the system-level stress-strain response of prismatic wires in torsion but also the resulting geometrically necessary dislocation density tensor fields. To establish a connection between explicit description of dislocations and the continuum assumed with crystal plasticity simulations, we ascertain the minimum length-scale at which meaningful dislocation density fields appear. Our results show that, for the case of torsion, the two material models can produce comparable spatial dislocation density distributions.

Keywords: Dislocation dynamics, crystal plasticity, dislocation density tensor

1. Introduction

The importance of accurate simulation of the plastic deformation of ductile metals in the design of structures and components to performance and failure criteria is well known. Plasticity models describe the influence of elastic and inelastic deformation on stress within a body that undergoes a specific displacement/loading path. These models are conventionally constructed with parameters, such as elastic constants, yield stress and work hardening, fitted to experimentally measured data.

Crystal plasticity (CP) is a particular form of a plasticity model that takes into account some details of the underlying crystal structure. In CP, characteristics of the motion of dislocations (nanometer scale line imperfections in the crystal structure of a material) influence the deformation experienced by the material. Early CP models connected plastic strain rate to Burgers vectors, velocity and line length density of inherent material dislocations (Bilby, 1960; Ortiz & Popov, 1982; Asaro, 1983). More recent models cast this response in a finite deformation framework, and decompose the plastic portion of the velocity gradient into separate contributions from various families of dislocations, each associated with a specific slip system (Kuchnicki et al., 2006; Lee et al., 2010; Zhao et al., 2016).

Given the nanoscale nature of the underlying defects that govern plastic deformation in metals, it is logical to consider whether simulation methods that explicitly model these defects, such as atomistic simulation and discrete dislocation dynamics (DD) see *e.g.* Mordehai et al. (2008) and Groh et al. (2009), can be used to improve the fidelity of CP constitutive relations. The connection between dislocation interaction and plasticity was pioneered by Orowan (1934), Polanyi (1934), and Taylor (1934a); Taylor (1934b). Since that time, a number of researchers have made connections between dislocation mechanisms and crystal plasticity, see Déprés et al. (2006); Zhou et al. (2010); Chandra et al. (2015).

A number of recent efforts have made comparisons of molecular dynamics results with continuum plasticity simulations. Horstemeyer et al. (2002); Horstemeyer et al. (2003) compared atomistic simulation of FCC copper and nickel in simple shear and torsion with both CP and macroscopic internal state variable theories. This comparison revealed qualitative similarities between the different methods, such as the number of shear bands, but also quantitative differences due to the presence of thermal vibrations that occur for atomistic simulation at room temperature. In particular, it was observed that a much narrower stress distribution arose for the finite element analyses than for that extracted from atomistics. Efforts have also been made by Vitek and Paidar (2008); Vitek (2011), Gröger et al. (2008a); Gröger et al. (2008b); Gröger et al. (2008c), and Weinberger et al. (2012); Lim et al. (2013); Hale et al. (2015) to directly use atomistic simulation results to parameterize microscale yield response and CP models in body-centered cubic (BCC) metals. Vitek and coworkers examined the non-Schmidt behavior of the slip systems of group VB and VIB systems using a Finnis-Sinclair inter-atomic potential. Gröger *et al.* used bond order potentials to model the behavior of screw dislocations in Mo and W, quantifying a stress/loading-direction dependency to the critical resolved shear stress and characterizing the resulting non-Schmid behavior. Weinberger et al. leveraged this approach to fit parameters of a single crystal yield constitutive model to zero-temperature molecular statics simulation results. They then combined this parameterization with experimental data to develop a temperature and stress dependent model for the evolving plastic strain rate. This approach is particularly well-suited for BCC metals, where lattice friction for individual dislocations represents a significant barrier to glide.

Dislocation dynamics has also been recently employed to make comparisons between continuum plasticity and lower-level simulation methods. Wang and Beyerlein (2011) built on the insights of Vitek et al. and Gröger et al. in developing a three-dimensional discrete DD model to characterize the relationship between dislocation glide behavior and macro-scale plastic slip in single crystal BCC Ta. For FCC metals, where interactions between dislocations and with other obstacles, *e.g.* precipitates, strongly dictate the speed of dislocation motion and resulting rate of plastic deformation, atomistic simulation is limited in its ability to simulate a significant quantity of dislocations. In this regime, DD can play a useful role in providing relevant characteristics used in the fitting of CP model parameters. Such a connection was exploited by Van der Giessen and Needleman (2003), who employed simple boundary value problems to compare the predictions of nonlocal plasticity theories with DD results. Groh et al. (2009) later used both atomistic simulation and DD in a hierarchy of modeling methods, with results from atomistics used to quantify individual dislocation mobilities for use in DD simulations, and DD results to quantify parameters related to work hardening within a CP framework. These researchers successfully used this combined approach to predict the mechanical response of an aluminum single crystal deformed under uniaxial compressive loading along the [421] crystal direction. The computed strain-stress response agreed well with experimental data that was not used in model calibration.

In addition to MD and DD, other methods such as phase field crystal methods using diffusive dynamics have also proven useful in simulating plastic behavior with near atomic detail, see, *e.g.* Berry et al. (2012) and Wang et al. (2016).

While these efforts have achieved some degree of success, they have been limited in how they use the data from the evolution of explicit dislocations to enhance the fidelity of the plasticity models considered. In general, only the global stress-strain response of a system containing a distribution of dislocations is used to parameterize the hardening behavior; no local information is used to connect the two models. In particular, a direct comparison of the dislocation density tensor field (DDT) resulting from the evolution of the dislocation distribution in the case of DD and from the plastic deformation in the case of CP has not been considered thus far. However, the inclusion of the DDT as an independent field or a parameter in the flow rules and yield criteria of single-crystal plasticity models has been explored, see Forest (1998); Menzel and Steinmann (2000); Gurtin (2002); Evers et al. (2004b); Evers et al. (2004a); Gupta et al. (2007); Gurtin et al. (2007); Gurtin (2008); Hansen et al. (2013); Leung et al. (2015); Lee et al. (2010); Aoyagi and Shizawa (2007), and can be used to model length-scale effects (Stölken and Evans, 1998; Evers et al., 2004b), predict different dislocation microstructures (Edmiston et al., 2013), and allows comparison with EBSD and X-ray micro-diffraction data (Larson et al., 2008; Kysar et al., 2007, 2010; Field et al., 2010; Field & Alankar, 2010).

A comparison of the dislocation density tensor field resulting from corresponding DD and CP simulations is the primary contribution of this work. An attendant development is the means of extracting the DDT field from both methods and the associated length-scale at which one may resolve the DD-based densities. After a review of the basic theory and the two simulation methods in Sec. 2 and Sec. 3, we compare corresponding simulations resulting from both methods in Sec. 4. To make the comparison meaningful we make the two models as consistent as possible given the different formulations. First, by using nominally *homogeneous* compression and shear loading simulations, we calibrate a simple, representative flow rule for the CP model based on the DD response in Sec. 4.1. To eliminate dynamic effects we omit inertia and consider loading rates where both the DD and CP results are relatively insensitive to changes of rate. Another key predicate for consistency is the appropriate scale for the comparison based on the averaging volume used to translate discrete dislocations in DD into a density field. We hypothesize that there exists a range of volumes where a field is resolved and remains practically indistinguishable for small changes in the averaging volume. This notion of an *intermediate asymptotic scale*, refer to Ulz et al. (2013), for the dislocation density field is crucial to this work. Then we compare the spatially-varying DDT field, as well as the global response, resulting from DD and CP simulations (Sec. 4.2) of torsion of wires with square cross section. The rationale behind this choice is that, on the one hand, torsion induces a deformation state which is *inhomogeneous* within the cross section, and therefore its plastic component can be associated with *geometrically necessary* dislocation microstructures. On the other hand, the deformation state is homogeneous along the twist axis, therefore allowing meaningful averages along this direction for all fields extracted from DD and CP. The simulation of inhomogeneous loading in DD is enabled by method of Po et al. (2014) and the MODEL code (Amodeo et al., 2015), which treats the boundary effects of a finite system with interacting dislocations (see also Devincre & Gatti, 2015). Others (Fleck & Hutchinson, 1997; Horstemeyer et al., 2002; Weinberger & Cai, 2010; Kaluza & Le, 2011; Senger et al., 2011) have chosen torsion as a boundary value problem particularly well-suited to examining plastic behavior and validating plasticity theories, especially ones with intrinsic scales. In particular, Senger et al. (2011) investigate the DD response to torsion for a case similar to the one we employ. We conclude with a discussion of our results and those of Senger et al., as well as insights on how aspects of CP models can be improved using DD simulation in the final section. To our knowledge, comparison of DD and CP response at the levels of both overall stress-strain response and microstructural fields has not been investigated before.

2. Theory

The dislocation density tensor (DDT, $\boldsymbol{\alpha}$) introduced by Nye (1953) is a measure of the geometrically necessary dislocation density in a region of a crystal. (The geometrically necessary portion of the total dislocation density is the part ensuring compatibility of the overall deformation from reference lattice to current configuration, as opposed to the statistically stored portion.) The DDT is a fundamental quantity connecting the individual dislocations at the micro-scale to the macroscopic elastic and plastic deformation gradient fields, which are the essential kinematic ingredients in the macroscopic phenomenological theory of plasticity, see, e.g., Lubliner (2008).

The dislocation density tensor α is defined by its relationship with the (total) Burgers vector **b** for oriented area **A**=A**N**

$$\mathbf{b}N(A) \equiv \oint_{\partial A_{\kappa_0}} \mathbf{F}_p d\mathbf{X} = \int_{A_{\kappa_0}} \underbrace{\nabla_\mathbf{X} \times \mathbf{F}_p}_{\alpha} d\mathbf{A} = \int_{A_{\kappa_\ell}} \nabla_\mathbf{X} \times \mathbf{F}_p \left(\mathbf{F}_p^*\right)^{-1} d\mathbf{a}_\ell \qquad (1)$$

where κ_0 is the dislocation-free reference configuration and κ_ℓ is the intermediate (or *lattice*) configuration where the loop closure **b** for area A is measured. This definition relies on Stokes theorem and the usual multiplicative decomposition, originally due to Lee (1969), of the deformation gradient **F**,

$$\mathbf{F} = \mathbf{F}_e \mathbf{F}_p \qquad (2)$$

where the deformation gradient $\mathbf{F} = \nabla_\mathbf{X} \chi$ is the derivative of the motion $\mathbf{x} = \chi(\mathbf{X}, t)$ with respect to the reference coordinates **X**, refer to Figure 1. As denoted in Equation (1), the DDT α, being the curl of the plastic deformation

$$\alpha = \nabla_\mathbf{X} \times \mathbf{F}_p \qquad (3)$$

maps (oriented) elements of area $d\mathbf{A}$ in the reference configuration κ_0 to infinitesimal Burgers vectors $d\mathbf{b}$ in the incompatible lattice configuration. This definition of α is supported by topological interpretations of dislocation kinematics using Cartan's modern differential geometry, see, *e.g.*, Acharya and Bassani (2000); Clayton et al. (2005); Hochrainer (2013).

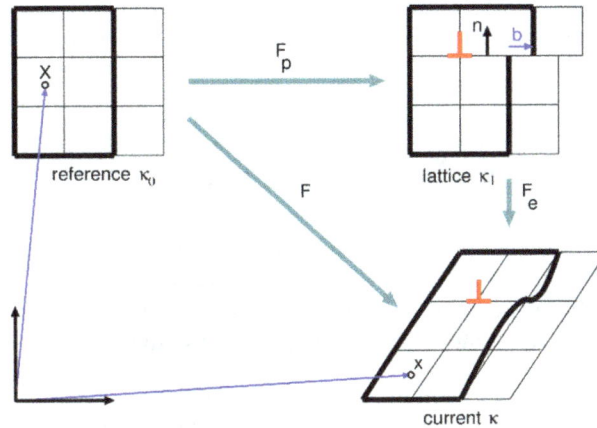

Figure 1. Reference κ_0, intermediate/lattice κ_ℓ, and current κ configurations. The Burgers circuit and associated area are outlined with a bold black line. The thin black lines indicate material lines. In this depiction the dislocation line tangent, ξ, is out of the plane, ($\xi = \partial_S \mathbf{x}_\ell(s)$ with s being the arc-length coordinate for line position $\mathbf{x}_\ell$). The slip plane normal $\mathbf{n}$, Burgers vector $\mathbf{b}$ and ξ form an orthogonal triad in the lattice configuration κ_ℓ. For the edge dislocation shown the slip vector $\mathbf{s}$ is aligned with $\mathbf{b}$

The primary kinematic assumption of crystal plasticity is the rate of plastic deformation is of the form:

$$\mathbf{L}_p = \sum_a \dot{\gamma}_a \mathbf{P}_a \Leftrightarrow \dot{\mathbf{F}}_p = \left(\sum_a \dot{\gamma}_a \mathbf{P}_a \right) \mathbf{F}_p \qquad (4)$$

where $\dot{\gamma}_a$ is the slip rate on plane a and the Schmid tensors $\mathbf{P}_a = \mathbf{s}_a \otimes \mathbf{n}_a$ reside in the intermediate configuration and are associated with the crystallographic planes in which glissile dislocations reside. (Here, $\mathbf{s}_a$ is the slip direction and $\mathbf{n}_a$ is the slip plane normal.) This leads to an overall deformation rate given by

$$\mathbf{L} = \dot{\mathbf{F}} \mathbf{F}^{-1} = \mathbf{L}_e + \mathbf{F}_e \mathbf{L}_p \mathbf{F}_e^{-1} \qquad (5)$$

in the finite deformation context. Figure 1 shows one such slip plane for an edge dislocation with the slip direction $\mathbf{s}$ aligned with the Burgers vector $\mathbf{b}$. In truth, Figure 1 is a very much simplified picture of plasticity with the one dislocation shown representing a family of parallel dislocation lines as Nye envisaged; however, the essential irreversibility of macro-plasticity is a manifestation of both dislocation motion and entanglement.

Under the assumption of linearized kinematics ($\mathbf{x} \approx \mathbf{X}$, small strain and rotation), the total deformation gradient can be written as

$$\mathbf{F} \approx \mathbf{I} + \boldsymbol{\beta}_e + \boldsymbol{\beta}_p \qquad (6)$$

where $\boldsymbol{\beta}_e = \mathbf{F}_e - \mathbf{I}$ and $\boldsymbol{\beta}_p = \mathbf{F}_p - \mathbf{I}$ are the elastic and plastic distortions, respectively. In particular, the process of imposing a relative displacement $\mathbf{b}_\ell(\mathbf{x})$ to the two sides of a cut in surface $\mathbf{A}_\ell$ (refer to Figure 1) results in the plastic distortion tensor, as in Mura (1987)

$$\boldsymbol{\beta}_p(\mathbf{z}) = - \int_{A_\ell} \delta(\mathbf{z} - \mathbf{x}) \mathbf{b}_\ell(\mathbf{x}) \otimes d\mathbf{A}(\mathbf{x}) \qquad (7)$$

If the relative displacement is constant over $\mathbf{A}_\ell$, the defect is a Volterra dislocation and it models a crystal dislocation with Burgers vector $\mathbf{b}_\ell$. In this case, the dislocation density tensor is concentrated on the closed line bounding $\mathbf{A}_\ell$:

$$\breve{\boldsymbol{\alpha}}(\mathbf{z}) = \boldsymbol{\nabla}_{\mathbf{x}} \times \boldsymbol{\beta}_p = \oint_{\partial A_\ell} \delta(\mathbf{z} - \mathbf{x}) \mathbf{b}_\ell \otimes d\mathbf{x} \qquad (8)$$

where $\boldsymbol{\nabla}_{\mathbf{x}} \times \mathbf{F}_p = \boldsymbol{\nabla}_{\mathbf{x}} \times \boldsymbol{\beta}_p$.

Particularly germane to the present study is volume average of the dislocation density tensor (8). The average of Equation (8) over a region Ω with volume V centered at $\mathbf{y}$ can easily be obtained by virtue of the sifting property of the Dirac-δ, that is

$$\boldsymbol{\alpha}(\mathbf{y}) = \frac{1}{V} \int_{\Omega(\mathbf{y})} \breve{\boldsymbol{\alpha}}(\mathbf{z}) \ dV(\mathbf{z}) = \frac{1}{V} \sum_{\ell \in \Omega} \mathbf{b}_\ell \otimes \int_\ell d\mathbf{x} \tag{9}$$

where the line integral now extends over the portion of the dislocation line inside Ω. Here, $\breve{\boldsymbol{\alpha}}(\mathbf{z})$ is at the scale of dislocations embedded in lattice and the field $\boldsymbol{\alpha}(\mathbf{y})$ is at a larger length scale where continuous fields can be obtained. This definition classifies dislocation lines into geometrically necessary or statistically stored depending on whether line is open or closed within the volume Ω. In previous studies (Mandadapu et al., 2013) we derived a generalization of Equation (9),

$$\boldsymbol{\alpha}(\mathbf{y}) = \int_\kappa d^3x \Delta(\mathbf{x} - \mathbf{y}) \underbrace{\left[\sum_\ell \mathbf{b}_\ell \otimes \int_\ell d\mathbf{z}\delta(\mathbf{z} - \mathbf{x}) \right]}_{\breve{\alpha}(\mathbf{x})} = \sum_\ell \mathbf{b}_\ell \otimes \int_\ell \Delta(\mathbf{x} - \mathbf{y})d\mathbf{x} \tag{10}$$

$$\approx \sum_\ell \mathbf{b}_\ell \otimes \left[\sum_{(ab)\in\ell} \mathbf{x}_{ab} \underbrace{\int_0^1 \Delta(\lambda\mathbf{x}_{ab} + \mathbf{x}_b - \mathbf{y})d\lambda}_{\Phi_{ab}(y)} \right] = \sum_\ell \mathbf{b}_\ell \otimes \left[\sum_{(ab)\in\ell} \mathbf{x}_{ab} \Phi_{ab}(\mathbf{y}) \right] \tag{11}$$

which uses bell-shaped smoothing kernels $\Delta(\mathbf{x} - \mathbf{y})$ that are not necessarily piecewise constant. This definition is evaluated using a quadrature based on straight segments $\mathbf{x}_{ab} = \mathbf{x}_a - \mathbf{x}_b$ of length $\ell_{ab} = \left| \frac{ds}{d\lambda} \right|$ with tangents $\boldsymbol{\xi}_{ab} = \mathbf{x}_{ab}/\ell_{ab}$. In a similar fashion the local line density $\rho(\mathbf{y})$ can be estimated:

$$\rho(\mathbf{y}) = \sum_\ell \int_\ell \Delta(\mathbf{x} - \mathbf{y}) \ \|d\mathbf{x}\| \approx \sum_\ell \sum_{(ab)\in\ell} |\mathbf{x}_{ab}| \Phi_{ab}(\mathbf{y}). \tag{12}$$

Given that the kernel function has compact support, we can use its radius of influence to estimate the range of length scales for the estimated dislocation density to be smooth and yet resolved the variations in the system. In fact, we will estimate the minimum of the range of intermediate asymptotic scales in order to maximize the resolution of trends in the dislocation density tensor field. In averaging via Equation (10) to find the minimum number of dislocation lines in a given volume to obtain smooth density field and to compare with CP results, we are also making the conjecture that this is the scale of CP.

It bears mentioning that in our previous work (Mandadapu et al., 2013) we demonstrated that the definition of α, Equation (9), satisfies the fundamental extensive property of the Burgers vector, namely that the resultant Burgers vector of a Burgers circuit is the sum of Burgers vectors of all individual dislocation lines crossing the enclosed surface. We also showed that Nye's original definition of dislocation density and Equation (9) become equivalent when certain conditions regarding the spacing between the dislocations and curvature of dislocations are obeyed. Consistency, for example

$$\int \rho(\mathbf{y})dV \approx \sum_I \rho_I w_I \approx \sum_I \sum_\ell \sum_{(ab)\in\ell} \|\mathbf{x}_{ab}\| \Phi_{ab}(\mathbf{y}_I)w_I$$

$$= \sum_\ell \sum_{(ab)\in\ell} \|\mathbf{x}_{ab}\| \sum_I \Phi_{ab}(\mathbf{y}_I)w_I$$

$$= \sum_\ell \sum_{(ab)\in\ell} \|\mathbf{x}_{ab}\| = \bar{\rho}V, \tag{13}$$

where $\bar{\rho}$ is the average density, implies the requirement that

$$\sum_I \Phi_{ab}(\mathbf{y}_I)w_I = \sum_I \int_0^1 \Delta(\lambda\mathbf{x}_{ab} + \mathbf{x}_b - \mathbf{y}_I)d\lambda\, w_I$$

$$= \int_0^1 \sum_I \frac{1}{V_I} N_I(\lambda\mathbf{x}_{ab} + \mathbf{x}_b - \mathbf{y}_I)\, d\lambda\, w_I$$

$$= \int_0^1 \sum_I N_I(\lambda\mathbf{x}_{ab} + \mathbf{x}_b - \mathbf{y}_I)\, d\lambda = \int_0^1 d\lambda = 1 \qquad (14)$$

given (a) the localization $\Delta(\mathbf{y}_I) = \frac{1}{V_I}N_I$ is based on a partition of unity $N_I(\mathbf{x})|\sum_I N_I(\mathbf{x}) = 1 \quad \forall \mathbf{x}$, (b) exact integration of the line integral, and (c) the integration weights equal to kernel volumes $w_I = V_I$. For a piece-wise constant basis N_I with compact support on a cubical region, exact integration is simply

$$\int_0^1 N_I(\lambda\mathbf{x}_{ab} + \mathbf{x}_b - \mathbf{y}_I)d\lambda = \int_{\lambda_0}^{\lambda_1} d\lambda = \lambda_1 - \lambda_0 \qquad (15)$$

given the two intersection points $\mathbf{x}_{0|1} = \lambda_{0|1}\mathbf{x}_{ab} + \mathbf{x}_b$ of the segment with the boundary of the cube where $\lambda_0 < \lambda_1 \in [0,1]$. If either end of the segment does not intersect the cube and instead is in the interior of the cube then $\lambda_{0|1} = 0|1$.

3. Method

We employed the discrete dislocation dynamics (DD) method developed by Po et al. (2014), as implemented in the computer program MODEL (Amodeo et al., 2015). A brief explanation of the method is given here; a full account of the method can be found in Po et al. (2014). The crystal plasticity model was implemented in the finite deformation code Albany (Salinger et al., 2015).

Given the aim of this work to compare and correlate DD and CP, we have maximized the commonalities of the two modeling techniques. For both models, we use: (a) isotropic elasticity with the slip planes of a cubic lattice (which allows the use of Green's functions in DD), (b) acceleration is ignored in the balance of linear momentum, however there are dynamic/viscous effects in each model, (c) the momentum balance is solved on finite element meshes with corresponding boundary conditions. In this mode, both the response of CP, governed by its flow rule, and that of DD, governed by its mobility equation, are relatively insensitive to loading rate.

The formula (10) allows to construct a continuum dislocation density field from DD simulations, which can be directly compared to the corresponding quantity defined in Sec. 3.2 in the CP framework.

3.1 Dislocation Dynamics

In general, DD approaches treat dislocation lines explicitly, with lines decomposed into discrete edges (segments) bounded by vertices. Given the formulas in Sec. 2, the dislocation density field $\alpha(\mathbf{x})$ can be computed for a network of dislocations; however, their evolution from a given initial configuration must be computed from a closure equation yielding the dislocation velocity $\mathbf{w}$ as a function of the local stress state. Such an equation of motion for the discrete dislocation configuration can be obtained from thermodynamic considerations (Po et al., 2014). For FCC metals and by neglecting inertia terms, the equation of motion can be written as:

$$-B\mathbf{w} + (\sigma\mathbf{b}) \times \boldsymbol{\xi} = \mathbf{0} \qquad (16)$$

Equation (16) expresses the balance of the total configurational force per unit line of dislocation, and is comprised of: a viscous drag contribution (Note 1), $-B\mathbf{w}$, and a mechanical contribution, $(\sigma\mathbf{b}) \times \boldsymbol{\xi}$, the Peach-Koehler force. In contrast to CP, the assumption of linearized kinematics adopted in DD allows to write the (small-strain) stress field σ appearing in the Peach-Koehler force as the sum of two terms: (a) a dislocation-dislocation interaction term which can be computed accurately using the Peach-Koehler stress equation (Peach & Koehler, 1950) and (b) a correction term which accounts for the mechanical boundary conditions imposed to the simulation domain.

The DD formulation used in this work solves Equation (16) in weak/variational form for the dislocation velocity field using the finite element method, refer to Po et al. (2014). The numerical implementation is based on discretization of dislocation lines into segments connecting pairs of nodes. Segment shape functions and nodal degrees of freedom are then used for the unknown dislocation velocity field $\mathbf{w}$, so that Equation (16), upon assembly over the dislocation network, is transformed into a discrete system of equations for the nodal velocities. Finally, the nodal velocities are used to update the nodal positions and evolve the dislocation configuration in time.

3.2 Crystal Plasticity

In our CP model we use Kirchhoff-St. Venant elasticity $\mathbf{S} = C\mathbf{E}_e$ with elastic Lagrange strain $\mathbf{E}_e = \frac{1}{2}(\mathbf{F}_e^T \mathbf{F}_e - \mathbf{I})$.

This elastic constitutive relation reduces to $\sigma = \frac{1}{\det \mathbf{F}_e}\mathbf{F}_e\mathbf{S}(\mathbf{E}_e)\mathbf{F}_e^T \approx C\frac{1}{2}(\boldsymbol{\beta}_e + \boldsymbol{\beta}_e^T)$ in the limit of small strain and

rotation. The elastic modulus C with cubic symmetry is

$$C = \sum_{i,j \neq i} C_{11}\mathbf{e}_{ii}\otimes\mathbf{e}_{ii} + C_{12}\mathbf{e}_{ii}\otimes\mathbf{e}_{jj} + C_{44}\mathbf{e}_{ij}\otimes\mathbf{e}_{ij} \tag{17}$$

where $\mathbf{e}_{ij}$ is the dyad $\mathbf{e}_i\otimes\mathbf{e}_j$, which we reduce to isotropy by setting $C_{44} = G$, $C_{11} = 2G\frac{1-v}{1-2v}$, $C_{12} = 2G\frac{2v}{1-2v}$, for correspondence with the DD simulations.

For the plastic behavior, the kinematic assumption in Equation (4) can be seen as a plastic flow rule of the form found in Ortiz and Stainier (1999). A particularly simple hardening rule for the rate of slip $\dot{\gamma}_a$ for slip system a is

$$\dot{\gamma}_a = \dot{\gamma}_a(\tau_a) = C_a\left|\frac{\tau_a}{\tau_a^{\text{crit}} + H_a\gamma_a}\right|^{1/m} \text{sgn}(\tau_a) \tag{18}$$

subject to a τ_a^{crit} threshold and (isotropic) hardening modulus H_a. This simple functional form is appropriate for the present study of DDT fields since it is in wide-spread use, has few parameters to calibrate and its phenomenological behavior is easy to interpret. This is in contrast to models which add non-local, gradient and coupled slip effects or include dislocation density as an evolving phenomenological parameter or a physical field, see, *e.g.*, Bargmann et al. (2010). The exact form of these modifications of the basic CP equations is still a matter of debate and we give a discussion of these models in light of our results in Sec. 5. Here $\tau_a = \mathbf{P}_a \cdot \boldsymbol{\Sigma}$ is the shear stress resolved on the slip plane with normal $\mathbf{n}_a$ in the intermediate/lattice configuration and the Mandel stress $\boldsymbol{\Sigma}$ is simply $\boldsymbol{\Sigma} = (\mathbf{I} + 2\mathbf{E}_e)\mathbf{S}(\mathbf{E}_e)$, refer to Cuitino and Ortiz (1993). The implicit update algorithm largely follows Ortiz and Stainier (1999).

Given a solution to momentum balance, we recover a dislocation density field by first using a local, element-wise L_2 projection to move $\mathbf{F}_p$ data at the integration points to the element nodes. Then using the finite element basis functions N_I to interpolate $\mathbf{F}_p$ on *each element*, which allows differentiation with respect the reference coordinates $\mathbf{X}$, we obtain:

$$\boldsymbol{\alpha} = \nabla_{\mathbf{x}} \times \mathbf{F}_p = \boldsymbol{\epsilon} \cdot \nabla_{\mathbf{x}}\mathbf{F}_p = \sum_I \boldsymbol{\epsilon}\mathbf{F}_p(\mathbf{X}_I)\nabla_{\mathbf{x}}N_I = \sum_{I,a,A,B,C} \epsilon_{ABC}F_{aA}(\mathbf{X}_I)N_{I,B}\mathbf{e}_a\otimes\mathbf{E}_C \tag{19}$$

where I indexes nodes, $\boldsymbol{\epsilon}$ is the permutation tensor, $\mathbf{e}$ and $\mathbf{E}$ are basis vectors in the spatial and reference frames (respectively), and we have dropped the subscript p for clarity. Chain rule provides the spatial gradient $N_{I,B}$ in terms of the partial derivatives of the basis N_I with respect to the local coordinates ζ_a

$$N_{I,B} = \sum_a N_{I,a}(\partial_\zeta\mathbf{X})_{aB}^{-1} = \sum_a N_{I,a}\left(\sum_J \mathbf{X}_J\otimes\partial_\zeta N_J\right)_{aB}^{-1} \tag{20}$$

where $\partial_\zeta\mathbf{X}$ is the Jacobian of the local ζ-to-referential $\mathbf{X}$ coordinate map. Lastly, we average the local node values to obtain a unique value for every node and hence a continuous plastic deformation curl field. Alternately a global L_2 projection could be used.

4. Results

For our comparison studies we chose face-centered cubic single crystals of Cu. . Its 12 slip systems are defined by the Schmid tensors $\{\boldsymbol{s}_a \otimes \boldsymbol{n}_a\} = \{\langle 110\rangle\otimes\langle 111\rangle\}$. As mentioned, we employ an isotropic elasticity model with effective Poisson ratio $v=0.34$ and shear modulus $G=48$ GPa. Given that MODEL is a dimensionless code, we adopt the same normalization in our results. Lengths are normalized to the material's Burgers vector $b=|\mathbf{b}| = 2.556$ Å, stress-like quantities are normalized by G, the drag coefficient is set at $B=10^{-4}$ Pa-s, and time is normalized by $\tau = B/G = 2.08$ fs.

After preliminary studies, over a wide range initial (fully gissile) dislocation densities and loading rates, we chose an initial dislocation density of $\rho_0 = 10^{13}$ m^{-2} = 6.533×10^{-7} b^{-2}, consisting of prismatic dislocation loops with $\langle 110\rangle$-type Burgers vectors lying in $\{111\}$ planes, and loading rate of 10^{-11} dimensionless inverse-time units (corresponding to 4.8×10^3 sec^{-1}) to approximate the rate-independent loading regime. From the initial dislocation density we employ a length-scale $\ell = \frac{1}{\sqrt{\rho_0}} = 1.29 \times 10^3 b$ to non-dimensionalize α fields. Given that initially the dislocation network consists entirely of loops, the total dislocation density tensor starts at zero which also gives some support to assuming that $\mathbf{F}_p = \mathbf{I}$ in the initial state of the material modelled with CP. (To allow potentially unphysical arrangement of dislocations to relax, we equilibrated the systems for 10,000 steps under zero applied load. Although minor fluctuations in the loop geometries did occur during this simulated time, in general no major changes in the configurations were observed.) Note, Equation (16) introduces a time-scale to DD and Equation (18) introduces time-scale to CP, yet neither has inertia so scaling time can lead to self-similar solutions dependent only on loading and not the rate of loading. For DD, this limit is achieved when the dislocation interactions and collisions are sufficiently resolved. For CP, time-scale and loading rate invariance is apparent only as $m \to \infty$ and also for small strains with a linear flow rule, as in Steinmann and Stein (1996) and Miehe and Schröder (2001). It should be noted that, despite the absence of inertial effects, the states generated by both the DD and CP models are dynamic. Using identical system size and loading, we examine the response of DD and CP for strains up to 1%, well below the limit of validity of the linear elastic model in DD.

In Sec. 4.1 we use homogeneous compression and shear loading results from DD to calibrate the CP flow rule and to estimate a minimum asymptotic scale. This calibration makes an implicit connection between the initial dislocation distribution in the DD systems and the hardening parameters of the CP model. Then, in Sec. 4.2 we compare the dislocation density fields and global response resulting from inhomogeneous loading due to torsion. For all comparisons an ensemble of 10 replicas of the DD systems are used to alleviate sensitivity to initial dislocation arrangement.

4.1 Calibration with Homogeneous Loading Simulations

With the given elastic constants and symmetries, we use compression and (simple) shear loading DD and CP simulations to calibrate the hardening parameters of the flow rule specified in Equation (18). For these nominally homogeneous loading cases we employed a cube of dimensions $2000b \times 2000b \times 2000b$ oriented with the $\langle 100 \rangle$ crystal directions. In compression, the top face of the cube (the face with outward normal in the $+\mathbf{e}_3$ direction) is displaced in $\mathbf{e}_3$ and the bottom face x_3 (face with normal in the $-\mathbf{e}_3$ direction) is fixed; the $+x_1$ and $+x_2$ lateral faces are constrained in their normal directions to promote homogeneous deformation and allow for expansion. In shear, the top face is displaced in x_1 while fixed in x_2 and x_3; the bottom face is fixed in x_1, x_2 and x_3 and all lateral faces are fixed in x_2 and x_3.

Figure 2 shows the stress-strain response for compression. The initial slope of the DD response, $2.54G$, compares well with the expected Young's modulus of $2(1+v)G = 2.68G$. Likewise, Figure 3 shows the stress-strain response for simple shear loading. In this case, the initial slope is $0.92G$ near the expected value $1.00G$. These results imply that the initial distribution of dislocations minimally affects the material's elastic response. After approximately 0.001 strain in compression and 0.004 strain in shear the total line density starts to rise rapidly which corresponds to a distinct decrease in the slope of the stress-strain response. Using direct fits to the DD response we determined that the slope of this plastic regime was $0.12G$ in the compression case and $0.42G$ in the shear case and that the onset of plastic response was at 0.00125 and 0.0031 strain, respectively. These values were used as initial guesses in calibrating the CP model. Figure 2 and 3 also show that the calibrated CP stress-strain response compares well to that of DD using the hardening parameters: $C_a = 1$ (inverse time units), $\tau_a^{\text{crit}} = 0.11$ GPa $= 0.0022\,G$, and $H_a = 30.5$ GPa $= 0.635\,G$, for all slip systems $a \in [1,12]$. We employed the exponent $m=4$ which was the highest value that offered numerical stability (The issue of instability upon approaching the rate independent limit $m \to \infty$ is well-known, as discussed by Steinmann and Stein, 1996; Miehe & Schröder, 2001; and Borja, 2013).

Lastly, by using a kernel estimator, Equation (10), positioned near the center of a $4000b \times 4000b \times 4000b$ system we were able to estimate the asymptotic region for the dislocation density tensor field using the compression case. Figure 4 shows that three relevant components of α converge to near constant, non-trivial values for kernel radii above $\approx 500\,b$, which is also well below the system size used for this study.

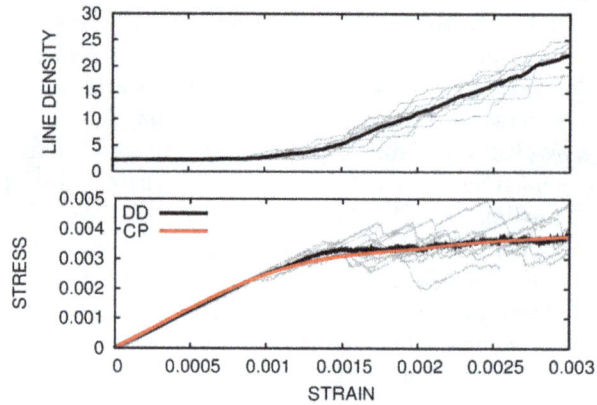

Figure 2. Compression: (top) dislocation line density and (bottom) 11-stress versus 11-strain. Note the response of the individual DD simulations are shown in gray, line density is normalized by ρ_0 and stress by G

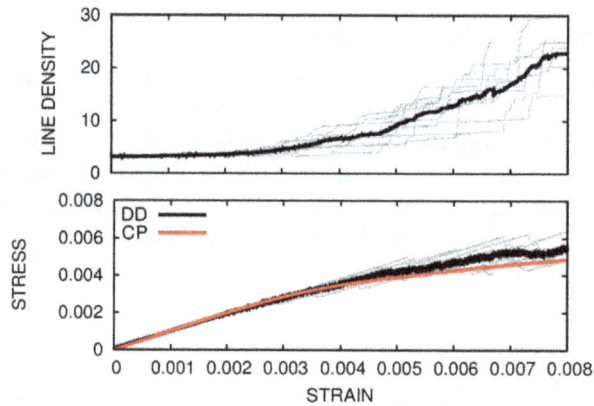

Figure 3. Shear: (top) dislocation line density and (bottom) 12-stress versus 12-strain . Note the response of the individual DD simulations are shown in gray, line density is normalized by ρ_0 and stress by G

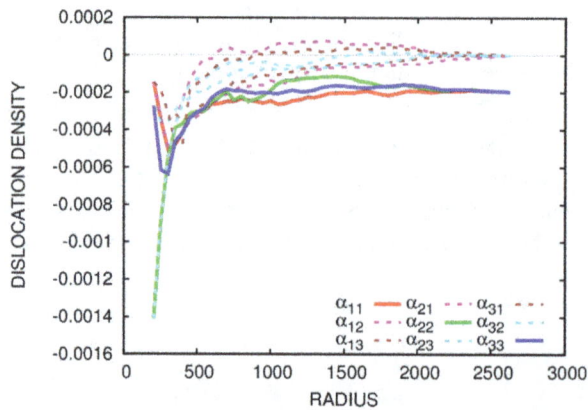

Figure 4. Dislocation density tensor components *vs.* kernel radius for 3.25 percent compressive strain at an interior point in the system. The apparent minimum asymptotic scale is approximately $500b$ (1/8 of the system size). Note that the values of α are non-dimensionalized by $\ell = 1/\sqrt{\rho}$

4.2 Comparison of Inhomogeneous Loading Simulations

For comparing the predictions of the DD and CP models we simulated the torsion of a $L_1 \times L_2 \times L_3 = 2000b \times 2000b \times 4000b$ rectangular wire. The torsion was effected by a twisting displacement of the top relative to the bottom faces, with the bottom face fully fixed and the lateral faces traction-free. Both $\langle 100 \rangle$ and $\langle 110 \rangle$ crystallographic orientations of the lateral faces of the wire are explored. Since the elastic response is isotropic, only slip plane orientation relative to the loading distinguishes these two cases. The general solution for the stress and displacement fields in an isotropic elastic wire due to torsion are Timoshenko and Goodier (1951):

$$\sigma = \frac{1}{2} G\Theta \left((w_{,1} - x_2)(\mathbf{e}_1 \otimes \mathbf{e}_3 + \mathbf{e}_3 \otimes \mathbf{e}_1) + (w_{,2} + x_1)(\mathbf{e}_2 \otimes \mathbf{e}_3 + \mathbf{e}_3 \otimes \mathbf{e}_2) \right) \tag{21}$$

and

$$\mathbf{u} = -\Theta x_3 (x_2 \mathbf{e}_1 - x_1 \mathbf{e}_2) + \Theta w \mathbf{e}_3 \tag{22}$$

in terms of a warping function $w = w(x_1, x_2)$. Given the warping function characteristic of our prismatic wire with square cross-section and traction-free sides, the torsional modulus K relating the torsional moment $M\mathbf{e}_3 = \int \mathbf{x} \times \sigma d\mathbf{A}$ to the twist per axial distance Θ/L_3 is

$$K = GL_1^4 \left(\frac{1}{3} - \frac{64}{\pi^3} \sum_{n=0}^{\infty} \frac{\tanh(\pi(2n+1)/2)}{(2n+1)^5} \right) \approx 0.1406 GL_1^4 \tag{23}$$

where $L_1 = L_2$ is the side length of the square cross-section.

The torque-twist response for both models are shown in Figure 5. The correspondence between the calibrated CP model and the DD response for the $\langle 100 \rangle$ orientation of the wire is similar to that shown in Figure 3 for shear. For the $\langle 110 \rangle$ oriented wire, however, the DD response shows a delayed onset of significant plasticity, as evidenced both by the torque-twist response as well as the line density evolution, whereas the CP torque-twist for this orientation is nearly the same as for $\langle 100 \rangle$. The quasi-linear increase in dislocation density with twist angle was also observed in Senger et al. (2011) and is in agreement with Fleck and Hutchinson (1997). Figure 6 shows the evolution of the dislocation network with twist for the $\langle 100 \rangle$ case. The dislocations are primarily screw dislocations on different glide planes. They reach their shown locations after a series of cross slip events from a few initial common sources. Clearly, between twist Θ=0.02 radians and 0.04 radians, the initial random network multiplied through Frank-Read mechanisms (no nucleation model is employed, as in Senger et al. (2011)) and aligned with the diagonals of the square cross-section. This regime also corresponds to the apparent onset of plasticity shown in Figure 5.

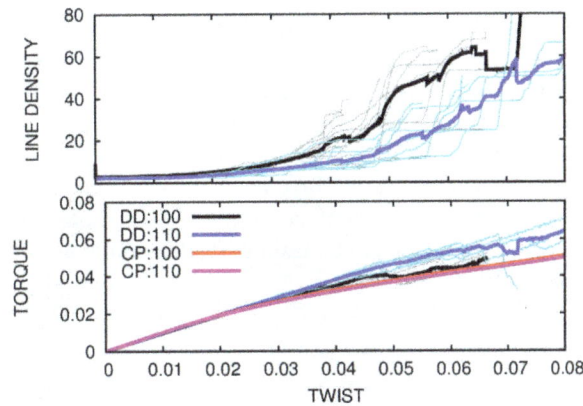

Figure 5. Torsion: (top) normalized dislocation line length normalized by the line density ρ, and (bottom) torque normalized by the torsion modulus K versus twist angle (in radians)

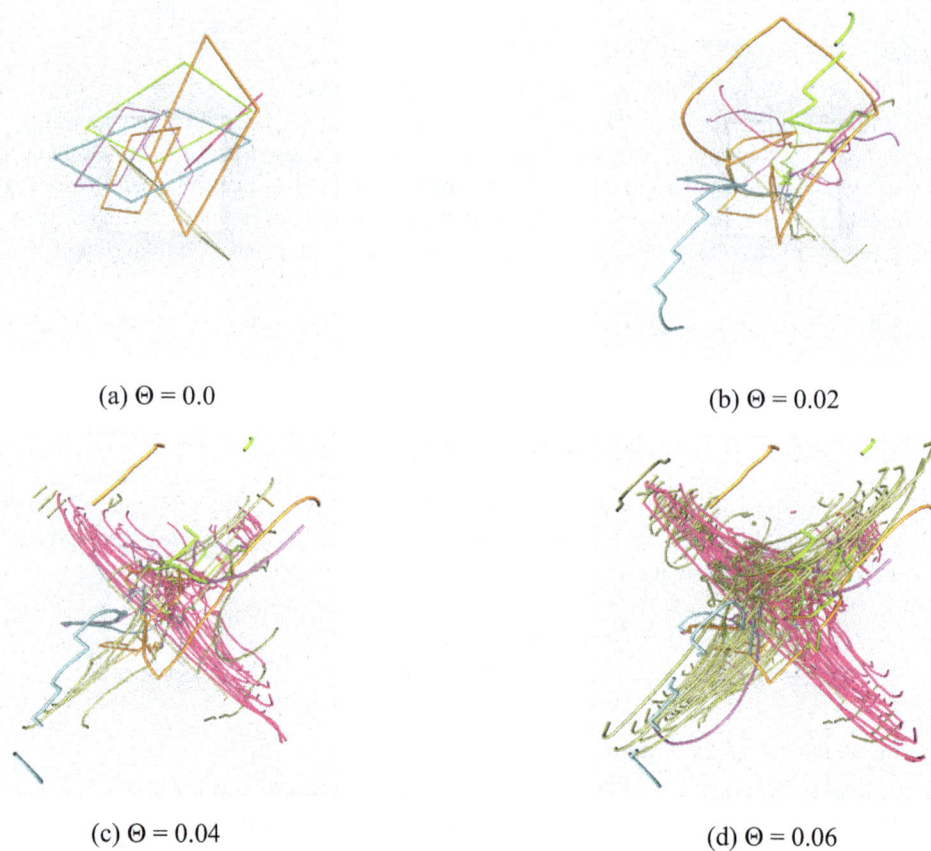

(a) $\Theta = 0.0$

(b) $\Theta = 0.02$

(c) $\Theta = 0.04$

(d) $\Theta = 0.06$

Figure 6. Torsion with orientation $\langle 100 \rangle$: evolution of the dislocation network initially composed of a random distribution of glissile prismatic loops viewed along the twist axis. Colors indicate dislocations lines with different Burgers vectors and the jogs in the dislocation lines are due to cross-slip events

Figure 7a shows the dislocation network and resulting line density at distinctly plastic state, twist angle $\Theta = 0.06$ radians for both the $\langle 100 \rangle$ and $\langle 110 \rangle$ cases. In the $\langle 100 \rangle$ case, dislocation lines spanning the diagonals of the wire's square cross-section pile up; whereas, in the $\langle 110 \rangle$ case shown, the majority of the lines are vertical and span the opposing face mid-lines while a minority span one pair of the diagonal corners as in the $\langle 100 \rangle$ case, *cf.* Senger et al. (2011). Clearly from the line networks themselves and the resulting line density, the dislocation lines in the $\langle 110 \rangle$ case are more dispersed spatially and less numerous in total than in the $\langle 100 \rangle$ case. (Note these networks are still dense relative to the nominally homogeneous cases where there is no tendency for dislocations to accumulate in any particular region of the body.) In DD, dislocation lines are generated in high stress regions near the corners of the wire in torsion and migrate to the low/zero stress basins shown in green in Figure 7b, given the Peach-Koehler forces in Equation (16). This hypothesis is corroborated by the findings in Weinberger and Cai (2010) and Senger et al. (2011). For the two cases, the orientations of the slip planes relative to the wire are different, but elastic stress field is same, up until yield, due to the isotropic moduli assumption. The resolved shear stress (RSS) is plotted in Figure 7b for both cases. In the $\langle 100 \rangle$ case the diagonal basin in RSS is dominant and in the $\langle 110 \rangle$ case the vertical basin is dominant but broader than in the $\langle 100 \rangle$.

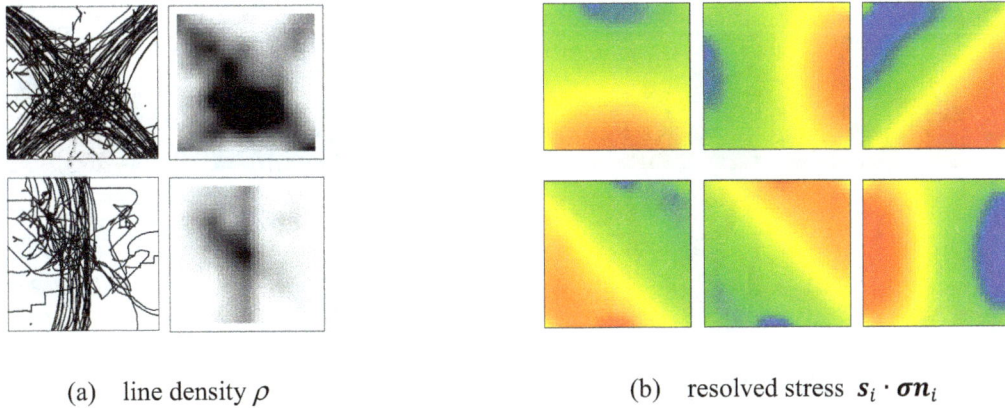

(a)   line density ρ (b)   resolved stress $\boldsymbol{s}_i \cdot \boldsymbol{\sigma n}_i$

Figure 7. Torsion: (a) dislocation lines (left), line densities (grayscale, same scale for both panels) and (b) resolved shear stress $\boldsymbol{s} \cdot \boldsymbol{\sigma n}$ on plane with normal $\boldsymbol{n}=(1,1,1)$ projected onto the x_1-x_2 plane for $\boldsymbol{s}=(1,1,1)$, $(-1,1,0)$, and $(1,-1,0)$ (rainbow, same scale for all panels). The $\langle 100 \rangle$-orientation is shown on the top row and the $\langle 110 \rangle$-orientation on the bottom row

Since our system size and total line density are at the threshold of being able to be resolved given the estimated minimum asymptotic scale, we average the DD data through the x_3-axis since slices of the solution are nominally equivalent along the twist axis. Figure 8 and Figure 9 show the two dimensional projection of the in-plane components of the dislocation density tensor for the $\langle 100 \rangle$ and $\langle 110 \rangle$ orientations of the wire, respectively. In general, the α fields generated with DD and CP are only comparable up to a divergence-free/solenoidal field $i.e.$ $\mathbf{F}_p$ gradient of scalar function does not contribute to $\boldsymbol{\alpha}$ (which follows directly from a Helmholtz decomposition of $\mathbf{F}_p$). Also, the solutions have nominal 90 degree rotation and inversion symmetry. The fields resulting from both DD and CP have comparable magnitudes. And yet, the $\boldsymbol{\alpha}$ fields generated by DD have significant components in the center of wire, unlike CP. In the $\langle 100 \rangle$ orientation case, the magnitudes of $\boldsymbol{\alpha}$ components are similar; but in the $\langle 110 \rangle$ case the weak pattern correspondence breaks down due to more dispersed nature of dislocation line density. As mentioned, in DD lines are generated at high stress regions at the lateral edges of the prismatic wire and migrate to the RSS basins either diagonal or crossing the midpoints of the wire's cross-section. In contrast, in a CP simulation, a background of mobile, geometrically necessary dislocations are assumed to be present everywhere and plastic slip occurs in regions of high stress. In CP there is no mechanism for explicit generation of dislocations nor compatibility of plastic slip induced by the motion of the implicit dislocation network. Hence, the dislocation density field produced by CP only responds to the regions of high RSS, and there is no driving force for migration/flow of dislocation density to low RSS basins.

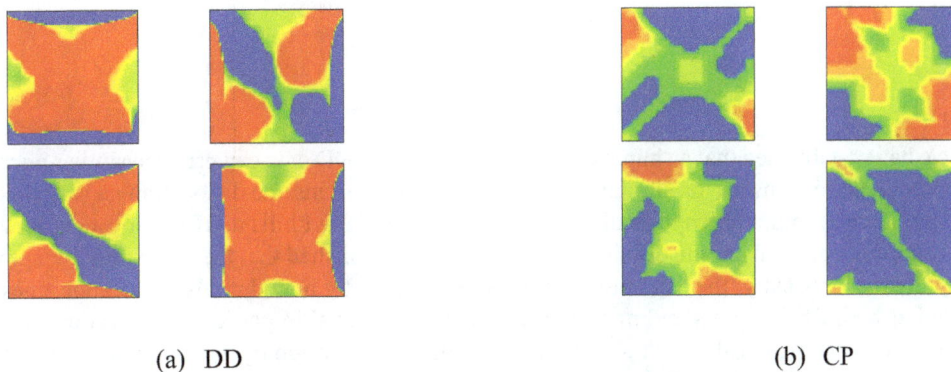

(a)   DD   (b)   CP

Figure 8. Torsion with orientation $\langle 100 \rangle$: comparison of α_{11} (upper left), α_{12} (upper right), α_{21} (lower left), α_{22} (lower right) resulting from (a) DD and (b) CP simulations at twist = 0.006 radians. The same scale is used in all panels where red corresponds to $\alpha = 0.00258/\ell$ and blue to $\alpha = -0.00258/\ell$

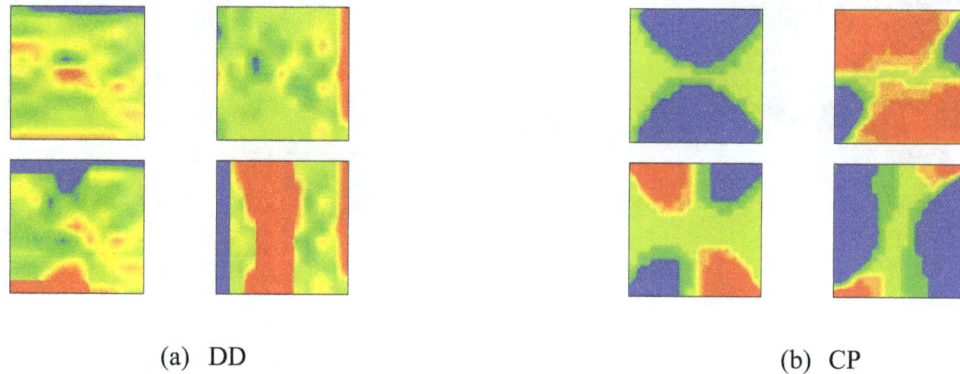

(a) DD   (b) CP

Figure 9. Torsion with orientation $\langle 110 \rangle$: comparison of α_{11} (upper left), α_{12} (upper right), α_{21} (lower left), α_{22} (lower right) resulting from (a) DD and (b) CP simulations at twist = 0.006 radians. The same scale is used in all panels where red corresponds to $\alpha = 0.00258/\ell$ and blue to $\alpha = -0.00258/\ell$

As a last comparison, Figure 10 shows the sensitivity of the dislocation density field for the $\langle 100 \rangle$ case predicted by CP to the flow rule (18) exponent m. Clearly the significant dislocation density approaches the center of the cross-section and the "×" pattern becomes more prominent and sharper with increased m showing better correspondence with the DD results. An explanation for this better correspondence is discussed in the concluding section. Due to numerical issues with the form the simple flow rule (18) it was not possible to increase m to further improve the correspondence with the DD results.

(a) m = 1   (b) m = 4

Figure 10. Torsion: dislocation density α_{11} for flow rule exponent $m = 1,4$ (same color scale in both panels)

5. Discussion

In this work, we have established that a simple CP model calibrated to DD data can produce similar microstructural fields as well as overall response. This similarity is predicated on resolving the dislocation density field at a scale above the intermediate asymptotic length scale intrinsic to the material with its native dislocation population. For the material we examined, Cu with line density $\rho = 10^{13}$ m^{-2}, we established that the minimum asymptotic scale is approximately 500 b. Physically, the asymptotic scale is determined by how densely dislocation lines pack after being generated at high RSS regions and migrating to low RSS regions. In particular, we estimated the scale at which continuum dislocation density fields should appear with nominally homogeneous simulations where there is little difference between high and low RSS regions. This estimate proved robust in constructing dislocation density fields in the inhomogeneous loading cases we considered. For the finite systems we examined, it appears that the dislocation pile-ups on particular slip planes reach a characteristic spacing due to the fact that forces between dislocations scale as b^2, while forces due to the applied stress scale as b, the Burgers vector of the material (refer to Chap. 21 in Hirth & Lothe, 1992). Hence, we conjecture that the asymptotic scale for the dislocation density is strongly dependent on material and less dependent on the particular loading case, (see Senger et al., 2011) similar patterns for a crystallographic orientation not examined in this work.)

By examining Equation (10) and Equation (12), the connection between the dislocation line density ρ and dislocation density tensor $\boldsymbol{\alpha}$ can be seen. Clearly ρ indicates the general magnitude of $\boldsymbol{\alpha}$, *i.e.* there are no $\boldsymbol{\alpha}$ contributions where ρ is zero; but, $\boldsymbol{\alpha}$ adds vectorial information from the line direction as well as from the Burgers vector, see Equation (10) In contrast, with a CP model the dislocation density α results from the curl of the plastic deformation gradient through the CP flow rule, Equation (18). So, unlike the explicit representation of dislocation lines in DD, CP, through its flow rule, conflates generation and migration and only creates mobile slip in high stress/deformation gradient regions. Moreover it simulates the hardening due to entanglement of dislocation lines with the flow rule.

A loose correspondence can be made between the CP flow rule and the dynamics of the dislocation lines that governs DD evolution. The Peach-Koehler forces driving the dislocation dynamics are analogous to the resolved shear stresses driving plastic evolution in the flow rule so that the dislocation line velocity is proportional to the Peach-Koehler force, and the plastic slip rate is proportional to the local resolved shear stress. From our results, is it clear that, although the global response of corresponding DD and CP models can be made the same, the microstructural dislocation density tensor field is much more sensitive to details of the calibrated flow rule. Although algorithmic stability issues prevented us from exploring the response of flow rules with high exponents, we were able to show that increased exponent tends to improve the qualitative similarity of the in-plane α fields for a prismatic wire in torsion where there is a strong tendency for the dislocation network to align with geometric features. It stands to reason that this improvement will saturate before the $m \rightarrow \infty$ limit is reached, where plastic slip, and by analogy dislocation velocity, is insensitive to RSS magnitude. In fact, it is hard to imagine how a flow rule could be constructed to perfectly represent the α patterns seen with DD unless some compatibility of the plastic slip representing dislocation motion is enforced. This modeling direction is currently being developed within the framework of *continuum dislocation theory*, as in Acharya and Roy (2006); Xia and El-Azab (2015); Hochrainer et al. (2014); Le (2016), where dislocation densities are independent field quantities evolving according to transport equations, and plastic slip is computed from dislocation fluxes.

In general, the continuum field, α given by Equations (3) and (10), captures the (geometrically necessary) dislocation density and is a quantitative measure useful in understanding the microstructures that result from large strains in the plastic regime. This tensor can be used in the yield function of macroscopic phenomenological theories of plasticity to reproduce realistic dislocation microstructures. Such models form alternatives to the crystal-plasticity formulations based on active slip systems. For example, Edmiston et al. (2013) include the DDT in the yield function, and develop a rate-independent macroscopic theory based on material symmetry. As the authors demonstrate, inclusion of DDT in the yield function introduces an additional length-scale into the macroscopic theory. Using this length-scale as a parameter, Edmiston et al. (2013) have shown the formation of dislocation microstructures with spatially localized DDT similar to our work. For similar structures, the value of the length-scale in the theory are also similar to the range of intermediate asymptotic length scales that we estimated are needed to obtain a field description of the DDT from the DD simulations. This suggests that performing DD simulations to evaluate the DDT tensor may serve as a practical tool for calibrating and validating the phenomenological theories of plasticity. Also the correspondence between DD and CP could perhaps be improved by using a flow rule with explicit dependence on the DDT computed from the plastic deformation gradient. An analysis along these lines is left for future work.

6. Conclusion

With Cu simulations of the same size and loading rate, we were able to compare the response of DD and CP models in torsion after calibration of the CP model with system-level DD data to fix the hardening parameters. The comparison of the dislocation density field involved novel field estimators applied to the DD data and we found an asymptotic scale on the order of hundreds of Burgers vectors for Cu with the given initial dislocation density. Below this limit, which is still challenging for DD simulations, comparison would be meaningless. At the scales we chose, the correspondence in the results of the two models is both qualitative, in that the dislocation density fields have the same pattern, and loosely quantitative, given that the magnitudes of the two resulting fields are on-par. The primary discrepancy appears in regions where there is no apparent driving force for slip, *i.e.* in the center of the torsion specimens. Here, CP predicts insignificant dislocation densities, whereas the DD model show that dislocations gravitate to region of low shear stress.

Acknowledgements

The authors would like to acknowledge helpful discussions with J. Ostien, A. Mota, F. Abdeljawad, H. Lim, and R. Sills (Sandia), and support from Sandia National Laboratories' Advanced Simulation and Computing/Physics & Engineering Models (ASC/P&EM) program. Sandia National Laboratories is a multi-program laboratory managed and operated by Sandia Corporation, a wholly owned subsidiary of Lockheed Martin Corporation, for the U.S. Department of Energy's National Nuclear Security Administration under contract DE-AC04-94AL85000.

References

Acharya, A., & Bassani, J. (2000). Lattice incompatibility and a gradient theory of crystal plasticity. *Journal of the Mechanics and Physics of Solids*, 48(8), 1565–1595. http://dx.doi.org/10.1016/s0022-5096(99)00075-7

Acharya, A., & Roy, A. (2006). Size effects and idealized dislocation microstructure at small scales: Predictions of a Phenomenological model of Mesoscopic Field Dislocation Mechanics: Part I. *J Mech Phys Solids*, 54(8), 1687–1710. http://dx.doi.org/10.1016/j.jmps.2006.01.009

Amodeo, R., Azab, A. E., Wang, Z., Sun, L., Hsi Tong, S., Huang, J.-M., … Ghoniem, N. (2015). MODEL: the Mechanics of Defect Evolution Library. https://bitbucket.org/model/model/wiki/Home.

Aoyagi, Y., & Shizawa, K. (2007). Multiscale crystal plasticity modeling based on geometrically necessary crystal defects and simulation on fine-graining for polycrystal. *Int J Plasticity*, 23(6), 1022–1040. http://dx.doi.org/10.1016/j.ijplas.2006.10.009

Asaro, R. J. (1983). Crystal plasticity. *Journal of applied mechanics*, 50(4b), 921–934. http://dx.doi.org/10.1115/1.3167205

Babu, B., & Lindgren, L.-E. (2013). Dislocation density based model for plastic deformation and globularization of Ti-6Al-4V. *International Journal of Plasticity*, 50:94–108. http://dx.doi.org/10.1016/j.ijplas.2013.04.003

Bargmann, S., Ekh, M., Runesson, K., & Svendsen, B. (2010). Modeling of polycrystals with gradient crystal plasticity: A comparison of strategies. *Philosophical Magazine*, 90(10), 1263–1288. http://dx.doi.org/10.1080/14786430903334332

Berry, J., Provatas, N., Rottler, J., & Sinclair, C. W. (2012). Defect stability in phase-field crystal models: Stacking faults and partial dislocations. *Physical Review B*, 86(22), 224112. http://dx.doi.org/10.1103/physrevb.86.224112

Bilby, B. (1960). Continuous distributions of dislocations. *Progress in solid mechanics*, 1(1), 329–398.

Borja, R. I. (2013). *Plasticity: Modeling & Computation*. Springer.

Chandra, S., Samal, M., Chavan, V., & Patel, R. (2015). Multiscale modeling of plasticity in a copper single crystal deformed at high strain rates. *Plasticity and Mechanics of Defects*, 1(1). http://dx.doi.org/10.1515/pmd-2015-0001

Clayton, J. D., Bammann, D. J., & McDowell, D. L. (2005). A geometric framework for the kinematics of crystals with defects. *Philosophical Magazine*, 85(33-35), 3983–4010. http://dx.doi.org/10.1080/14786430500363312

Cuitino, A. M., & Ortiz, M. (1993). Computational modelling of single crystals. *Modelling and Simulation in Materials Science and Engineering*, 1(3), 225. http://dx.doi.org/10.1088/0965-0393/1/3/001

Déprés, C., Robertson, C., & Fivel, M. C. (2006). Low-strain fatigue in 316l steel surface grains: a three dimension discrete dislocation dynamics modelling of the early cycles. part 2: Persistent slip markings and micro-crack nucleation. *Philosophical Magazine*, 86(1), 79–97. http://dx.doi.org/10.1080/14786430500341250

Devincre, B., & Gatti, R. (2015). Physically justified models for crystal plasticity developed with dislocation dynamics simulations. *AerospaceLab*, (9), 1–7.

Edmiston, J., Steigmann, D., Johnson, G., and Barton, N. (2013). A model for elastic–viscoplastic deformations of crystalline solids based on material symmetry: Theory and plane-strain simulations. *International Journal of Engineering Science*, 63:10–22. http://dx.doi.org/10.1016/j.ijengsci.2012.10.001

Evers, L., Brekelmans, W., and Geers, M. (2004a). Non-local crystal plasticity model with intrinsic ssd and gnd effects. *Journal of the Mechanics and Physics of Solids*, 52(10), 2379–2401. http://dx.doi.org/10.1016/j.jmps.2004.03.007

Evers, L., Brekelmans, W., and Geers, M. (2004b). Scale dependent crystal plasticity framework with dislocation density and grain boundary effects. *International Journal of solids and structures*, 41(18), 5209–5230. http://dx.doi.org/10.1016/j.ijsolstr.2004.04.021

Field, D. P., & Alankar, A. (2010). Observation of Deformation and Lattice Rotation in a Cu Bicrystal. *Metall and Mat Trans A*, 42(3), 676–683. http://dx.doi.org/10.1007/s11661-010-0570-2

Field, D. P., Magid, K. R., Mastorakos, I. N., Florando, J. N., Lassila, D. H., and Morris, Jr., J. W. (2010). Mesoscale strain measurement in deformed crystals: A comparison of X-ray microdiffraction with electron backscatter diffraction. *Philosophical Magazine*, 90(11), 1451–1464. http://dx.doi.org/10.1080/14786430 903397297

Fleck, N., & Hutchinson, J. (1997). Strain gradient plasticity. *Advances in applied mechanics*, 33:296–361. http://dx.doi.org/10.1016/s0065-2156(08)70388-0

Forest, S. (1998). Modeling slip, kink and shear banding in classical and generalized single crystal plasticity. *Acta Materialia*, 46(9), 3265–3281. http://dx.doi.org/10.1016/s1359-6454(98)00012-3

Gao, Y., Zhuang, Z., Liu, Z., You, X., Zhao, X., and Zhang, Z. (2011). Investigations of pipe-diffusion-based dislocation climb by discrete dislocation dynamics. *International Journal of Plasticity*, 27(7), 1055–1071. http://dx.doi.org/10.1016/j.ijplas.2010.11.003

Gröger, R., Bailey, A. G., and Vitek, V. (2008a). Multiscale modeling of plastic deformation of molybdenum and tungsten: I. Atomistic studies of the core structure and glide of $1/2\langle111\rangle$ screw dislocations at 0 K. *Acta Materialia*, 56:5401–5411. http://dx.doi.org/10.1016/j.actamat.2008.07.018

Gröger, R., Racherla, V., Bassani, J. L., , and Vitek, V. (2008b). Multiscale modeling of plastic deformation of molybdenum and tungsten: II. Yield criterion for single crystals based on atomistic studies of glide of $1/2\langle111\rangle$ screw dislocations. *Acta Materialia*, 56:5412–5425. http://dx.doi.org/10.1016/j.actamat.2008. 07.037

Gröger, R., Racherla, V., Bassani, J. L., , and Vitek, V. (2008c). Multiscale modeling of plastic deformation of molybdenum and tungsten: III. Effects of temperature and plastic strain rate. *Acta Materialia*, 56:5426–5439. http://dx.doi.org/10.1016/j.actamat.2008.07.027

Groh, S., Marin, E., Horstemeyer, M., and Zbib, H. (2009). Multiscale modeling of the plasticity in an aluminum single crystal. *International Journal of Plasticity*, 25(8), 1456–1473. http://dx.doi.org/10.1016/j.ijplas.2008. 11.003

Gupta, A., Steigmann, D. J., and Stölken, J. S. (2007). On the evolution of plasticity and incompatibility. *Mathematics and Mechanics of Solids*, 12:583–610. http://dx.doi.org/10.1177/1081286506064721

Gurtin, M. E. (2002). A gradient theory of single-crystal viscoplasticity that accounts for geometrically necessary dislocations. *Journal of the Mechanics and Physics of Solids*, 50:5–32. http://dx.doi.org/10.1016/s0022-5096 (01)00104-1

Gurtin, M. E. (2008). A finite-deformation, gradient theory of single-crystal plasticity with free energy dependent on densities of geometrically necessary dislocations. *International Journal of Plasticity*, 24(4), 702–725. http://dx.doi.org/10.1016/j.ijplas.2007.07.014

Gurtin, M. E., Anand, L., and Lele, S. P. (2007). Gradient single-crystal plasticity with free energy dependent on dislocation densities. *Journal of the Mechanics and Physics of Solids*, 55(9), 1853–1878. http://dx.doi.org/10.1016/j.ijplas.2007.07.014

Haghighat, S. H., Eggeler, G., and Raabe, D. (2013). Effect of climb on dislocation mechanisms and creep rates in γ-strengthened Ni base superalloy single crystals: a discrete dislocation dynamics study. *Acta Materialia*, 61(10), 3709–3723. http://dx.doi.org/10.1016/j.actamat.2013.03.003

Hale, L. M., Lim, H., Zimmerman, J. A., Battaile, C. C., and Weinberger, C. R. (2015). Insights on activation enthalpy for non-Schmid slip in body-centered cubic metals. *Scripta Materialia*, 99:89–92. http://dx.doi.org/10.1016/j.scriptamat.2014.11.035

Hansen, B. L., Beyerlein, I. J., Bronkhorst, C. A., Cerreta, E. K., and Dennis-Koller, D. (2013). A dislocation-based multi-rate single crystal plasticity model. *Int J Plasticity*, 44:129–146. http://dx.doi.org/10.1016/j.ijplas.2012.12.006

Hirth, J. P., & Lothe, J. (1992). *Theory of Dislocations*. Krieger Publishing Company, Malabar, Florida, 2nd edition.

Hochrainer, T. (2013). Moving dislocations in finite plasticity: a topological approach. *ZAMM-Journal of Applied Mathematics and Mechanics/Zeitschrift für Angewandte Mathematik und Mechanik*, 93(4), 252–268. http://dx.doi.org/10.1002/zamm.201100159

Hochrainer, T., Sandfeld, S., Zaiser, M., and Gumbsch, P. (2014). Continuum dislocation dynamics: Towards a physical theory of crystal plasticity. *J Mech Phys Solids*, 63:167–178. http://dx.doi.org/10.1016/j.jmps.2013.09.012

Horstemeyer, M., Lim, J., Lu, W., Mosher, D., Baskes, M., Prantil, V., and Plimpton, S. (2002). Torsion/simple shear of single crystal copper. *Journal of Engineering materials and technology*, 124(3), 322–328. http://dx.doi.org/10.1115/1.1480407

Horstemeyer, M. F., Baskes, M. I., Prantil, V. C., Philliber, J., and Vonderheide, S. (2003). A multiscale analysis of fixed-end simple shear using molecular dynamics, crystal plasticity, and a macroscopic internal state variable theory. *Modelling and Simulation in Materials Science and Engineering*, 11:265–286. http://dx.doi.org/10.1088/0965-0393/11/3/301

Kaluza, M., & Le, K. (2011). On torsion of a single crystal rod. *International Journal of Plasticity*, 27(3), 460–469. http://dx.doi.org/10.1016/j.ijplas.2010.07.003

Kuchnicki, S., Cuitino, A., and Radovitzky, R. (2006). Efficient and robust constitutive integrators for single-crystal plasticity modeling. *International Journal of Plasticity*, 22(10), 1988–2011. http://dx.doi.org/10.1016/j.ijplas.2010.07.003

Kysar, J. W., Gan, Y. X., Morse, T. L., Chen, X., and Jones, M. E. (2007). High strain gradient plasticity associated with wedge indentation into face-centered cubic single crystals: Geometrically necessary dislocation densities. *J Mech Phys Solids*, 55(7), 1554–1573. http://dx.doi.org/10.1016/j.jmps.2006.09.009

Kysar, J. W., Saito, Y., Oztop, M. S., Lee, D., and Huh, W. T. (2010). Experimental lower bounds on geometrically necessary dislocation density. *Int J Plasticity*, 26(8), 1097–1123. http://dx.doi.org/10.1016/j.ijplas.2010.03.009

Larson, B. C., Tischler, J. Z., El-Azab, A., and Liu, W. (2008). Dislocation density tensor characterization of deformation using 3D X-ray microscopy. *J Eng Mater-T ASME 130(2), 021024*. http://dx.doi.org/10.1115/1.2884336

Le, K. C. (2016). Three-dimensional continuum dislocation theory. *International Journal of Plasticity*, 76:213–230. http://dx.doi.org/10.1016/j.ijplas.2015.07.008

Lee, E. H. (1969). Elastic-plastic deformation at finite strains. *Journal of Applied Mechanics*, 36(1), 1–6. http://dx.doi.org/10.1115/1.3564580

Lee, M. G., Lim, H., Adams, B. L., Hirth, J. P., and Wagoner, R. H. (2010). A dislocation density-based single crystal constitutive equation. *International Journal of Plasticity*, 26(7), 925–938. http://dx.doi.org/10.1016/j.ijplas.2009.11.004

Leung, H. S., Leung, P. S. S., Cheng, B., and Ngan, A. H. W. (2015). A new dislocation-density-function dynamics scheme for computational crystal plasticity by explicit consideration of dislocation elastic interactions. *Int J Plasticity*, 67:1–25. http://dx.doi.org/10.1016/j.ijplas.2014.09.009

Lim, H., Weinberger, C. R., Battaile, C. C., and Buchheit, T. E. (2013). Application of generalized non-Schmid yield law to low-temperature plasticity in bcc transition metals. *Modelling and Simulation in Materials Science and Engineering*, 21(4), 045015. http://dx.doi.org/10.1088/0965-0393/21/4/045015

Lubliner, J. (2008). *Plasticity Theory*. Dover Publications.

Mandadapu, K. K., Jones, R. E., and Zimmerman, J. A. (2013). On the microscopic definitions of the dislocation density tensor. *Mathematics and Mechanics of Solids*, pages 1–14. http://dx.doi.org/10.1177/1081286513486792

Menzel, A., & Steinmann, P. (2000). On the continuum formulation of higher gradient plasticity for single and polycrystals. *Journal of the Mechanics and Physics of Solids*, 48(8), 1777–1796. http://dx.doi.org/10.1016/s0022-5096(99)00024-1

Miehe, C., & Schröder, J. (2001). A comparative study of stress update algorithms for rate-independent and rate-dependent crystal plasticity. *International Journal for Numerical Methods in Engineering*, 50(2), 273–298. http://dx.doi.org/10.1002/1097-0207(20010120)50:2<273::aid-nme17>3.3.co;2-h

Mordehai, D., Clouet, E., Fivel, M., and Verdier, M. (2008). Introducing dislocation climb by bulk diffusion in discrete dislocation dynamics. *Philosophical Magazine*, 88(6), 899–925. http://dx.doi.org/10.1080/14786430 801992850

Mura, T. (1987). *Micromechanics of Defects in Solids*. Kluwer Academic Publishers, Dordrecht, The Netherlands, 2nd edition.

Nye, J. F. (1953). Some geometrical relations in dislocated crystals. *Acta Metall.*, 1:153–162. http://dx.doi.org/10.1016/0001-6160(53)90054-6

Orowan, E. (1934). Zur kristallplastizitat II. *Z. Phys.*, 89:614.

Ortiz, M., & Popov, E. (1982). A statistical theory of polycrystalline plasticity. *Proceedings of the Royal Society of London A: Mathematical, Physical and Engineering Sciences*, 379(1777), 439–458. http://dx.doi.org/10. 1098/rspa.1982.0025

Ortiz, M., & Stainier, L. (1999). The variational formulation of viscoplastic constitutive updates. *Computer methods in applied mechanics and engineering*, 171(3), 419–444. http://dx.doi.org/10.1016/s0045-7825 (98)00219-9

Peach, M., & Koehler, J. (1950). The Forces Exerted on Dislocations and the Stress Fields Produced by Them. *Physical Review*, 80(3), 436–439. http://dx.doi.org/10.1103/physrev.80.436

Po, G., Mohamed, M. S., Crosby, T., Erel, C., El-Azab, A., and Ghoniem, N. (2014). Recent progress in discrete dislocation dynamics and its applications to micro plasticity. *JOM*, 66(10), 2108–2120. http://dx.doi.org/10. 1007/s11837-014-1153-2

Polanyi, M. N. (1934). Uber eine art gitterstroung, die einen kristall plastisch machen konnte. *Z. Phys.*, 89:660.

Salinger, A., Hansen, G. A., and Ostien, J. (2015). Sandia National Laboratories' Albany multiphysics code. https://github.com/gahansen/Albany.

Senger, J., Weygand, D., Kraft, O., and Gumbsch, P. (2011). Dislocation microstructure evolution in cyclically twisted microsamples: a discrete dislocation dynamics simulation. *Modelling and Simulation in Materials Science and Engineering*, 19(7), 074004. http://dx.doi.org/10.1088/0965-0393/19/7/074004

Steinmann, P., & Stein, E. (1996). On the numerical treatment and analysis of finite deformation ductile single crystal plasticity. *Computer Methods in Applied Mechanics and Engineering*, 129(3), 235–254. http://dx.doi.org/10.1016/0045-7825(95)00913-2

Stölken, J. S., & Evans, A. G. (1998). A microbend test method for measuring the plasticity length scale. *Acta Materialia*, 46:5109–5115. http://dx.doi.org/10.1016/s1359-6454(98)00153-0

Taylor, G. I. (1934a). The mechanism of plastic deformation of crystals. Part I. Theoretical. *Proceedings of the Royal Society of London. Series A, Containing Papers of a Mathematical and Physical Character*, pages 362–387.

Taylor, G. I. (1934b). The mechanism of plastic deformation of crystals. Part II. Comparison with observations. *Proceedings of the Royal Society of London. Series A, Containing Papers of a Mathematical and Physical Character*, pages 388–404.

Timoshenko, S., & Goodier, J. (1951). *Theory of elasticity*. McGraw-Hill, New York, NY.

Ulz, M. H., Mandadapu, K. K., and Papadopoulos, P. (2013). On the estimation of spatial averaging volume for determining stress using atomistic methods. *Modelling and Simulation in Materials Science and Engineering*, 21(1), 15010–15024. http://dx.doi.org/10.1088/0965-0393/21/1/015010

Van der Giessen, E., & Needleman, A. (2003). GNDs in nonlocal plasticity theories: lessons from discrete dislocation simulations. *Scripta Materialia*, 48:127–132. http://dx.doi.org/10.1016/s1359-6462(02)00332-9

Vitek, V. (2011). Atomic level computer modelling of crystal defects with emphasis on dislocations: Past, present and future. *Progress in Materials Science*, 56(6), 577–585. http://dx.doi.org/10.1016/j.pmatsci.2011.01.002

Vitek, V., & Paidar, V. (2008). Non-planar dislocation cores: a ubiquitous phenomenon affecting mechanical properties of crystalline materials. *Dislocations in solids*, 14:439–514. http://dx.doi.org/10.1016/s1572-4859 (07)00007-1

Wang, L., Liu, Z., and Zhuang, Z. (2016). Developing micro-scale crystal plasticity model based on phase field theory for modeling dislocations in heteroepitaxial structures. *International Journal of Plasticity*. http://dx.doi.org/10.1016/j.ijplas.2016.01.010

Wang, Z., & Beyerlein, I. (2011). An atomistically-informed dislocation dynamics model for the plastic anisotropy and tension–compression asymmetry of bcc metals. *International Journal of Plasticity*, 27(10), 1471–1484. http://dx.doi.org/10.1016/j.ijplas.2010.08.011

Weinberger, C. R., Battaile, C. C., Buchheit, T. E., and Holm, E. A. (2012). Incorporating atomistic data of lattice friction into bcc crystal plasticity models. *International Journal of Plasticity*, 37:16–30. http://dx.doi.org/10.1016/j.ijplas.2012.03.012

Weinberger, C. R., & Cai, W. (2010). Plasticity of metal wires in torsion: Molecular dynamics and dislocation dynamics simulations. *Journal of the Mechanics and Physics of Solids*, 58(7), 1011–1025. http://dx.doi.org/10.1016/j.jmps.2010.04.010

Xia, S., & El-Azab, A. (2015). Computational modelling of mesoscale dislocation patterning and plastic deformation of single crystals. *Model Simul Mater Sc*, pages 1–26. http://dx.doi.org/10.1088/0965-0393/23/5/055009

Zhao, P., Low, T. S. E., Wang, Y., and Niezgoda, S. R. (2016). An integrated full-field model of concurrent plastic deformation and microstructure evolution: Application to 3d simulation of dynamic recrystallization in polycrystalline copper. *International Journal of Plasticity*. http://dx.doi.org/10.1016/j.ijplas.2015.12.010

Zhou, C., Biner, S. B., and LeSar, R. (2010). Discrete dislocation dynamics simulations of plasticity at small scales. *Acta Materialia*, 58(5), 1565–1577. http://dx.doi.org/10.1016/j.actamat.2009.11.001

Note

Note 1. In general, the drag coefficient B is a function of velocity itself, *i.e.* the relation between drag force and dislocation velocity is non-linear; however, for the simulations presented here we assume that B is a constant material coefficient. Also of note is the fact that the drag coefficient is in general a second-order tensor exhibiting anisotropy depending on the crystal structure, and that it may also depend on the local dislocation character (*i.e.* edge versus screw), and on the type of dislocation motion (*i.e.* glide versus climb). As described in Po et al. (2014), the method used here utilizes Lagrange multipliers to constrain climb processes, which are often only active at very high temperatures. Approaches that do incorporate climb include the works by Gao et al. (2011), Haghighat et al. (2013), and Babu and Lindgren (2013).

5

Effect of Coke Granulometry on the Properties of Carbon Anodes based on Experimental Study and ANN Analysis

Arunima Sarkar[1], Duygu Kocaefe[1], Yasar Kocaefe[1], Dipankar Bhattacharyay[1], Brigitte Morais[1] & Patrick Coulombe[2]

[1] Department of Applied Sciences, University of Quebec at Chicoutimi, Québec, Canada

[2] Aluminerie Alouette Inc., 400, Chemin de la Pointe-Noire, C.P. 1650, Sept-Îles, Québec, Canada

Correspondence: Duygu Kocaefe, Department of Applied Sciences, University of Quebec at Chicoutimi, Québec, Canada. E-mail: Duygu_Kocaefe@uqac.ca

Abstract

The available anode-quality petroleum coke is not sufficient to cover the need created by the increase in the world aluminum production. Understanding the consequences of varying calcined coke quality is necessary to possibly compensate for the reduction in coke quality and adjust the anode paste recipe in the subsequent use of coke in order to obtain economically viable production of aluminum. Different fractions of coke particles were mixed to optimize the anode recipe; however, it was laborious to find experimentally the suitable percentage of each fraction in anode paste which would give good anode properties. In this study, Artificial Neural Network (ANN) model was developed for adjusting the granulometry of the raw materials for anode production. Tapped bulk density of dry aggregates was used to predict the anode paste recipe using the ANN method. A new anode recipe (by adjusting the medium fraction in the paste) was proposed based on the predictions of an ANN model, which resulted in improved anode properties.

Keywords: carbon anodes, coke, pitch, ANN, paste recipe

1. Introduction

During the anode fabrication, calcined petroleum coke, recycled material (butts, green and baked rejects), which is also called dry aggregate, and coal tar pitch are blended in certain percentages and mixed in kneader. The composition of this mixture is known as "anode recipe" or the "anode paste recipe". Considering the differences in raw material quality in different shipments, adjusting the anode recipe becomes a very important task for every manufacturer to ensure good anode quality. In order to facilitate selection of anode recipe, an Artificial Neural Network (ANN) model was developed to meet the specific needs of the aluminum production industry. The bulk density of dry aggregates is a good indicator of particle packing and volume occupied by dry aggregate fractions for different recipes. Particle size distribution not only affects the anode density but it also influences the mechanical strength, pore size distribution, electrical resistivity, air permeability, reactivity and even chemical composition of anodes (Farr-Wharton, 1980; Smith, 1991; Hulse, 2000; Adams, 2004; Figueiredo, 2005). In this study, tapped bulk density of dry aggregates was used to predict the anode paste recipes using ANN.

Fractions of different size particles are usually used in a recipe designed to give a dense mix where the voids between the coarse particles are filled with medium size particles and voids between these are filled by small size particles. An optimal quantity of coarse, medium and finer fractions is required for anode formulation to obtain good anode quality (Bowers, 2008). But, it is difficult to study experimentally the effect of an individual fraction sizes on a specific anode property since the change in one fraction is followed simultaneously by variations in other fractions. Also, an experimental study is laborious, time consuming, costly and involves extensive handing to find the appropriate percentage of each fraction to be used in anode paste recipe which would give good quality anode. For such cases, a trained and validated ANN model has proven to be a highly efficient tool, which can handle multiple-input conditions. ANN can give a fast response and can be upgraded continuously as more data become available. Artificial neural network is used for predicting the values of dependent parameters for which no mathematical relation is available (Parthiban, 2007) or even though some mathematical relationship is available, it is hard to find the numerical parameters (Milewski, 2009).

ANN models are widely used in various research fields including quality control (Bahlmann, 1999; Pang, 2004; Fruhwirth, 2007; Parthiban, 2007; Saengrung, 2007; Shang, 2008; Piuleac, 2010; Bhagavatula, 2012), prediction of compositions and properties of metallic and nonmetallic compounds (Wang, 2008; Asadi-Eydivand, 2014; Mohanty, 2014), aluminum reduction cell (Meghlaoui, 1998; Biedler, 2002; Boadu, 2010). However, there are only a few studies available (Berezin, 2002; Bhattacharyay, 2013; Bhattacharyay, 2015) in literature, which are directly related to carbon anodes used for the production of primary aluminum.

2. Method

2.1 Materials

Two different petroleum cokes from different suppliers and one coal tar pitch were used for laboratory scale anode production. Table 1 and Table 2 illustrate the properties of cokes and pitch.

Table 1. Physical and chemical properties of coke

Properties	RDC 1	RDC 2
Bulk Density* (g/cc)	0.89	0.901
Real Density(g/cc)	2.06	2.072
CO_2 Reactivity (%)	9	8
Ash Content*** (%)	0.2	0.15
Moisture Content (%)	0.1	0.19
Na (wt%)	0.007	0.0059
Si (wt%)	0.01	0.0095
P (wt%)	0.0006	-
S (wt%)	2.75	0.73
Ca (wt%)	0.01	0.004
V (wt%)	0.031	0.024
Fe (wt%)	0.02	0.01
Ni (wt%)	0.02	0.019

*Measured by ASTM D4292-10,**Measured by ISO 1014:1985, ***Measured by dry basis.

Table 2. Physical and chemical properties of pitch

Properties	Pitch
Ash at 900°C (%m/m)	0.12
β Resin (%m/m)	22.2
Density at 20°C (g/ml)	1.320
Quinoline insoluble (%m/m)	6.9
Toluene insoluble (%m/m)	29.1
Coking Value (%m/m)	59.1
Softening Point (°C)	119.6
Dynamic Viscosity 170°C (mPa.s)	1390
Surface Tension[a] (dyne/cm) at 170°C	39.33

2.2 Methodology

Anode production is a challenging task to perform and includes various complex production steps and circumstances. A set of 19 anodes were prepared at the pilot-scale anode production laboratory of the University of Quebec at Chicoutimi (UQAC). Sieved coke fractions, recycled anode butt, ball mill product (BMP) and filter dust (FD) were weighed according to ANN specified recipe and blended in an intensive mixer together with coal tar pitch. The paste temperature was maintained at 170°C during mixing. Green anodes were formed in a vibrocompactor using predetermined conditions. Then, these anodes were baked under the conditions similar to those used in industry. When the maximum baking temperature was reached, anodes were subjected to soaking for 8 h at this temperature.

2.3 Characterizations of Anodes

The anode properties were measured to ensure that they conform to general specifications, especially to produce high density, low resistivity and low reactivity anodes. The following tests and measurements were carried out:

air/CO_2 reactivity using the thermogravimetric (TGA) analysis (ASTM D6559 - 00a (2010) and ASTM D6558 - 00a (2010), density and electrical resistivity of whole anode using the device/method developed at UQAC, density and electrical resistivity of cylindrical anode cores using ASTM D6120-97(resistivity) and ASTM D5502-00 (density) standards. The compressive and flexural strengths were measured by ASTM C695-91 (2005) and ISO CD 12986, respectively.

3. Results

3.1 Development of ANN Model

Two customized feed-forward artificial neural network (ANN) models with back-propagation training were developed using Matlab 2014. For networks, one input layer, two hidden layers connected in series, and one output layer were selected. Various transfer functions such as logsig, tansig, purelin are associated with the hidden layers. The logsig function can be represented as logsig(n) = 1/(1 + exp(-n)). Similarly, tansig function can be represented as tansig(n) = 2/(1+exp(-2*n))-1. Purelin is a linear function represented as purelin(n) = n. The transfer functions process the input to a layer such that the output can be easily classified into groups of similar data, which is important for an efficient prediction. The networks were trained based on the measurement of error in prediction. The errors were measured in terms of mean squared error (mse) and mean average error (mae). All the densities are normalized by using following equation.

$$\text{Normalized value} = \frac{\text{Value to be Normalized - Minimum Value}}{\text{Maximum Value - Minimum Value}} \qquad (1)$$

In the first model, measured tapped bulk density of each fraction and the corresponding granulometry of different dry aggregate recipes were fed to model as input to predict tapped bulk density of the specific recipe. For this model, 46 sets of aggregate bulk density data were used to train the network and five random data, which were not used in training, were used for validation. 14 laboratory anodes were produced using different recipes. These recipes were chosen using the predictions of the first ANN model which indicated the percentage of each fraction to be used in order to obtain high aggregate density. To see if the high aggregate density will result in high quality anodes, the green densities of these anodes were measured. In the second model, tapped bulk density of an anode recipe and granulometry were used as input to predict the ratio of dry aggregate to green anode density. Ten of the fourteen data were utilized to train the database and four random recipes were chosen for validation. Afterwards, two models were combined to predict the green anode density for a chosen recipe. In order to minimize the errors in training and testing calculations, the values predicted by ANN were plotted against the experimental results for numerous test data sets. The coefficient of determination for linear regression was used as the criteria for the quality of the network predictions. The closer the value of the coefficient of determination to unity is, the better the model's ability for prediction is. It can be seen from the Figure 1 that the R^2 value for training and test sets were found as 0.966 and 0.975, respectively, for dry aggregate density (first model).

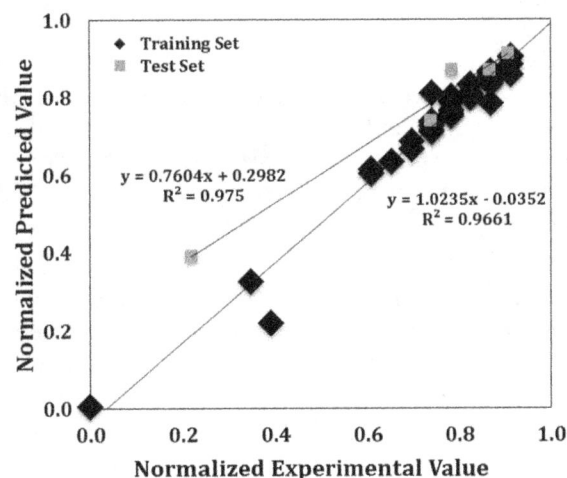

Figure 1. Normalized predicted and experimental dry aggregate densities using the first ANN model

Figure 2 shows the predicted vs experimental ratios of green anode density to dry aggregate density for the second model. The model had R^2 value of 0.982 for training data set. The accuracy of the prediction is further indicated by the R^2 (0.986) value of the test set.

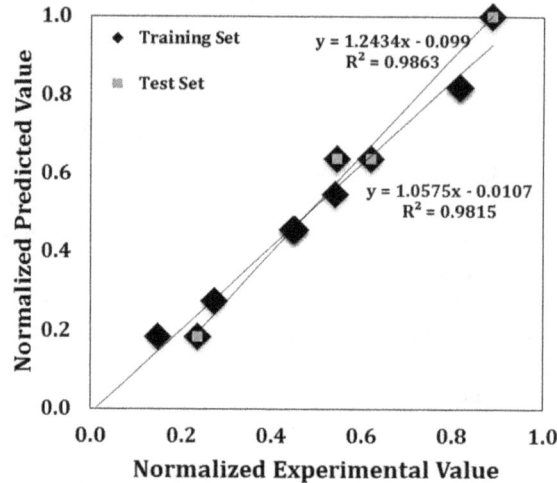

Figure 2. Normalized predicted and experimental values of the ratio of green anode density to dry aggregate density using the second ANN model

The ANN model developed could be utilized to determine the granulometry required to obtain a desired tapped density of an aggregate and green anode density which is produced using this aggregate recipe. It was possible to vary the percentages of all the different size fractions which could yield a desired range of dry aggregate or green anode density. However, this might lead to a large number of combinations with different size fractions.

As it was discussed earlier, each fraction has an effect on anode properties. The model was used to demonstrate the effect of medium fractions and recycled anode butt on dry aggregate density and green anode density. To demonstrate the application of the model here, only the percentages of two particle sizes (described as coarse and medium in decreasing order in size) of fresh coke were varied keeping all other parameters constant (Figure 3). Then, the aggregate densities and anode densities corresponding to the granulometry used were predicted. This model was trained for one type of coke only. Also the pitch percentage was kept constant at 15%. The percentages of all fractions are normalized based on the Equation 1. Effect of medium particles on dry aggregate density and green anode density is illustrated in Figure 3. The results show that an optimum percentage of medium fractions were required to obtain high green anode density. High green anode density did not necessarily correspond to high dry aggregate density, rather usually low value of dry aggregate density resulted in high values of green anode density. It can be noted that when the coarse fractions were in the lower end of the range and when medium fraction percentage decreased, dry aggregate density increased. The green anode density reached an optimum at a certain medium fraction percentage (Figure 3 (a)). When the coarse fractions were in medium range, it can be seen from Figure 3 (b) that there was an optimum composition for maximum green anode density, even if the variation in dry aggregate density was not very large. For the anode with a high amount of coarse fraction, the green anode density increased with decreasing medium fraction percentage whereas dry aggregate density showed the completely opposite trend (Figure 3 (c)). The results also showed that, at a higher level of medium fraction percentage and at a certain point, the dry aggregate and green anode densities were the same and showed opposite trend on both sides of this point.

Similar approach was taken to analyze the effect of recycled anode butt percentage on dry aggregate and green anode density. It could be seen from Figure 4 that predicted dry aggregate density and green anode density increased with increasing butt content up to certain level. Afterwards, it decreased. These results indicate that it is important to select an optimum range for each fraction to improve the anode density.

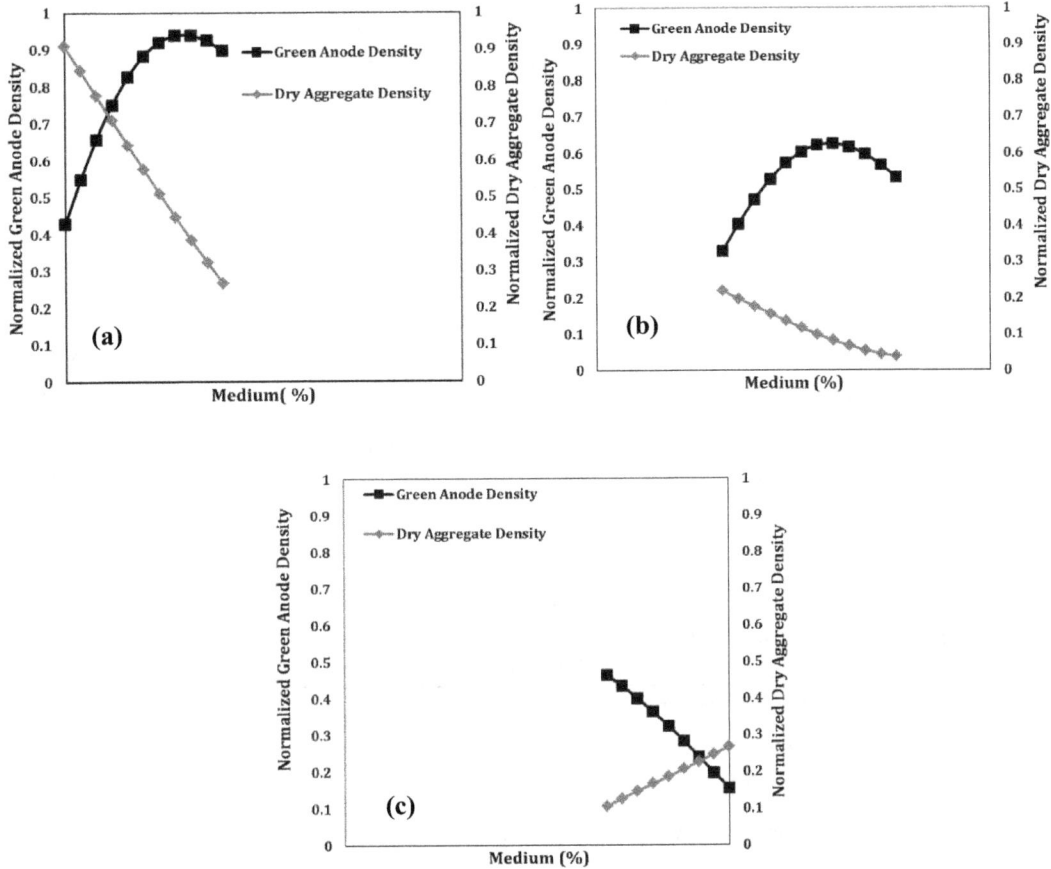

Figure 3. Effect of medium fractions on normalized dry aggregate density and normalized green anode density (a) low amount of coarse fraction (b) medium amount of coarse fraction (c) high amount of coarse fraction

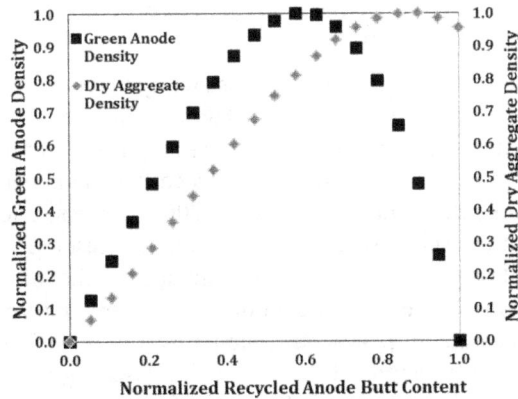

Figure 4. Effect of normalized recycled butt on normalized dry aggregate density and normalized green anode density

3.2 Effect of Particle Size Distribution on Anode Properties

To establish the impact of particle size distribution on anode density, two sets of experiments were carried out. Several recipes were prepared using various combinations of recycled anode butt as well as the coarse, medium, and fine fractions of fresh coke. Characterization of anodes produced from different recipes is discussed in this section. All recipes with their corresponding anode numbers are presented in Table 3.

Table 3. Anode recipes and their corresponding anode number and properties

Anode No	Butt(%)	Coarse*	Medium*	Fine*	Type of coke	Green Anode Density(g/cc)	Green Elec.Res. ($\mu\Omega$m)	Baked Anode Density(g/cc)	Baked Elec.Res. ($\mu\Omega$m)
1**[1]	25	0.0	1.0	0.0	CPC 1	1.587	5566	1.543	59.04
2**	25	0.0	1.0	0.0	CPC 1	1.586	6357	1.528	59.86
3[1]	25	0.3	0.6	0.3	CPC 1	1.638	3837	1.572	54.88
4	25	0.3	0.6	0.3	CPC 1	1.629	3951	1.561	54.29
5	25	0.3	0.6	0.3	CPC 1	1.632	3933	1.563	54.22
6***	25	0.3	0.6	0.3	CPC 1	1.585	6647	1.535	57.93
7	25	0.6	0.0	1.0	CPC 1	1.568	5868	1.478	65.05
8	25	0.3	0.8	0.0	CPC 1	1.603	4539	1.546	60.02
9	25	0.3	0.6	0.4	CPC 1	1.587	6349	NA	NA
10	25	0.3	0.4	0.6	CPC 1	1.593	5090	1.512	65.35
11	25	0.7	0.2	0.6	CPC 1	1.590	4256	NA	NA
12	0	1.0	0.0	0.6	CPC 1	1.574	5840	NA	NA
13	25	1.0	0.0	0.6	CPC 1	1.623	4043	1.546	57.23
14	25	0.3	0.5	0.5	CPC 1	1.625	4487	1.552	58.11
15	0	0.0	1.0	0.0	CPC 1	1.576	4505	1.511	54.01
16	15	0.0	1.0	0.0	CPC 1	1.577	6357	1.506	59.09
17	35	0.0	1.0	0.0	CPC 1	1.568	6409	1.496	59.74
18**	25	0.0	1.0	0.0	CPC 2	1.594	5909	1.546	60.41
19	25	0.3	0.6	0.3	CPC 2	1.626	4258	1.571	55.05

[1] The anodes 1 and 3 were produced earlier, and the vibro speed was not measured.

*All the coke fractions are normalized by column wise using Equation 1.

** Standard recipe

***14% pitch was used where the other anodes were made with 15% pitch

CPC 1: Calcined petroleum coke 1, CPC 2: Calcined petroleum coke 2.

3.2.1 Green and Baked Density of Anodes

To establish the effect of fraction sizing on green anode density, a total of 19 pilot-scale anodes were prepared from two different sources of petroleum coke using different recipes including a standard recipe (Anode 1 and 2, Table 3. Aggregate size distributions, specially medium and coarse fractions, were varied. The anode making conditions were kept the same for all the anodes.

The recipes with different recycled butt contents showed consistently that the green and baked densities are higher at 25% butt content compared to those at lower and higher butt contents (Table 3). In all recipes, 15% pitch was used except for anode 6. This result is in accordance with the results given in literature (Fischer, 1991). For anode 17, the butt content was increased to 35%, which slightly reduced the anode density. In general, recycled anode butt is more wettable compared to petroleum coke (Adams, 2004. Adams, 2004). However, it also has lower surface porosity. It is possible that the anode was overpitched due to presence of excess butt, hence, pitch layer around the butt particles (as it has less porosity) was thicker compared to that of the petroleum coke. This resulted in an increase in anode volume which in turn decreased the density. In addition, the shape of the recycled butt particles is completely different than that of petroleum coke (Bhattacharyay, 2014). It is possible that this might have affected the packing behavior of the aggregate causing a decrease in green anode density. It is also possible that higher butt content changes the binder demand (Prouix, 1993) which, in turn, affects density if the binder content is not adjusted accordingly.

Green and baked densities of anodes produced from different recipes with constant butt and pitch contents are illustrated in Table 3 and in 3-D plots in Figures 5, respectively. In 3-D plots, symbol sizes increases and color code changes with increasing density values. The axes represents the normalized coarse, medium, and fine fractions used in the recipe. Anode 1 and 2 represent the typical standard recipe. Aiming to improve the anode density, trials were carried out to adjust the medium and coarse fractions in the recipe based on ANN results. Decrease of medium fraction around 40% compared to that of the standard recipe (Table 3 and Figure 5) improved the green and baked anode densities when 25% butt was maintained in the recipe. It is a well-known fact that an increase in fine fractions increases the anode density but it can also affect the anode mechanical properties and binder demand. In this study, the ultra-fine fractions (BMP and FD) were kept similar to that of the standard recipe. Finally, an improved anode recipe (anodes 3, 4, and 5) was found to result in higher green and baked anode densities compared to those of the standard recipe. Green density of anode increased from 1.586 g/cc to 1.638 g/cc and baked density increased from 1.543 g/cc to 1.572 g/cc. Two repetitive measurements were done to confirm the improvement of the anode properties. An increase in density was observed for an anode produced using a different coke with similar recipe (see anode 19 in Table 3 and Figure 5). Recipe corresponding to anode 14 also revealed a promising trend. The results showed that lower amount of the medium fractions in the paste recipe gave comparatively better green anode density. Hulse (Hulse 2000) also found similar trend and mentioned in her thesis that lower intermediate fraction could improve the dry aggregate density yielding higher anode density. The results also showed that complete removal of medium fraction sharply reduced the anode quality (anodes 7 and 12) except for anode 13. It can be seen that for anode 7, density was reduced even though it contained high amount of fines. It was possible that density was reduced because inter-particle spaces were not well filled as the medium fraction was completely missing. Also it was possible that higher amount of fines increased the binder demand and coke particles were not wetted sufficiently by pitch due to under-pitching. As it can be seen, the recipe of anode 12 and 13 were similar with exception of their butt content. The density of anode 12 was lower as this anode did not contain any recycled anode butt. This result was in good agreement with the previous findings of the present study (Sarkar 2015). This showed that suitable amount of recycled anode butt, medium, coarse, and fines fractions were required to have good packing and good anode properties. Thus, it is evident that the presence of lower medium fractions with optimal recycled butt, coarse, and fine particles can improve the anode density. As it can be seen in Table 3, the pitch level had pronounced influence on the green and baked density (anodes 3 and 6). As noted, the anode containing 14% pitch (anode 6) had inferior quality than that with 15% (anode 3) pitch. The baked anode densities also followed same trend as that of the green anode densities.

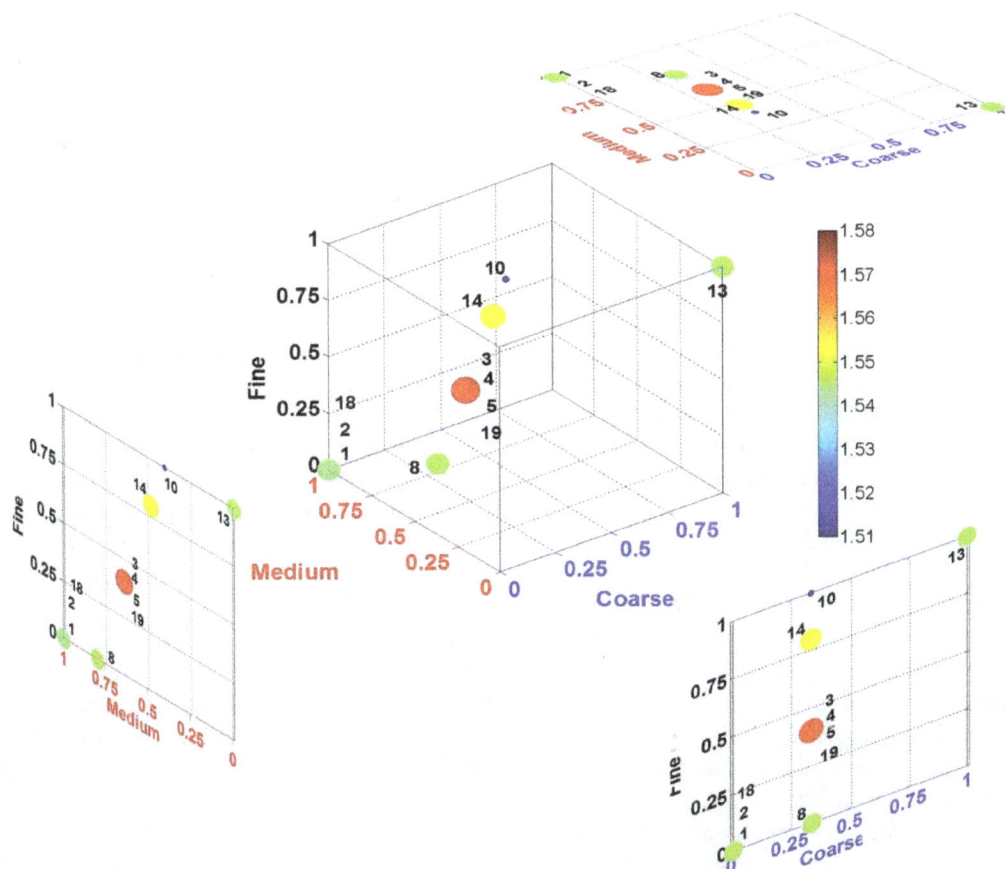

Figure 5. Baked anode density of different recipes (symbol size increases with increasing density)

3.2.2 Electrical Resistivity of Green and Baked Anodes

The electrical resistivity (ER) measurement is essential to determine the anode quality. The ER gives an indication of the structural integrity through the current conducting properties of the anode. Electrical resistivity of anode is closely related to anode density. Generally higher density anodes should yield lower ER as dense anode contains lower porosity. However, this is not always the case as highly dense anodes can have more crack formation during baking which results in higher ER.

Several physico-chemical transitions occur in anodes at different stages of heat treatment. The principal chemical reaction is dehydrogenative polymerization of aromatic compounds. Initially, monomeric PAHs undergo addition reaction to produce biaryls and oligo-aryls. Afterwards, bi- and oligo-aryls go through intramolecular dehydrocyclization reactions to form pericondenzed aromatic systems which continue to grow larger by cross linking (Zande, 1997; Engvoll, 2001). At this stage, evolution of mesophase occurs and these spherical molecules go through polymerization reactions which generate long chains of molecules, consequently, reduce specific electrical resistivity of anodes (Brooks, 1968; Engvoll, 2001). On the contrary, production of bigger mesophase spheres in higher quantity could hinder the pitch penetration into pores. In addition, bigger molecules could generate more cracks. Also, this kind of polymerization reactions are reversible in nature and upon reaching to their decomposition temperature, the reversible transition begins. This might lead to an increase in the anode electrical resistivity (ER). Therefore, it is hard to find a clear-cut relation with baked density and ER. The ER of the green and baked anodes was measured using custom made equipment developed at UQAC and results are presented in Table 3 and Figure 6.

The results showed that anode produced without butt had the lowest electrical resistivity. ER increased if butt was added to the aggregate recipe. However, it slightly changed with further increase in butt content. This result is in contrast with the work reported by Belitkus (Belitskus, 1981). Anode 17 exhibited lowest density that resulted highest value of ER. Anode 2 had somewhat higher ER compared to anode 1 (both were made from the same recipe and butt content). Highest green and baked density of anode 1 (standard recipe) gave lowest ER among all the recipes studied with different butt content. The decrease in baked anode resistivity observed for anode 15 compared to that of the standard anode could be attributed to the absence of butt particles. Since the mechanical properties of butts and fresh coke are different, the stress created at the positions where butt is present might cause new cracks to form.

Table 3 gives the ER of the anodes for pitch contents. The ER was reduced with increased pitch level to 15% from 14%. This result was in good agreement with the findings of Figueiredo et al. (2005). Previous investigation demonstrated that for higher pitch content, density was higher which was directly reflected on electrical resistivity results. Of course this is only true if the anodes are not overpitched.

The ER values of the green and baked anodes produced from different paste recipes are in Table 3 and Figure 6 respectively. The changes in ER of anodes with different recipes generally followed the opposite of the trend found for the density. In most cases, highly dense green and baked anodes displayed lower ER. For some baked anodes, this trend was not followed. This can be explained with the mesosphere formation which was described earlier. Decrease in medium fraction with adjusted coarse, fine and butt contents led to a general decrease in ER of green anodes. The best recipe (anodes 3, 4, and 5) found among the recipes tried during this study resulted in highest green and baked anode densities (table 3 and Figure 6) as well as lowest ER for both green and baked anodes. The 3-D plot (Figure 6) showed that anode electrical resistivity of baked anodes produced from the best recipe (anodes 3, 4 and 5) reduced significantly from 59 μΩm to 54 μΩm compared to standard recipe (anodes 1 and 2). Anode produced from CPC2 also displayed similar tendency and anode ER of baked anodes formed from the best recipe (anode 19) reduced from 60 μΩm to 55 μΩm compared to standard recipe (anode 18) (Figure 6). Anodes 12 and 13 were produced from similar recipe but green ER of anode 12 was higher (Table 3). Since the recipe of anode 13 resulted in much higher density, its resistivity was lower. Anode 14, which had a higher density value, exhibited lower ER in green stage; but, after baking, the ER value was just slightly lower than the standard recipe. Similar trend was observed for anode 13. Alternatively, anode 7 and 10 which had comparatively lower ER before baking gave higher ER after baking compared to the standard recipe (Table 3). Thermochemical changes and formation of mesophase during baking could be the probable reason behind this behavior as described in different literatures (Brooks 1968, Zande 1997, Engvoll 2001). Also, cracks forming during baking can affect ER. Maximum values of ER were measured for anodes 2 and 9 before baking (Table 3). Similarly, anode 10 had the maximum value of ER after baking (Figure 6).

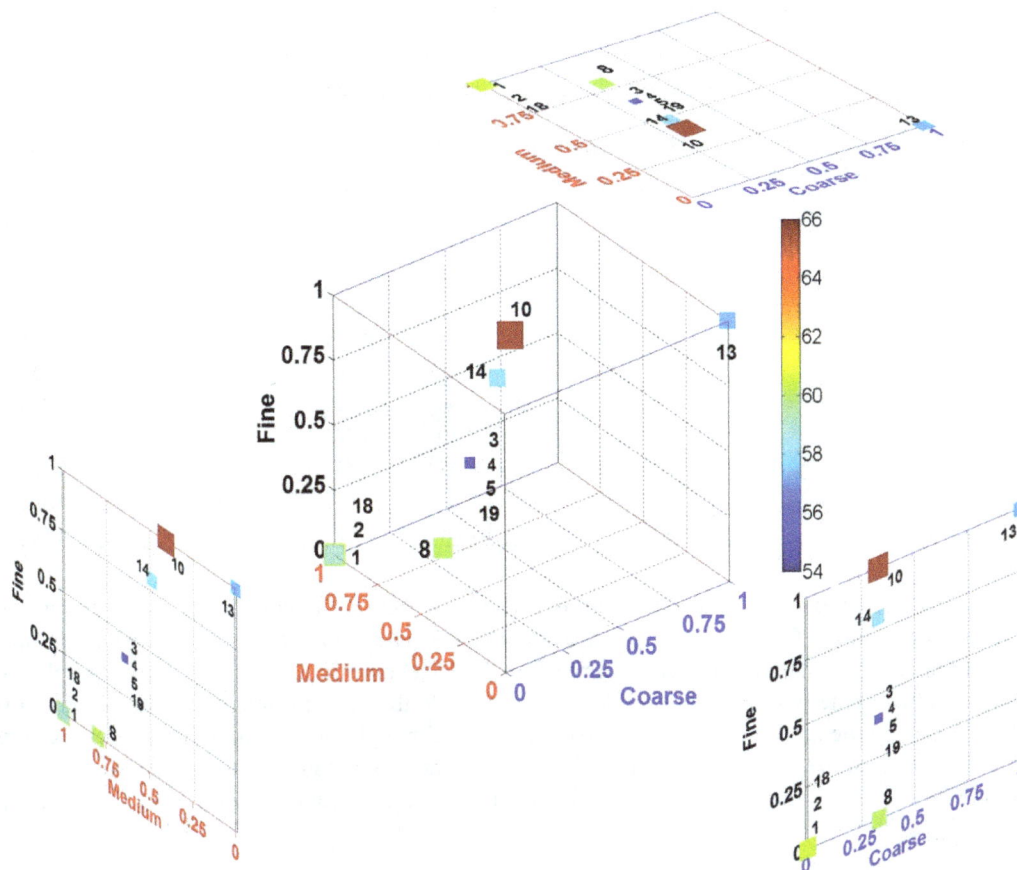

Figure 6. Electrical resistivity of baked anodes produced from different recipes (symbol size increases with increasing electrical resistivity)

3.2.3 Air Reactivity

During the electrolysis, although the anodes are covered with a mixture of alumina and crushed bath material for protection, air can still penetrate through the cover material and react with the anodes leading to dusting. The air reactivity and dusting data obtained using ASTM D6559-00 a method are given in Table 4; Figures 7 illustrates the trends observed for air reactivity of anodes, made with two different types of coke and with different recipes including different butt and pitch contents. In this section, only the air reactivities of high density baked anodes compared to that of the standard recipe with the objective of finding out how the reactivity was affected when the anode quality was improved (high density, low ER) using ANN. It can be seen from Table 4 and Figure 7 that there were differences in reactivities of the anodes produced from the same recipe such as anodes 1 and 2 (both produced from the standard recipe) as well as anodes 3 and 4 (both produced from the best recipe).

Table 4 shows the effect of butt content on anode total air reactivity and dusting. The results corresponded well with those reported by Belitskus (Belitskus, 1981). The recycled butt additions seemed to increase the overall anode air reactivity due to its sodium content and the presence of other impurities such as potassium, calcium, aluminum, iron, sulfur, nickel, silicon, fluoride etc. (Schmidt-Hatting, 1991; Suriyapraphadilok, 2005, Fischer, 2013). The reactivity was minimum when anode did not contain any butt. The air reactivity was more affected by butt content than the CO_2 reactivity, which will be presented in the next section. The results also showed that at 35% butt content, reactivity suddenly dropped, but dusting increased continuously with increasing butt content. This could be correlated with density of the anodes. Air could react with impurities, aromatic and aliphatic carbon on the surface of the anode. Higher density represents a higher amount of carbon on the surface which in turn increases the air reactivity. As it was previously shown, anode containing 25% of butt exhibited highest green and baked density. Therefore, this could be the possible reason for higher air reactivity.

Table 4. Air reactivity and dusting data of the anodes produced from different recipes

Anode No	Butt (%)	Coarse	Medium	Fine	Pitch (%)	Total Rate (mg/cm²h)	Dusting Rate (mg/cm²h)	%/min	Sample Baked Density (g/cc) (core 3)
1	25	0.0	1.0	0.0	15	66.46	5.83	0.088	1.541
2	25	0.0	1.0	0.0	15	97.69	5.81	0.135	1.525
3	25	0.3	0.6	0.3	15	73.70	5.60	0.095	1.543
4	25	0.3	0.6	0.3	15	108.09	3.06	0.142	1.555
6	25	0.3	0.6	0.3	14	62.32	3.38	0.082	1.541
7	25	0.0	0.6	0.0	15	56.53	6.86	0.078	1.502
8	25	0.3	0.8	0.0	15	108.85	4.52	0.146	1.537
13	25	1.0	0.0	0.6	15	70.74	4.47	0.094	1.550
14	25	0.3	0.5	0.5	15	116.28	5.66	0.152	1.553
15	0.0	0.0	1.0	0.0	15	43.44	2.52	0.059	1.517
16	15	0.0	1.0	0.0	15	74.50	4.74	0.100	1.522
17	35	0.0	1.0	0.0	15	77.87	7.56	0.103	1.524
18	25	0.0	1.0	0.0	15	127.33	3.57	0.165	1.553
19	25	0.3	0.6	0.3	15	121.50	3.46	0.157	1.565

The air reactivity of the anodes seemed to be strongly affected by the anode recipe including different pitch level (Table 4). The anode (anode 6) with lower pitch content (14%) had a lower overall air reactivity and dusting compared with the anode containing higher amount of pitch (15%) (anode 2). Table 4 shows that the amount of dust produced due to air reactivity from the anode was sensitive to the binder pitch level. The higher air reactivity was probably caused by the mechanisms of selective oxidation although the oxidation of coke was also very fast. Binder matrix is more reactive than coke causing selective air burn of the binder matrix (Lhuissier, 2007, Lhuissier 2009, Sulaiman 2012). The higher amount of pitch resulted in increased amount of dust and air reactivity.

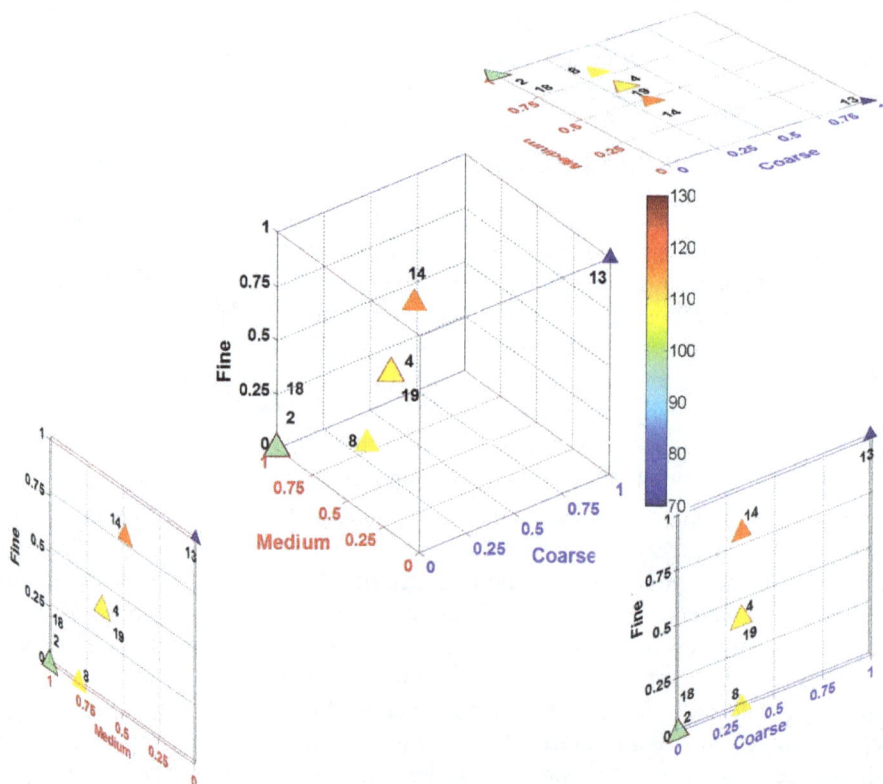

Figure 7. Total air reactivity of anodes prepared with different recipes (symbol size increases with increasing air reactivity)

Table 4 and Figure 7 present the total air reactivity measurements and dusting during air reactivity for different anode paste recipes. The 3-D plot (Figure 7) indicates that anode air reactivity increased with decreasing medium particles content (anode 4, 8, 14, 19). This is again due to the increase in the anode density with decreasing medium fractions in the anode paste recipe. As discussed earlier, highly dense anodes led to higher air reactivity due to increased contact surface. The best recipe (anode 4) of this study had higher air reactivity but lower dusting compared to the standard recipe (anode 2). In most of the recipes studied, highly dense anodes (anode 14) had higher overall air reactivity and dusting. Figure 7 shows that anode 13, which did not contain any medium fraction, had lowest air reactivity. Figure 7 also illustrates that anode prepared from CPC 2 coke (anode 18 and 19) had higher air reactivity compared to anode produced from CPC 1 coke (anode 2 and 4) for the same recipe due to higher sodium content and higher real density for CPC 2 coke (see Table 1). It can be also noted from the Table 4 that dusting did not follow any trend with the air reactivity. Anode 4 showed the lowest and anode 14 indicated the highest dusting values within the range of study.

3.2.4 CO_2 Reactivity

Carbon dioxide is produced during the alumina reduction which can react with the anode carbon to produce carbon monoxide. Similar to air reactivity, CO_2 reactivity also leads to dusting due to the partial disintegration of the anodes. In order to investigate the effect of anode paste recipe on the CO_2 reactivity and dusting, anode samples were tested using ASTM D6558 - 00a (2010) method.

Table 5. CO_2 reactivity data of the anodes produced from different recipes

Anode No	Butt (%)	Coarse	Medium	Fine	Pitch (%)	Total Rate (mg/cm²h)	Dusting Rate (mg/cm²h)	%/min	Sample Baked Density (g/cc) (core 1)
1	25	0.0	1.0	0.0	15	20.25	2.24	0.027	1.535
2	25	0.0	1.0	0.0	15	15.10	1.66	0.021	1.521
3	25	0.3	0.6	0.3	15	20.62	1.48	0.026	1.545
4	25	0.3	0.6	0.3	15	15.72	0.59	0.021	1.549
6	25	0.3	0.6	0.3	14	17.95	1.48	0.023	1.536
7	25	0.0	0.6	0.0	15	23.49	3.48	0.031	1.518
8	25	0.3	0.8	0.0	15	17.43	1.23	0.023	1.530
13	25	1.0	0.0	0.6	15	17.75	1.47	0.023	1.539
14	25	0.3	0.5	0.5	15	16.89	0.98	0.022	1.554
15	0.0	0.0	1.0	0.0	15	15.31	0.88	0.021	1.514
16	15	0.0	1.0	0.0	15	16.27	1.22	0.022	1.524
17	35	0.0	1.0	0.0	15	25.13	4.11	0.033	1.517
18	25	0.0	1.0	0.0	15	25.44	4.50	0.035	1.556
19	25	0.3	0.6	0.3	15	24.76	3.22	0.032	1.569

The effect of butt content on anode CO_2 reactivity can be seen in Table 5. Recycled anode butt contains many impurities. These include sodium, potassium, calcium, aluminum, iron, sulfur, nickel, silicon, fluoride and others (Schmidt-Hatting 1991, Suriyapraphadilok 2005, Fischer 2013). As expected, anode CO_2 reactivity increased with increasing butt content due to the presence of impurities which can catalyze the reaction. However, anode CO_2 reactivity is not only dependent on the level of impurities but also on anode density. All anodes with different butt levels displayed a similar trend with respect to density. Higher density of anodes exhibited lower CO_2 reactivity. Though anode 2 has 25% butt content, it exhibited lower CO_2 reactivity due to its higher density. CO_2 reactivity is controlled by diffusion where it is solely determined by the ability of this gas to penetrate though the open pores and react with the carbon body of the anodes. Higher density corresponds to lower porosity which in turn helps reduce the CO_2 reactivity. Table 5 presents that anode dust formation, which increased with increasing butt content due to increased impurity content.

In addition to anode density and impurity content, CO_2 reactivity of anodes depends on several other factors such as the type and proportion of raw materials in the anode recipe (amount of pitch, coke and butt) and graphitization level (Lavigne 1993, Hume 1999, Tkac 2007). CO_2 reactivity and dust also decreased as the level of pitch was increased (anode 4 and 6; Table 4). The CO_2 reactivity and dusting results for other recipes are presented in Figures

8 and Table 5. Figure 8 shows that anode CO_2 reactivity reduces with the reduction in medium fractions in the paste recipe compared to that of the standard recipe (anode 2). This result is directly related to the anode density. A trend was previously observed that a reduction (anodes 4, 8) or complete elimination (anode 13) of medium fractions from recipe with adjusted amounts of coarse, fine and butt improved the anode density (Figures 8 and Table 3). It can be also seen from Figures 8 and Table 5 that total CO_2 reactivity of anode 4 (best recipe) was similar to anode 2 (standard recipe), but the dust rate was much lower for anode 4 (best recipe) due to higher density of anode 4 compared to anode 2, which helped reduce the CO_2 gas diffusion into anode. Another recipe (anode 14) also exhibited significantly lower dusting compared to that of the standard recipe due to its higher density. Anode produced from CPC 2 coke (anodes 18 and 19, Figures 8 and Table 5), had comparatively higher rate of reactivity and dusting due to lower concentration of sulfur. It was found in literature that sulfur inhibited the catalytic activity of metal impurities (Hume, 1999; Tran, 2009; Hume, 2013).

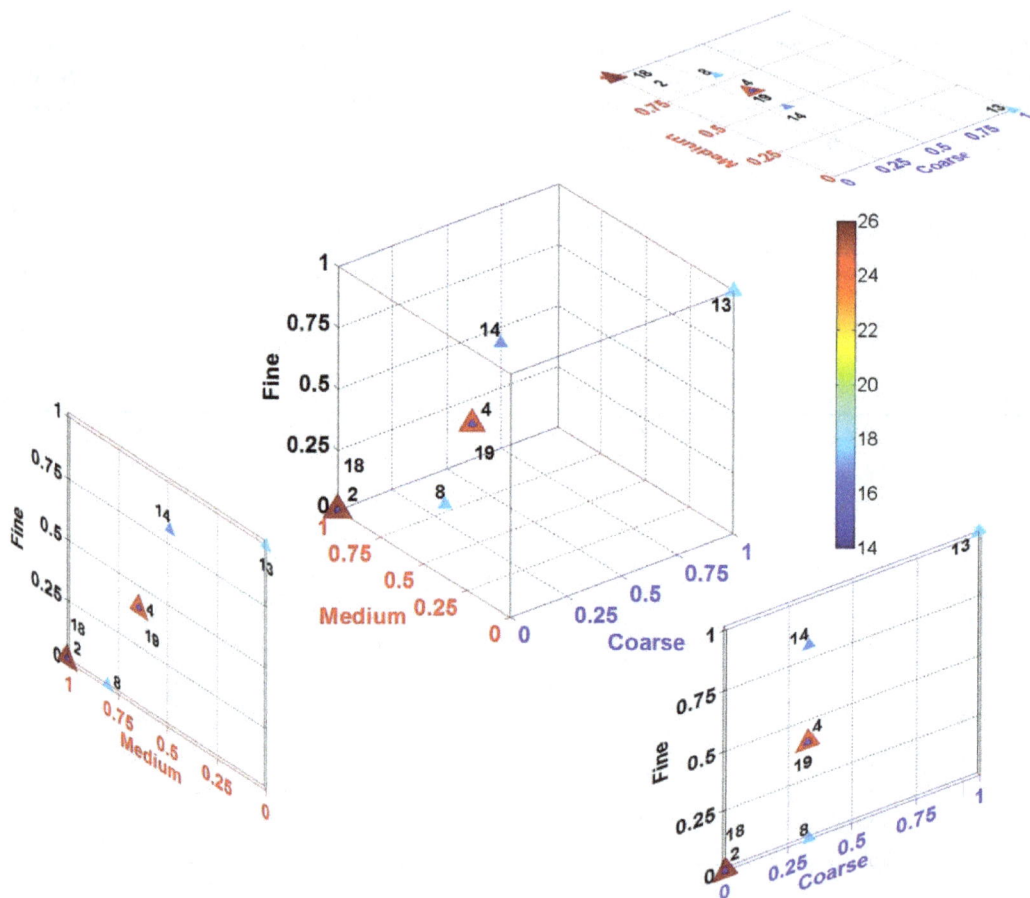

Figure 8. Total CO_2 reactivity of anodes prepared with different recipes (symbol size increases with higher values)

The carbon anodes are consumed during the electrolysis process and have to be replaced every 2-4 weeks depending on the size and density of the anode and the operating conditions of the electrolytic cell. Therefore, 7 h standard CO_2 reactivity test does not represent the actual reactivity and dusting in the electrolysis cell as anodes are exposed to the bath for longer times. For a better understanding, a 21-hour CO_2 reactivity test was carried out for standard (anodes 2 and 18) and best recipes (anodes 4 and 19) for both cokes. Anodes 2 and 4 were produced from CPC 1 coke and anodes 18 and 19 were produced from CPC 2 coke. The results of the 21-hour CO_2 test are shown in Figure 9. Results demonstrate that for best recipe (anodes 4 and 19), total CO_2 rates were reduced and compared to those of the standard recipe (anodes 2 and 18). It could be also seen from Figure 9 (a) that the final rate was significantly reduced for best recipe (anodes 4 and 19) for both cokes. Dusting rate was always better for adjusted best recipe regardless of the type of coke (anodes 4 and 19). The reason for this behavior could be higher density of the best recipe compared to that of the standard recipe. Also, anode produced from CPC 2 coke exhibited higher CO_2 reactivity and dusting due to lower sulfur content of this coke.

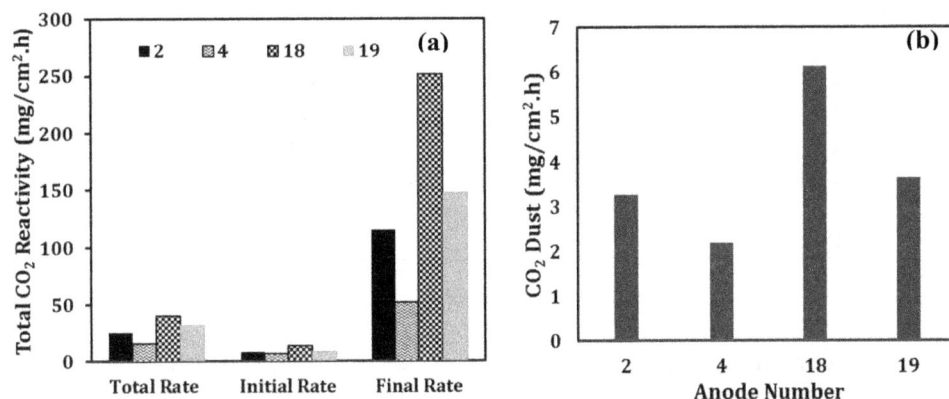

Figure 9. (a) Total CO_2 reactivity (21 h) and (b) CO_2 dust (21 h) of standard and best recipes

3.2.5 Uniaxial Compressive Strength

The compressive strength, carried out at room temperature, indicates the capacity of a material to withstand axially directed compressive forces. When the limit of the compressive strength is reached, the materials are crushed. The effect of the recipe on physical properties and reactivity was studied and the results are presented in the previous sections. However, the granulometry also affects the mechanical properties. Therefore, the compressive strength of baked anodes was measured according to ASTM C695-91 (2005) standard. The modulus of elasticity (Young's modulus) was determined experimentally from the slope of the stress-strain curve obtained during compression tests. A low Young's modulus means a small slope and therefore represents a material with high elasticity. The measurement of mechanical properties was carried out to ensure that the recipe proposed in this study was mechanically feasible. The compressive strength and Young modulus of the samples are tabulated in Table 6. All the Young modulus data are presented in dimensionless form by dividing the actual value by the highest value.

Table 6. Compressive strength of anode sample

Sample No (Anode)	Compressive Strength (MPa)	Young Modulus[a]	Specification
1	36.7	0.86	Std-CPC1
3	30.5	0.82	Best-CPC1
2	35.7	0.92	Std-CPC1
4	32.6	0.69	Best-CPC1
5	42.5	1.00	Best-CPC1
5	40.0	-	Best-CPC1
18	37.6	0.93	Std-CPC2
19	37.4	0.91	Best-CPC2

[a] Dimensionless value.

The compressive strength values were found to be within an acceptable range. Anodes 3 and 4 had slightly lower compressive strength compared to that of the standard but anode 5 exhibited higher value. Anode 19 produced from coke CPC 2 showed similar compressive strength with standard recipe, anode 18. Similar trend was observed with the Young modulus. Anodes 3 and 4 had marginally lower Young modulus compared to the standard. Anode 19 produced from coke CPC 2 displayed similar Young modulus with standard recipe (anode 18). A probable cause of this lower value of compressive strength for anode 3 was micro cracks. These results indicated that it was possible to produce anodes using the best recipe found during this study with a compressive strength within an acceptable range of 30-65 MPa (Perruchoud 2004).

3.2.6 Flexural Strength (Bending)

The flexural strengths of baked anodes produced from best and standard recipes were compared (Table 7) in order to carry out final screening. The test results showed that the anode from best recipe had similar flexural strength to that of standard recipe and varied within the range of 9-11 MPa. The flexural strength values of best recipe lied within in the practical range of industrially accepted anode flexural strength of 4–14 MPa (Perruchoud 2004).

Table 7. Flexural strength (Bending) of anodes

Anode No.	Bending stress (MPa)	Specification
1	11.3	Std-CPC1
3	8.7	Best-CPC1
2	10.7	Std-CPC1
4	11.0	Best-CPC1
18	10.7	Std-CPC2
19	9.8	Best-CPC2

4. Conclusions

Extensive pilot scale work on determining the effect of coke properties, particle size distribution, butt content and binder level on anode properties was done. A new recipe was developed which led to better anode properties compared to those of the standard recipe. The results showed that reducing the medium fraction to with adjusted amount of coarse, fine and recycled butt can improve the overall anode quality. This might make their application possible in industry. Information obtained from the ANN analysis might be useful for plants since controlling the recipe is not an easy task when the raw material quality is changing continuously. The ANN model developed might be used to improve the anode density by readjusting the aggregate granulometry when there is a change in raw material properties during operation.

Acknowledgments

The technical and financial support of Aluminerie Alouette Inc. as well as the financial support of the Natural Sciences and Engineering Research Council of Canada (NSERC), Dévelopment économique Sept-Îles, Université du Québec à Chicoutimi (UQAC), and the Foundation of Université du Québec à Chicoutimi (FUQAC) is greatly appreciated.

References

Adams, A. N. (2004). *Characterization of the pitch wetting and penetration behavior of petroleum coke and recycled butts in pre-baked carbon anode*. PhD, Pennsylvania State University.

Adams, A. N., & Schobert, H. H. (2004). Characterization of the surface properties of anode raw materials. *Light Metals.*, 495-498.

Asadi-Eydivand, M., Solati-Hashjin, M., Farzadi, A., & Osman, N. A. A. (2014). Artificial neural network approach to estimate the composition of chemically synthesized biphasic calcium phosphate powders. *Ceramics International, 40*(8 PART A), 12439-12448. http://dx.doi.org/10.1016/j.ceramint.2014.04.095

Bahlmann, C., Heidemann, G., & Ritter, H. (1999). Artificial neural networks for automated quality control of textile seams. *Pattern Recognition, 32*(6), 1049-1060. http://dx.doi.org/10.1016/S0031-3203(98)00128-9

Belitskus, D. (1981). Effect of carbon recycle materials on properties of bench scale prebaked anodes for aluminum smelting. *Metallurgical Transactions B, 12*(1), 135-139. http://dx.doi.org/10.1007/BF02674766

Berezin, A. I., Polaykov, P. V., Rodnov, O. O., Klykov, V. A., & Krylov, V. L. (2002). Improvement of green anodes quality on the basis of the neural network model of the carbon plant workshop. *Light Metals*, 605-608.

Bhagavatula, Y. S., Bhagavatula, M. T., & Dhathathreyan, K. S. (2012). Application of artificial neural network in performance prediction of PEM fuel cell. *International Journal of Energy Research, 36*(13), 1215-1225. http://dx.doi.org/10.1002/er.1870

Bhattacharyay, D., Kocaefe, D., Kocaefe, Y., & Morais, B. (2015). An artificial neural network model for predicting the CO2 reactivity of carbon anodes used in the primary aluminum production. *Neural Computing and Applications*. http://dx.doi.org/10.1007/s00521-015-2093-7

Bhattacharyay, D., Kocaefe, D., Kocaefe, Y., Morais, B., & Gagnon, M. (2013). Application of the artificial neural network (ANN) in predicting anode properties. *Light Metals*, 1189-1194.

Bhattacharyay, D., Kocaefe, D., Kocaefe, Y., Sarkar, A., Morais, B., & Chabot, J. (2014). Characterization of dry aggregates in carbon anodes by image analysis. *Light Metals.*, 1111-1114. http://dx.doi.org/10.1002/9781118888438.ch185

Biedler, P., Banta, L., Dai, C., Love, R., Tommey, C., & Berkow, J. (2002). Development of a state observer for an aluminum reduction cell. *Light Metals*, 1091-1098.

Boadu, K. D., & Omani, F. K. (2010). Adaptive control of feed in the Hall-Héroult cell using a neural network. *JOM, 62*(2), 32-36. http://dx.doi.org/10.1007/s11837-010-0028-4

Bowers, R., Ningileri, S., Palmlund, D. C., Vitchus, B., & Cannova, F. (2008). New analytical methods to determine calcined coke porosity, shape, and size. *Light Metals.,* 875-880.

Brooks, J. D., & Taylor, G. H. (1968). *Chemistry and Physics of Carbon.* New York.

Engvoll, A. M. (2001). *Reactivity of anode raw materials and anodes for production of aluminium.* PhD, NTNU.

Farr-Wharton, R., Welch, B. J., Hannah, R. C., Dorin, R., & Gardner, H. J. (1980). Chemical and electrochemical oxidation of heterogeneous carbon anodes. *Electrochimica Acta, 25*(2), 217-221. http://dx.doi.org/10.1016/0013-4686(80)80046-6

Figueiredo, F. E. O., Kato, C. R., Nascimento, A. S., Marques, A. O. F., & Miotto, P. (2005). Finer fines in anode formulation. *Light Metals.,* 665-668.

Fischer, W. K., & Perruchoud, R. C. (2013). Interdependence Between Properties of Anode Butts and Quality of Prebaked Anodes. *Essential Readings in Light Metals: Electrode Technology for Aluminum Production, 4,* 267-270. http://dx.doi.org/10.1002/9781118647745.ch34

Fischer, W., & Perruchoud, R. (1991). Interdependence between properties of anode butts and quality of prebaked anodes. *Light Metals,* 721-724.

Fruhwirth, R. K., Filzwieser, A., Pesl, J., & Steinlechner, S. (2007). Computational intelligence - Neural computation in a copper refinery. *Proceedings - European Metallurgical Conference, EMC 2007, 1,* 13-29.

Hulse, K. L. (2000). *Anode Manufacture: Raw Materials, Formulation and Processing Parameters.* Sierre, Switzerland, R&D Carbon Ltd.

Hume, S. M. (1999). *Anode reactivity: influence of raw material properties.* Sierre Suisse., R & D Carbon Ltd.

Hume, S. M., Fischer, W. K., Perruchoud, R. C., Metson, J. B.,Terry, R., & Baker, K. (2013). Influence of petroleum coke sulphur content on the sodium sensitivity of carbon anodes. *Essential Readings in Light Metals: Electrode Technology for Aluminum Production, 4,* 123-129. http://dx.doi.org/10.1002/9781118647745.ch17

Lavigne, L., & Castonguay, L. (1993). Prediction of anode performance from calcined coke properties. *Light Metals.,* 569-575.

Lhuissier, J. (2007). *About the use of under-calcined coke for the production of low reactivity anodes.* International pitch and calcined petroleum coke conference-Industrial quimica del nalon and Jacobs consultancy.

Lhuissier, J., Bezamanifary, L., Gendre, M., & Chollier, M. J. (2009). Use of under-calcined coke for the production of low reactivity anodes. *Light Metals.,* 979-983.

Meghlaoui, A., Bui, R. T., Thibault, J., Tikasz, L., & Santerre, R. (1998). Predictive control of aluminum electrolytic cells using neural networks. *Metallurgical and Materials Transactions B: Process Metallurgy and Materials Processing Science, 29*(5), 1007-1019. http://dx.doi.org/10.1007/s11663-998-0069-z

Milewski, J., & Świrski, K. (2009). Modelling the SOFC behaviours by artificial neural network. *International Journal of Hydrogen Energy, 34*(13), 5546-5553. http://dx.doi.org/10.1016/j.ijhydene.2009.04.068

Mohanty, C., & Jena, B. K. (2014). Optimization of aluminium die casting process using artificial neural network. *International Journal of Emerging Technology and Advanced Engineering, 4*(7), 146-149.

Pang, G., Xu, W., Zhai, X., & Zhou, J. (2004). Forecast and control of anode shape in electrochemical machining using neural network. *Lecture Notes in Computer Science (including subseries Lecture Notes in Artificial Intelligence and Lecture Notes in Bioinformatics), 3174,* 262-268. http://dx.doi.org/10.1007/978-3-540-28648-6_41

Parthiban, T., Ravi, R., & Kalaiselvi, N. (2007). Exploration of artificial neural network [ANN] to predict the electrochemical characteristics of lithium-ion cells. *Electrochimica Acta, 53*(4), 1877-1882. http://dx.doi.org/10.1016/j.electacta.2007.08.049

Perruchoud, R. C., Meier, M. W., & Fischer, W. K. (2004). Survey on worldwide prebaked anode quality. *Light Metals.,* 573-578.

Piuleac, C. G., Rodrigo, M. A., Ca-izares, P., Curteanu, S., & Sáez, C. (2010). Ten steps modeling of electrolysis processes by using neural networks. *Environmental Modelling and Software, 25*(1), 74-81. http://dx.doi.org/10.1016/j.envsoft.2009.07.012

Prouix, A. L. (1993). Optimum binder content for prebaked anodes. *Light Metals.,* 657-661.

Saengrung, A., Abtahi, A., & Zilouchian, A. (2007). Neural network model for a commercial PEM fuel cell system. *Journal of Power Sources, 172*(2), 749-759. http://dx.doi.org/10.1016/j.jpowsour.2007.05.039

Sarkar, A. (2015). *Effect of coke properties on anode properties.* PhD, University of Quebec at Chicoutimi.

Schmidt-Hatting, W., Kooijman, A. A., & Perruchoud, R. (1991). Investigation of the quality of recycled anode butts. *Light Metals,* 705-720.

Shang, G. Q., & Sun, C. H. (2008). Application of BP neural network for predicting anode accuracy in ECM. *2008 International Symposium on Information Science and Engineering, ISISE 2008.* http://dx.doi.org/10.1109/ISISE.2008.55

Smith, M., Perruchoud, R., Fischer, W., & Welch, B. (1991). Evaluation of the effect of dust granulometry on the properties of binder matrix bench scale electrodes. *Light Metals,* 651-655.

Sulaiman, D., & Garg, R. (2012). Use of under calcined coke to produce baked anodes for aluminium reduction lines. *Light Metals.,* 1147-1151.

Suriyapraphadilok, U., Halleck, P., Grader, A., & Andresen, J. M. (2005). Physical, chemical and X-ray Computed Tomography characterization of anode butt cores. *Light Metals,* 617-621.

Tkac, M. (2007). *Porosity development in composite carbon materials during heat treatment.* Philosophiae doctor (PhD), Norwegian University of Science and Technology.

Tran, K. N., Berkovich, A. J., Tomsett, A., & Bhatia, S. K. (2009). Influence of sulfur and metal microconstituents on the reactivity of carbon anodes. *Energy and Fuels, 23*(4), 1909-1924. http://dx.doi.org/10.1021/ef8009519

Wang, L., Apelian, D., Makhlouf, M., & Huang, W. (2008). Predicting compositions and properties of aluminum die casting alloys using artificial neural network. *Metallurgical Science and Technology, 26*(1), 16-21.

Zande, M. (1997). *Introduction to Carbon Technologies.* University of Alicante, Alicante.

6

Molecular Dynamics Simulations of CdTe / CdS Heteroepitaxy - Effect of Substrate Orientation

Jose J. Chavez[1], Xiao W. Zhou[2], Sergio F. Almeida[1], Rodolfo Aguirre[1] & David Zubia[1]

[1] Department of Electrical and Computer Engineering, the University of Texas at El Paso, Texas 79968, USA

[2] Mechanics of Materials Department, Sandia National Laboratories, Livermore, California 94550, USA

Correspondence: Jose J. Chavez, Department of Electrical and Computer Engineering, University of Texas at El Paso, TX 79968 USA. Email: jjchavez5@miners.utep.edu

Abstract

Molecular dynamics simulations were used to catalogue atomic scale structures of CdTe films grown on eight wurtzite (wz) and zinc-blende (zb) CdS surfaces. Polytypism, grain boundaries, dislocations and other film defects were detected. Dislocation lines were distributed in three distinct ways. For the growths on the wz {0001} and zb {111} surfaces, dislocations were found throughout the epilayers and formed a network at the interface. The dislocations within the films grown on the wz {$\bar{1}$100}, wz {11$\bar{2}$0}, zb {$\bar{1}$10}, zb {010}, and zb {$\frac{1}{10}$ 1 $\frac{1}{10}$} surfaces formed an interface network and also threaded from the interface towards the film's surface. In contrast, the growth on the zb {11$\bar{2}$} surface only had dislocations localized to the interface. This film exhibited a different orientation from the substrate to reduce the lattice mismatch strain energies, and therefore, its misfit dislocation density. Our study indicates that the substrate orientation could be utilized to modify the morphology of dislocation networks in lattice mismatched multi-layered systems.

Keywords: defects, dislocations, CdTe, substrates, surfaces, cadmium compounds, thin films

1. Introduction

Cadmium telluride (CdTe) thin films are widely used in photovoltaic (Colegrove et al., 2012) (McCandless and Sites 2003) (Okamoto, Yamada, & Konagai 2001) and other optoelectronic applications (Liang et al., 2012) (Brill et al., 2005) (Heiss et al., 2006) due to its low production cost, a desired direct band gap of ~1.5 eV, and an excellent absorption coefficient of 6×10^4 m^{-1} at 600 nm. A variety of methods have been applied to synthesize CdTe films such as molecular beam epitaxy (MBE) (Kim et al., 2004) (Han, Kang, & Kim 1999), physical vapour deposition (PVD) (Moutinho et al., 2008), high vacuum evaporation (HVE) (Terheggen et al., 2003), and close space sublimation (CSS) (Paudel, Xiao, & Yan 2014) (Yan, Al-Jassim and Jones 2001). These efforts indicated that due to a large lattice mismatch with the underlying substrate (e.g., CdS), CdTe films always contain a high density of performance limiting lattice defects (Moseley et al., 2014), including point defects, stacking faults, dislocations, and grain boundaries.

Past efforts (Aguirre et al., 2014) (Cruz-Campa et al., 2012) (Zubia et al., 2007) have focused on reducing defects during the fabrication of the epilayers, although the fundamental mechanisms of defect formation are not well understood. Experimental studies have shown that careful preparation of the substrate surface termination (Myers et al., 1983) and the orientation of the substrate (Smith et al., 2000) (Sarney & Brill, 2004) (Terheggen et al., 2003) can greatly influence the microstructure and the quality of the heteroepitaxial film. However, the high-resolution characterization techniques used to obtain these results are destructive, expensive, and time consuming. For example, transmission electron microscopy (TEM) (Li et al., 2013) (Yan, Al-Jassim, & Jones, 2001) is effective at revealing two-dimensional (2-D) microstructures with atomic resolution, but cannot provide three-dimensional (3-D) information that determines material properties and defect evolution. Moreover, atomic probe tomography (Kelley & Miller, 2007) (Miller & Forbes, 2009) provides rich 3-D compositional information, but is unable to resolve lattice and defect structure with atomic resolution. Alternatively, atomistic simulations such as molecular dynamics (MD) (Alder & Wainwright, 1959) (Rapaport, 2006) allow virtual experiments that isolate external factors and offer a 3-D analysis of the heteroepitaxial microstructures. More importantly, MD simulations can rapidly generate a large database of atomic scale structures of films as a function of substrate characteristics including all commonly encountered substrate orientations. A similar database may take many years to develop using experiments alone.

In recent years, the present authors have developed a robust MD method capable of simulating the growth of II-VI compounds (Zhou & Ward, 2013). Based on this MD method, we have discovered and corrected issues of the traditional misfit dislocation theory, designed defect free, nanostructured CdTe/CdS solar cells (Zhou et al., 2015), and catalogued the atomic scale structures of CdS homoepitaxial films on commonly encountered CdS surfaces without the lattice mismatch effect (Almeida et al., 2016). In the present paper, we further extend the study to catalogue the atomic scale structures of heteroepitaxial CdTe films on commonly encountered wz and zb CdS surfaces with the lattice mismatch effect. A separate study will also be performed to explore the mechanisms of the observed phenomena. Our ultimate objective is to create a convenient handbook characterizing substrate orientation effects on atomic scale structure of films that can help material scientists interpret experimental observations and reduce defects in multi-layered films via selection of substrate orientations.

2. Molecular Dynamics Simulation Details

The simulations were performed with the Large-Scale Atomic/Molecular Massively Parallel Simulator (LAMMPS) code (Plimpton, 1995) employing a Zn-Cd-Hg-S-Se-Te Stillinger-Weber potential (Zhou et al., 2013) capable of predicting the crystal structures and defects observed experimentally. The CdTe growth was performed on eight different wurtzite (wz) and zinc blende (zb) CdS substrate surfaces. The wz substrates included; $\{0001\}$, $\{\bar{1}100\}$, and $\{11\bar{2}0\}$ surfaces. The zb substrates included; $\{111\}$, $\{11\bar{2}\}$, $\{\bar{1}10\}$, $\{010\}$, and $\left\{\frac{1}{10} \; 1 \; \frac{1}{10}\right\}$ surfaces. Here the $\left\{\frac{1}{10} \; 1 \; \frac{1}{10}\right\}$ surface essentially mimics a ~8° miscut from the $\{010\}$ surface. The CdS lattice constants were $a_{wz} = 4.134$ Å and $c_{wz} = 6.752$ Å for the wz crystals, and $a_{zb} = 5.847$ Å for the zb crystals. The lattice parameters for wz CdTe were $a_{wz} = 4.590$ Å and $c_{wz} = 7.495$ Å, while the zb lattice parameter was $a_{zb} = 6.491$ Å. These lattice constant values were calculated from time averaged MD simulations of bulk single crystals at 300 K and are in excellent agreement (less than 1% difference) with experimental values (Donnay and Ondik 1973) and molecular statics calculations (Zhou et al., 2013).

The deposition parameters were selected to ensure crystalline growth based on previous results (Zhou, Johnson, & Wadley, 1997) (Zhou & Wadley, 2000) (Zhou & Wadley, 2001) and a complete description of the vapour deposition simulations methods has been discussed previously (Zhou et al., 2012). Briefly, CdTe growth was simulated at a deposition temperature of 1200 K under isothermal conditions by employing the Nose-Hoover algorithm (Hoover, 1985). The total deposition time was ~19.6 ns with a deposition rate of ~0.48 nm/ns. Periodic boundary conditions were employed in the x- and z- directions while a free boundary condition was employed in the y- (growth) direction. The adatoms had an incident energy of 5.0 eV and a stoichiometric vapour ratio of Cd:Te = 1:1. During simulations, positions of atoms in the bottom two monolayers of the substrates were fixed to prevent crystal drift due to adatom momentum transfer. It should be noted that the heteroepitaxy growth mechanism differs from that of homoepitaxy due to additional effects from lattice mismatch and chemistry. These effects are taken into account by a high-fidelity interatomic potential. The Stillinger-Weber potential employed in this work ensures the lowest energy configuration of the equilibrium compounds in addition to capturing their lattice constants and cohesive energies.

Following MD simulations of growth, a three-step post-deposition treatment was used to reduce thermal noises and improve the clarity of the data. First, an MD annealing (without adding more adatoms) was performed for ~8 ns to relax the structures. The MD simulation was then continued for another ~8 ns where the temperature was linearly cooled down to 50 K. Finally, a molecular statics energy minimization simulation was carried out to relax the structures at 0 K.

The effect of system size was explored and larger systems (between ~200 Å and ~600 Å in the lateral dimensions) generally captured a better physical description of the material defects at the expense of an increased computing time. In contrast, simulations that are less computationally intensive (< ~100 Å per side) could capture only a few single dislocation segments in general. The heterostructure samples employed had a size of ~215 Å, ~120 Å, and ~160 Å respectively for the x, y, and z dimensions. This system size appropriately captured the dislocation networks studied in this work while consuming an appropriate amount of computing resources.

3. Analysis of MD Data

The Open Visualization Tool (Stukowski, 2010) (OVITO) version 2.5.1 was employed to analyse the MD data and to render the images used in this work. The common neighbour analysis algorithm for diamond structures provided by the tool was used for structural characterization of the simulated films. The dislocation extraction algorithm (DXA) (Stukowski & Albe, 2010) (Stukowski, Bulatov, & Arsenlis, 2012) was used for the identification of linear defects within the data. Two types of dislocation analyses were explored for this study: one assumes a perfect wz lattice and the other assumes a perfect zb lattice. The Burger's vectors assigned to the dislocation segments were given in Miller-Bravais indices for the wurtzite analysis and cubic Miller index format for the zinc blende analysis.

4. Results

4.1 Lattice Structure Percentage and Dislocation Density

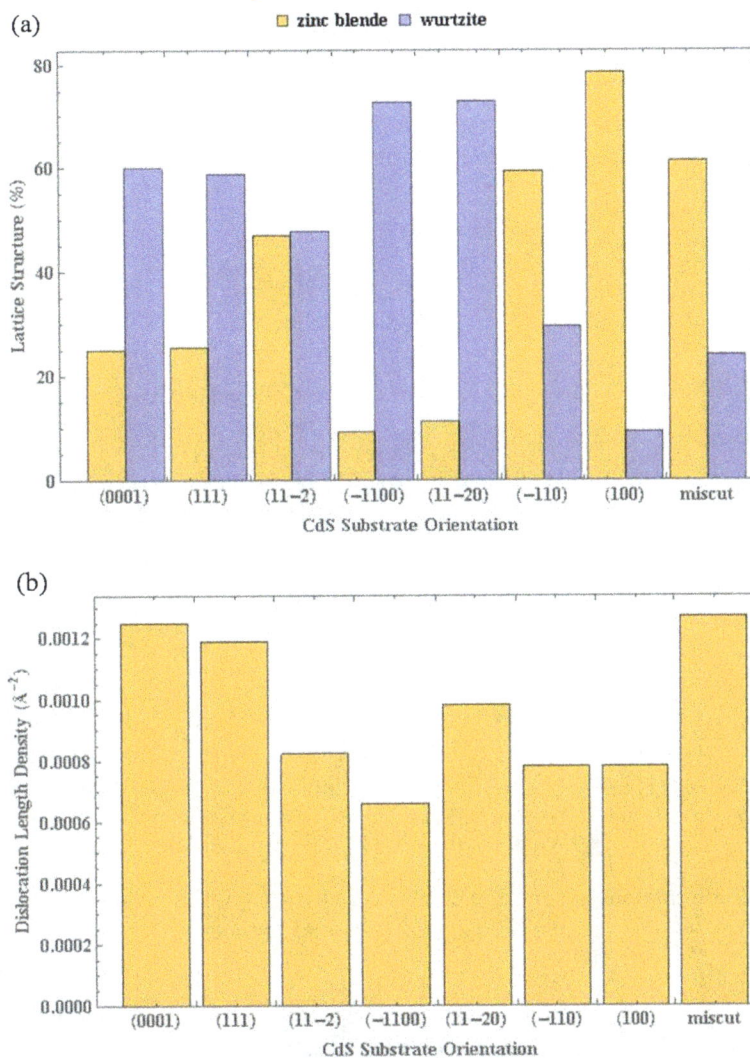

Figure 1. (a) Lattice structure percentage and (b) dislocation density for all simulated growths. Polytypism and misfit dislocations were observed in all cases

The lattice structure analyses results in Figure 1(a) show that each film contained multiple phase domains, having some amount of atoms associated to wurtzite and zinc blende structures. This polytypism is a common occurrence reported in CdTe films grown using MBE (Smith et al., 2000), as well as epitaxy using CSS (Al-Jassim et al., 2001) (Yan et al., 2000). Each simulated film also contained some regions where the structure did not match either the wz or zb lattice structures (unmatched percentages are not shown in the figure). These unmatched regions arise from other crystallographic imperfections such as point defects (vacancies, interstitials, etc.), grain boundaries, and dislocations. Visualizations of the atomic microstructure revealed that the epilayer grain orientation matched the substrate orientation in all cases except for the growth on the {11$\bar{2}$}. The mechanism to form a new growth orientation will be reported in detail elsewhere. Film dislocation densities near the interface (<~100 Å away from the substrate) were also analysed, and the results are shown in Figure 1(b) for all samples. The data indicates that films grown on the hexagonal surfaces ({0001} and {111}) are highly defected. The remaining six surfaces have rectangular geometries. Except for the $\{11\bar{2}0\}$ and miscut surfaces, four of the growths on these surfaces clearly had lower dislocation densities when compared to hexagonal surface growths. The lowest dislocation density occurred for the $\{\bar{1}100\}$ substrate. The trends observed in Figure 1 suggest that using rectangular rather than hexagonal substrate surfaces can reduce defect densities. The same trend was observed in MBE CdTe films grown on Si substrates (Terheggen et al., 2003). This was also observed in theoretical studies employing density functional theory (Yin et al., 2015) and in CdS homoepitaxy via MD (Almeida et al., 2016).

4.2 Burgers Vector and Dislocation Spatial Distributions

Analysis of the dislocation networks extracted using zinc blende DXA revealed three distinct spatial distribution categories associated with CdS substrate orientations; the $\{11\bar{2}\}$ orientation, orientations with rectangular surface geometry ($\{\bar{1}100\}$, $\{11\bar{2}0\}$, $\{\bar{1}10\}$, $\{010\}$, $\{\frac{1}{10}\,1\,\frac{1}{10}\}$), and orientations with hexagonal surface geometry ($\{0001\}$ and $\{111\}$). These three categories are represented in Figure 2 by the films grown on the $\{11\bar{2}\}$, $\{\bar{1}100\}$, and $\{0001\}$ oriented substrates. All the dislocations occurred in the grown films and none were found in the CdS substrates.

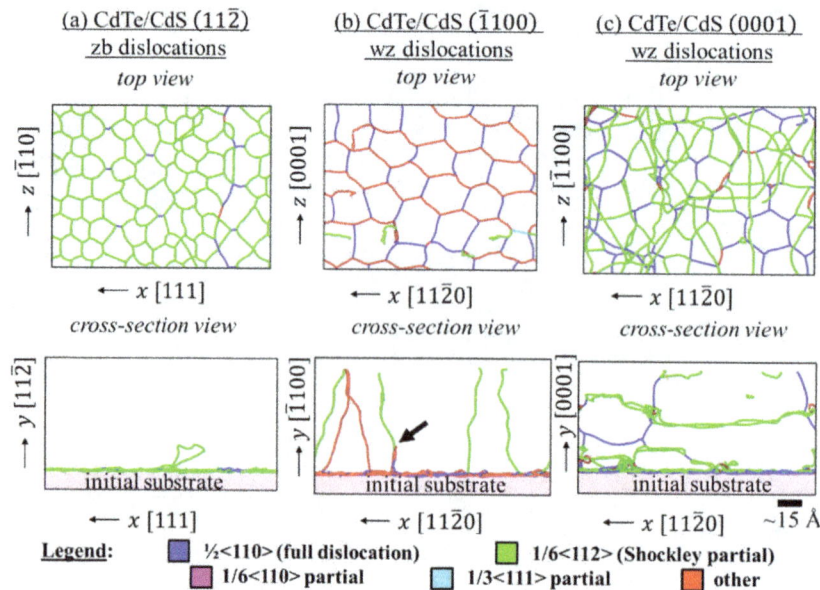

Figure 2. Dislocation configurations in (a) CdTe on zb ($11\bar{2}$) CdS, (b) CdTe on wz ($\bar{1}100$) CdS, and (c) CdTe on wz (0001) CdS substrate surfaces

The first distribution category consists of an interfacial dislocation network with virtually no dislocations in the bulk of the film as shown in Figure 2(a) suggesting a highly efficient mechanism at relieving lattice mismatch strain. The film grown on the $\{11\bar{2}\}$ substrate was the only one that showed this type of dislocation distribution. The Burgers vector of the interfacial dislocations are predominantly $\frac{1}{6}<112>$ Shockley partials (or $\frac{1}{3}<1100>$ for a wz DXA). A small number of $\frac{1}{3}<112>$ dislocations were also detected and are shown in red colour in Figure 2(a).

The second dislocation category is shown in Figure 2(b), where in addition to a dislocation network at the interface, a group of dislocations threaded from the interface to the surface. This type of dislocation distribution was observed in the films grown on the $\{\bar{1}100\}$, $\{11\bar{2}0\}$, $\{\bar{1}10\}$, $\{010\}$, and $\{\frac{1}{10}\,1\,\frac{1}{10}\}$ oriented CdS surfaces which possess a rectangular surface geometry. In this case, the Burgers vectors of the interfacial dislocations are roughly an equal mixture of $\frac{1}{2}<110>$ (or $\frac{1}{3}<1120>$) full dislocations and $\frac{1}{6}<114>$ (or $\frac{1}{6}<2203>$) shown in red colour in Figure 2(b). However, the threading dislocations in the cross-section view of Figure 2(b) are $\frac{1}{6}<112>$ Shockley partials and $\frac{1}{6}<114>$. These threading dislocations don't intersect and instead remain approximately parallel with respect to each other. In one instance (indicated by the black arrow), a vertical dislocation junction was formed between $\frac{1}{2}<110>$ and $\frac{1}{6}<114>$ dislocations near the interface, with one additional $\frac{1}{6}<112>$ segment threading towards the surface.

The third dislocation category is shown in Figure 2(c), where dislocations were present throughout the film. This type of dislocation distribution was observed in the films grown on surfaces with hexagonal surface geometry, {0001} and {111}. These films contained a mixture of $\frac{1}{6} < 112 >$ Shockley partial and $\frac{1}{2} < 110 >$ full dislocations both at the interface and in the bulk. The Burgers vectors of both $\frac{1}{6} < 112 >$ Shockley partials and $\frac{1}{2} < 110 >$ full dislocations lie on the closed-packed planes, and their formation is energetically favoured especially in materials with a low stacking fault energy that promote polytypism (Terheggen et al., 2003). Other types of dislocations were also detected but their densities were significantly lower.

4.3 Lattice Structure Morphology

The lattice structures of the CdTe films were further analysed. Figure 3 shows atomistic visualizations of the CdTe/{11$\bar{2}$} CdS heterostructure. Figure 3(a) is a cross-sectional view of a ~10 Å slice of the sample where the atoms are coloured according to species (Cd, S, and Te). The ternary CdS_yTe_{1-y} alloy was detected in a 10 Å thick region straddling the interface. The S atoms were observed to diffuse further into the epilayer in comparison to the Te into the substrate, similarly to experimental observations using a TEM equipped with an energy dispersive X-ray spectroscopy (EDS) detector (Terheggen et al., 2003) where low concentrations of S were detected to diffuse into the film but no Te diffusion into the substrate. This intermixing behaviour was a trend observed in all eight samples. Figure 3(b) is another cross-sectional view of the same data except that the atoms are now coloured according to their lattice structure. The light blue regions indicate a zinc blende structure and the red regions indicate a wurtzite structure. The orange atoms indicate atoms that could not be matched to either the zb or wz structures. This occurs in defected regions such as point defects (e.g., vacancies and interstitials), surfaces, grain boundaries, and dislocation cores. Comparison between Figures 3(a) and 3(b) clearly indicates the presence of polytypism in agreement with experimental reports (Smith et al., 2000) (Al-Jassim et al., 2001) (Yan et al., 2000) and stacking faults along a direction about ~22° from the interface plane. The majority of the grains have a zinc blende crystal structure.

Figure 3(c) shows a plan view lattice structure map near the interface as indicated by marker "C" in Figure 3(b). The orange atoms outline an interfacial dislocation network consistent with the dislocation network observed in Figure 2(a). In this region the film consisted of numerous grains with an average size of ~15 Å. Figure 3(d) shows a plan view lattice structure map about 80 Å away from the interface (marked as "B" in Figure 3(b)). In contrast to the region near the interface, the film in Figure 3(d) consisted of a single grain with point defects and alternating zb and wz regions lying along the x direction.

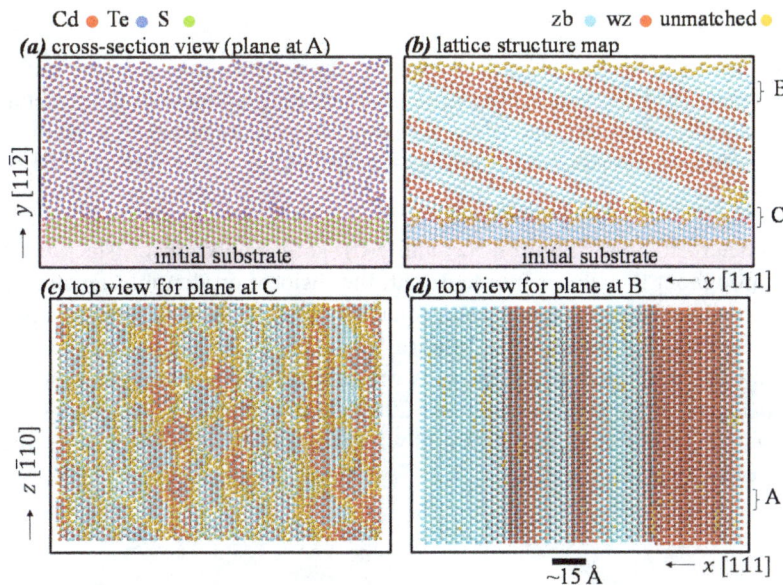

Figure 3. Visualization of the CdTe on zb (11$\bar{2}$) CdS: (a) front view where atoms are distinguished by species, (b) front view where atoms are distinguished by local structures as zinc blende (light blue), wurtzite (red), and "unmatched" (orange), (c) plan view of atom structures at a location near the interface, showing an average grain size of ~15 Å with zinc blende being the preferred structure, and (d) plan view of atom structures at a location ~80 Å away from the interface, showing polytypism without grain boundaries other than point defects

A similar analysis was performed for the CdTe/{$\bar{1}$100}CdS heterostructure in Figure 4. Figures 4(a) and 4(b) clearly indicate three different regions separated, for example, by the orange atoms in Figure 4(b). In this case, the orange atoms represent dislocation line cores which begin at the interface and extend to the surface. Point defects were detected within the grains (black arrow in Figure 4(b)). Figure 4(c) shows the dislocation network near the interface, in which the line defects connect to form geometries with an area of ~18 Å. This is similar to the phenomenon observed for the {11$\bar{2}$} growth in Figure 3. In contrast, Figure 4(d) indicates that at ~80 Å away from interface the film consisted of a single wurtzite grain with stacking faults, point defects, and the intersecting dislocation cores (indicated by the black arrows for two cases) from the threading dislocations.

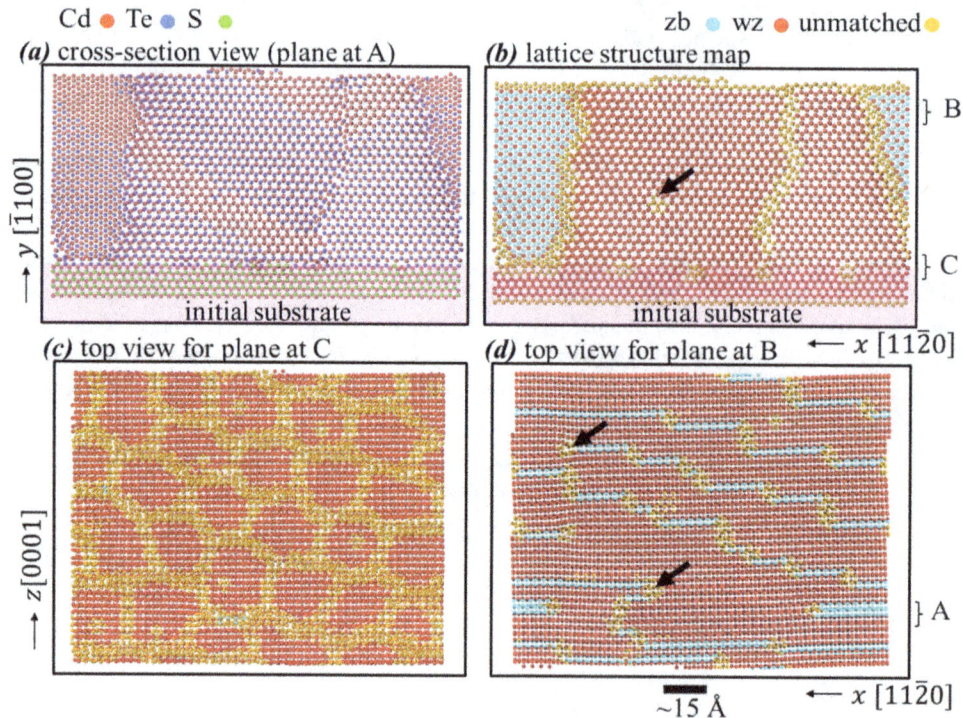

Figure 4. Visualization of the CdTe on wz ($\bar{1}$100) CdS using the same approach described in Figure 3: (a) front view of the species map; (b) front view of the structure map; (c) plan view of the structure map near the interface, showing a dislocation network; and (d) plan view of the structure map at a location ~80 Å away from the interface, showing polytypism, point defects, and dislocation cores (indicated by black arrows)

The CdTe/ {0001}CdS heterostructure (Figures 5(a) and 5(b)) shows polytypism, stacking faults, point defects, and dislocation cores throughout the film, consistent with the dislocation distribution showed in Figure 2(c). A polycrystalline network with hexagonal geometry and equal mixture of zb and wz grains is present at the interface with an average grain size of ~15 Å as shown in Figure 5(c). This is anticipated since the CdS substrate has a hexagonal surface geometry and experimental growths on this surface geometry commonly report polytypism (Smith et al., 2000) (Al-Jassim et al., 2001) (Yan et al., 2000). However only ~80 Å away from the interface, the hexagonal geometry is replaced by much larger zb and wz grains of ~80 Å nominal size and no remnant of the underlying hexagonal geometry (Figure 5(d)).

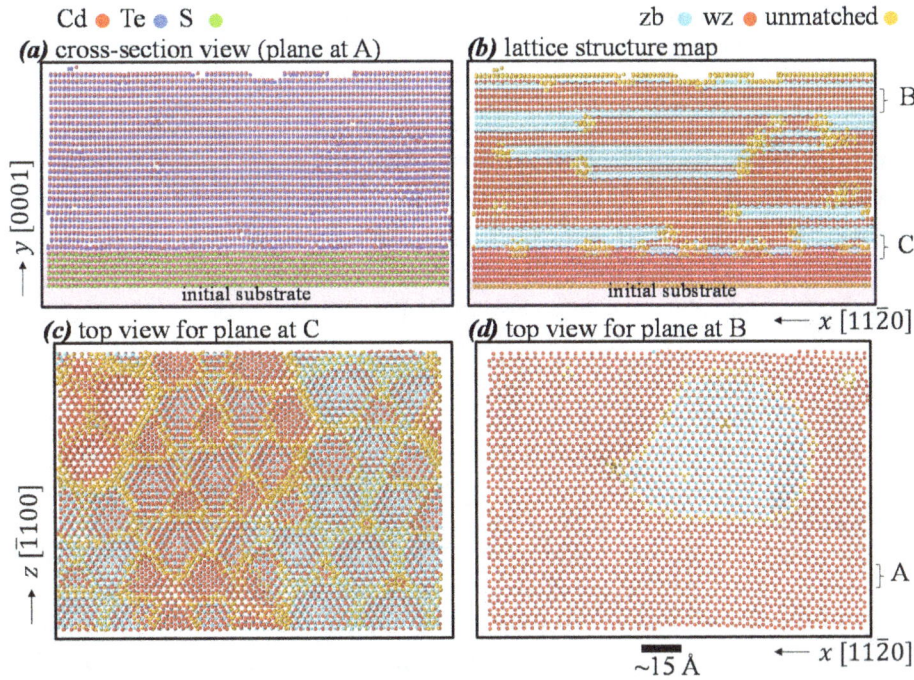

Figure 5. Visualization of the CdTe on wz (0001) CdS using the same approach described in Figure 3: (a) front view of the species map; (b) front view of the structure map showing polytypism, grain boundaries, and point defects; (c) plan view of the structure map near the interface, showing dislocation network, and an average grain size of ~15 Å with an even distribution of zinc blende and wurtzite grains; and (d) plan view of the structure map at a location ~80 Å away from the interface, showing polytypism in the form of a zinc blende grain surrounded by wurtzite material

5. Conclusions

Molecular dynamics simulations have been used to analyse CdTe films grown on eight commonly encountered CdS surfaces. The formation of the CdS_yTe_{1-y} alloy was observed at the interface due to multi-layer intermixing in all growths. The lattice structure analysis detected polytypism (coexistence of wurtzite and zinc blende domains) in all samples. Most epilayers had a dominant lattice structure that matched that of its initial substrate, except for the growths on the $\{11\bar{2}\}$ and $\{111\}$ surfaces. Lattice defects (point defects, stacking faults, dislocations and grain boundaries) were also found in all films. The film dislocation density was calculated and the data showed that growths on the hexagonal surfaces ($\{0001\}$ and $\{111\}$) are highly defected. In contrast, four of the growths on rectangular surfaces ($\{\bar{1}100\}$, $\{11\bar{2}\}$, $\{\bar{1}10\}$, and $\{010\}$) had lower dislocation densities. The exception were the films grown on the $\{11\bar{2}0\}$ and miscut surfaces. These results agree with findings from a surface stability study (Yin et al., 2015) in which density functional theory was employed to explore optimal growth conditions. The dislocation morphology was found to depend on substrate orientation and could be described in three categories. For films grown on the $\{11\bar{2}\}$ surface, dislocations were localized to the interface. An additional interesting phenomenon is that unlike other substrate surfaces, the film on the $\{11\bar{2}\}$ surface grew in a new orientation different from the substrate. This accounts for the lack of misfit dislocations because the new orientation enables the lattice mismatch strain to be released. For films grown on the $\{\bar{1}100\}$, $\{11\bar{2}0\}$, $\{\bar{1}10\}$, $\{010\}$, and $\{\frac{1}{10} 1 \frac{1}{10}\}$ rectangular surfaces, dislocations formed at the interface but also threaded out towards the surface. For the films grown on the wz $\{0001\}$ and $\{111\}$ hexagonal surfaces, the dislocations formed a network at the interface and extended throughout the films. Our study indicates that selecting an appropriate substrate orientation could offer some control over defect density and dislocation network morphology that is formed within the epilayer.

Acknowledgements

Sandia National Laboratories is a multiprogram laboratory managed and operated by Sandia Corporation, a wholly owned subsidiary of Lockheed Martin Corporation, for the US Department of Energy's National Nuclear Security Administration under contract DE-AC04-94AL85000. This work was supported by DOE project No. EE0005958,

National Science Foundation grants DGE-0903670 and CNS-1059430. This work used the Extreme Science and Engineering Discovery Environment (XSEDE), which is supported by NSF grant number ACI-1053575.

References

Aguirre, B. A., Zubia, D., Ordonez, R., Anwar, F., Prieto, H., Sanchez, C. A., ... & Mcclure, J. C. (2014). Selective growth of CdTe on nano-patterned CdS via close-space sublimation. *Journal of Electronic Materials, 43*(7), 2651-2657. http://dx.doi.org/10.1007/s11664-014-3104-7

Alder, B. J., & Wainwright, T. E. (1959). Studies in molecular dynamics. I. General method. *The Journal of Chemical Physics, 31*(2), 459-466. http://dx.doi.org/10.1063/1.1730376

Al-Jassim, M. M., Yan, Y., Moutinho, H. R., Romero, M. J., Dhere, R. D., & Jones, K. M. (2001). TEM, AFM, and cathodoluminescence characterization of CdTe thin films. *Thin Solid Films, 387*(1), 246-250. http://dx.doi.org/10.1016/S0040-6090(00)01707-7

Almeida, S., Chavez, J. J., Zhou, X. W., & Zubia, D. (2016). Effect of substrate orientation on CdS homoepitaxy by molecular dynamics. *Journal of Crystal Growth, 441*, 89-94. http://dx.doi.org/10.1016/j.jcrysgro.2016.02.006

Brill, G., Chen, Y., Amirtharaj, P. M., Sarney, W., Chandler-Horowitz, D., & Dhar, N. K. (2005). Molecular beam epitaxial growth and characterization of Cd-based II–VI wide-bandgap compounds on Si substrates. *Journal of electronic materials, 34*(5), 655-661. http://dx.doi.org/10.1007/s11664-005-0080-y

Colegrove, E., Banai, R., Blissett, C., Buurma, C., Ellsworth, J., Morley, M., ... & Scott, M. (2012). High-efficiency polycrystalline CdS/CdTe solar cells on buffered commercial TCO-coated glass. *Journal of electronic Materials, 41*(10), 2833-2837. http://dx.doi.org/10.1007/s11664-012-2100-z

Cruz-Campa, J. L., Zubia, D., Zhou, X., Aguirre, B. A., Ward, D., Sanchez, C. A., ... & Lu, P. (2012, June). Nanopatterning and bandgap grading to reduce defects in CdTe solar cells. In *Photovoltaic Specialists Conference (PVSC), 2012 38th IEEE* (pp. 000838-000842). IEEE. http://dx.doi.org/10.1109/PVSC.2012.6317734

Donnay, J, D. H., & Ondik, H. M. (1973). *Crystal Data Determinative Tables* (3rd ed., Vol. 2, inorganic compounds). U.S.A. : U. S. Department of Commerce, National Bureau of Standards, and Joint Committee on Power Diffraction Standards.

Han, M. S., Kang, T. W., & Kim, T. W. (1999). The dependence of the structural and optical properties on the CdTe epitaxial layer thicknesses in CdTe (111)/GaAs (100) heterostructures. *Applied surface science, 140*(1), 1-11. http://dx.doi.org/10.1016/S0169-4332(98)00363-8

Heiss, W., Groiss, H., Kaufmann, E., Hesser, G., Böberl, M., Springholz, G., ... & Yano, M. (2006). Centrosymmetric PbTe/CdTe quantum dots coherently embedded by epitaxial precipitation. *Applied physics letters, 88*(19), 192109. http://dx.doi.org/10.1063/1.2202107

Hoover, W. G. (1985). Canonical dynamics: equilibrium phase-space distributions. *Physical Review A, 31*(3), 1695. http://dx.doi.org/10.1103/PhysRevA.31.1695

Kelley, T F, & Michael, K. M. (2007). Invited review article: atom probe tomography. *Review of Scientific Instruments, 78*, 031101. http://dx.doi.org/10.1063/1.2709758.

Kim, T. W., Lee, H. S., Lee, J. Y., Ryu, Y. S., & Kang, T. W. (2004). Existence and atomic arrangement of microtwins in CdTe epilayers grown on GaAs (211) B substrates. *Solid state communications, 129*(8), 515-518. http://dx.doi.org/10.1016/j.ssc.2003.11.047

Li, C., Poplawsky, J., Wu, Y., Lupini, A. R., Mouti, A., Leonard, D. N., ... & Yan, Y. (2013). From atomic structure to photovoltaic properties in CdTe solar cells. *Ultramicroscopy, 134*, 113-125. http://dx.doi.org/10.1016/j.ultramic.2013.06.010

Liang, R., Xu, S., Yan, D., Shi, W., Tian, R., Yan, H., ... & Duan, X. (2012). CdTe Quantum Dots/Layered Double Hydroxide Ultrathin Films with Multicolor Light Emission via Layer‐by‐Layer Assembly. *Advanced Functional Materials, 22*(23), 4940-4948. http://dx.doi.org/10.1002/adfm.201201367

McCandless, B. E., & James, R. S. (2003). Cadmium Telluride Solar Cells. In *Handbook of Photovoltaic Science and Engineering* (pp. 617-662). John Wiley & Sons, Ltd.

Miller, M. K., & Forbes. R. G. (2009). Atom probe tomography. *Materials Characterization, 60*(6), 461-469. http://dx.doi.org/10.1016/j.matchar.2009.02.007

Moseley, J., Al-Jassim, M. M., Kuciauskas, D., Moutinho, H. R., Paudel, N., Guthrey, H. L., ... Ahrenkiel, R. K. (2014). Cathodoluminescence analysis of grain boundaries and grain interiors in thin-film CdTe. *Photovoltaics, IEEE Journal of, 4*(6), 1671-1679. http://dx.doi.org/10.1109/JPHOTOV.2014.2359732

Moutinho, H. R., Dhere, R. G., Romero, M. J., Jiang, C. S., To, B., & Al-Jassim, M. M. (2008). Electron backscatter diffraction of CdTe thin films: Effects of CdCl2 treatment. *Journal of Vacuum Science & Technology A, 26*(4), 1068-1073. http://dx.doi.org/10.1116/1.2841523

Myers, T. H., Schetzina, J. F., Magee, T. J., & Ormond, R. D. (1983). Growth of low dislocation density CdTe films on hydroplaned CdTe substrates by molecular beam epitaxy. *Journal of Vacuum Science & Technology A, 1*(3), 1598-1603. http://dx.doi.org/10.1116/1.572275

Okamoto, T., Yamada, A., & Konagai, M. (2001). Optical and electrical characterizations of highly efficient CdTe thin film solar cells. *Thin Solid Films, 387*(1), 6-10. http://dx.doi.org/10.1016/S0040-6090(00)01725-9

Paudel, N. R., Xiao, C., & Yan, Y. (2014). Close-space sublimation grown CdS window layers for CdS/CdTe thin-film solar cells. *Journal of Materials Science: Materials in Electronics, 25*(4), 1991-1998. http://dx.doi.org/10.1007/s10854-014-1834-1

Plimpton, S. (1995). Fast parallel algorithms for short-range molecular dynamics. *Journal of computational physics, 117*(1), 1-19. http://dx.doi.org/10.1006/jcph.1995.1039

Rapaport, D C. (2006). *The Art of Molecular Dynamics.* Cambridge: Cambridge University Press.

Sarney, W. L., & Brill, G. (2004). A microstructural study of the CdTe/ZnTe film morphology as related to the Si substrate orientation. *Solid-State Electronics, 48*(10), 1917-1920. http://dx.doi.org/10.1016/j.sse.2004.05.036

Smith, D. J., Tsen, S. C., Chandrasekhar, D., Crozier, P. A., Rujirawat, S., Brill, G., ... & Sivananthan, S. (2000). Growth and characterization of CdTe/Si heterostructures—effect of substrate orientation. *Materials Science and Engineering: B, 77*(1), 93-100. http://dx.doi.org/10.1016/S0921-5107(00)00480-3

Stukowski, A., & Albe, K. (2010). Extracting dislocations and non-dislocation crystal defects from atomistic simulation data. *Modelling and Simulation in Materials Science and Engineering, 18*(8), 085001. http://dx.doi.org/10.1088/0965-0393/18/8/085001

Stukowski, A. (2009). Visualization and analysis of atomistic simulation data with OVITO–the Open Visualization Tool. *Modelling and Simulation in Materials Science and Engineering, 18*(1), 015012. http://dx.doi.org/10.1088/0965-0393/18/1/015012

Stukowski, A., Bulatov, V. V., & Arsenlis, A. (2012). Automated identification and indexing of dislocations in crystal interfaces. *Modelling and Simulation in Materials Science and Engineering, 20*(8), 085007. http://dx.doi.org/10.1088/0965-0393/20/8/085007

Terheggen, M., Heinrich, H., Kostorz, G., Romeo, A., Baetzner, D., Tiwari, A. N., ... & Romeo, N. (2003). Structural and chemical interface characterization of CdTe solar cells by transmission electron microscopy. *Thin Solid Films, 431,* 262-266. http://dx.doi.org/10.1016/S0040-6090(03)00268-2

Yan, Y., Al-Jassim, M. M., & Jones, K. M. (2001). Characterization of extended defects in polycrystalline CdTe thin films grown by close-spaced sublimation. *Thin Solid Films, 389*(1), 75-77. http://dx.doi.org/10.1016/S0040-6090(01)00841-0

Yan, Y., Al-Jassim, M. M., Jones, K. M., Wei, S. H., & Zhang, S. B. (2000). Observation and first-principles calculation of buried wurtzite phases in zinc-blende CdTe thin films. *Applied Physics Letters, 77*(10), 1461-1463. http://dx.doi.org/10.1063/1.1308062

Yin, W. J., Yang, J. H., Zaunbrecher, K., Gessert, T., Barnes, T., Yan, Y., & Wei, S. H. (2015). Surface stability and the selection rules of substrate orientation for optimal growth of epitaxial II-VI semiconductors. *Applied Physics Letters, 107*(14), 141607. http://dx.doi.org/10.1063/1.4932374

Zhou, X. W., & Wadley, H. N. G. (2000). Atomistic simulation of the vapor deposition of Ni/Cu/Ni multilayers: Incident adatom angle effects. *Journal of Applied Physics, 87*(1), 553-563. http://dx.doi.org/10.1063/1.371899

Zhou, X. W., & Wadley, H. N. G. (2001). The low energy ion assisted control of interfacial structure: ion incident angle effects. *Surface science, 487*(1), 159-170. http://dx.doi.org/10.1016/S0039-6028(01)01088-3

Zhou, X. W., Ward, D. K., Martin, J. E., van Swol, F. B., Cruz-Campa, J. L., & Zubia, D. (2013). Stillinger-Weber potential for the II-VI elements Zn-Cd-Hg-S-Se-Te. *Physical Review B, 88*(8), 085309. http://dx.doi.org/10.1103/PhysRevB.88.085309

Zhou, X. W., Johnson, R. A., & Wadley, H. N. G. (1997). A molecular dynamics study of nickel vapor deposition: temperature, incident angle, and adatom energy effects. *Acta materialia, 45*(4), 1513-1524. http://dx.doi.org/10.1016/S1359-6454(96)00283-2

Zhou, X. W., Ward, D. K., Wong, B. M., Doty, F. P., Zimmerman, J. A., Nielson, G. N., ... & Zubia, D. (2012). High-fidelity simulations of CdTe vapor deposition from a bond-order potential-based molecular dynamics method. *Physical Review B, 85*(24), 245302. http://dx.doi.org/10.1103/PhysRevB.85.245302

Zhou, X. W., Ward, D. K., Doty, F. P., Zimmerman, J. A., Wong, B. M., Cruz ‐ Campa, J. L., ... & McClure, J. C. (2015). A prediction of dislocation ‐ free CdTe/CdS photovoltaic multilayers via nano ‐ patterning and composition grading. *Progress in Photovoltaics: Research and Applications, 23*(12), 1837-1846. http://dx.doi.org/10.1002/pip.2628

Zubia, D., López, C., Rodríguez, M., Escobedo, A., Oyer, S., Romo, L., ... & McClure, J. (2007). Ordered CdTe/CdS arrays for high-performance solar cells. *Journal of Electronic Materials, 36*(12), 1599-1603. http://dx.doi.org/10.1007/s11664-007-0276-4

7

Bamboo Particles-Polyvinyl Chloride Composites: Analysis of Particles Size Distribution and Composites Performance

Shahril Anuar Bahari[1] & Andreas Krause[2]

[1] Department of Wood Biology and Wood Products, Faculty of Forest Sciences and Forest Ecology, Georg-August-Universität Göttingen, Büsgenweg 4, 37077 Göttingen, Germany

[2] Centre for Wood Science, Hamburg University, Leuschnerstraße 91, 21031 Hamburg, Germany

Correspondence: Shahril Anuar Bahari, Department of Wood Biology and Wood Products, Faculty of Forest Sciences and Forest Ecology, Georg-August-Universität Göttingen, Büsgenweg 4, 37077 Göttingen, Germany. E-mail: sbinbah@gwdg.de

Abstract

Analysis of particles size distribution of Malaysian bamboo species (*Bambusa vulgaris* and *Schizostachyum brachycladum*) for polyvinyl chloride (PVC) composites production was conducted using dynamic image analysis (DIA). A wide distribution of bamboo particles length was recorded, varying from almost 0 to 1500-μm for both species. Inadequate amount of actual particles length distribution from each sieve size (75-μm and 1-mm) was also recorded. DIA observed an increase of aspect ratio from small to large particles, and fine particles were recorded to be slightly elongated than the large ones. However, the effects of bamboo particles size on the finished PVC composites performance were uncertain, considerable of numerous other factors that influence the performance. Only impact and water uptake properties of composites have been obviously affected by different particles size. Greater modulus value is observed in composites with high particles loading, though low impact strength and water resistance were recorded. The incorporation of high concentration of selected processing lubricants in the composites formulation helped to improve the impact and water resistance of the composites. Malaysian bamboo particles-PVC composites performance between different species was equivalent, demonstrated that both species displayed identical behaviour for composites production.

Keywords: bamboo, particles size distribution, polymer composites

1. Introduction

Bamboo is a plant for multi uses, not only for traditional, but also for industrial applications. It is a member of subfamily *Bambusoideae* of family *Graminae* (grasses) (Tewari, 1993). Its maturation period is very short, about 3 to 5 years (Wahab, Samsi, Ariffin, & Mustafa, 1997). Bamboo is a renewable resource due to its fast growing and wide availability especially in South East Asia, and it is also relatively cheaper than other wood resources (Gupta & Kumar, 2008). Bamboo is globally classified into 70 genera with over 1500 species (Tewari, 1993). In South East Asia countries such as Malaysia, Wong (1995) documented bamboo into 59 species, where 25 species are introduced or only known in cultivation and 34 are indigenous and found wild. Its wide availability and good properties encouraged some wood-based industries to exploit bamboo for commercial markets: e.g. production of duroplastic composite boards, furniture, papers and textiles made out of bamboo (Gupta & Kumar, 2008). The commercialization of bamboo-based products in Asia has tremendously increased the harvesting, manufacturing and marketing activities of bamboo (Hua & Kobayashi, 2004), therefore, make it possible to utilize its particles for wood polymer composite (WPC) production. Carus and Eder (2014) predicted that the production and use of WPC in the construction and extrusion sectors in Europe will be increased from 190000 t in 2012 to 400000 t in 2020. Therefore, it appears to be an opportunity in exploiting bamboo materials for WPC industry.

For this reason, introducing bamboo particles as a potential filler in WPC manufacturing will be also depending upon its particles behaviour. In principle, particles is commonly used as an extender/filler to reduce the polymer use, but sometimes it can also be used to modify the composite properties; e.g. increase strength and stiffness, scuff resistance, reduce tackiness, enhance electrical properties, and reduce material cost (Lutz & Dunkelberger, 1992). In general, bamboo particles are classified as irregularly shaped particles which make the strength of the resultant composite will decrease with higher filler loading due to the inability of the filler to support stresses transferred from the polymer matrix (Kassim, 1999). In this context, bamboo particles filled-thermoplastic composite properties varied significantly with particles loading. Some correlations of the properties of this composite with filler loading have been previously reported. According to Kassim (1999), the increased of filler loading from 10 to 50%, tensile modulus and flexural modulus were observed to increase by about 95.6% and 22.9% respectively, whereas tensile strength and flexural strength were decreased by 33.8 and 47.6% respectively. Based on a study by Ke and Jyh (2010), the highest bending strength was found in the bamboo particles-high density polyethylene (PE) composites with 20% or 30% bamboo particles with the maximum values of 23.7 MPa. As the flexural modulus of neat polypropylene (PP) was 1008 MPa, the PP composites mixed with 10%, 30%, and 50% bamboo fibre increased by 87.8%, 215.3%, and 383.2% respectively due to the stiffness increases of the composites (Lee, Chun, Doh, Kang, Lee, & Paik, 2009). According to Samal, Mohanty, and Nayak (2009), tensile modulus of PP-bamboo/glass fibre hybrid composites increased steadily with the increase of filler content from 10% – 40%. The increase was attributed to the increased wt% of the fibre loading within the matrix, leading to an efficient stress transfer from matrix to fiber (Samal et al., 2009). In term of impact strength, neat PP showed the values of 2.62 kg cm/cm^2, though the addition of 10% – 50% bamboo fibre to the net PP, the values was ranged from 2.94 – 3.13 kg cm/cm^2 (Lee et al., 2009). However, Ke and Jyh (2010) reported that bamboo particles-high density PE composites with 30% to 60% bamboo particles showed slight decreases in bending strength, and a drastic decrease when it was more than 60%. According to Bouza, Lasagabaster, Abad, and Barral (2008), this was attributed by the aggregation of woody materials in the composition. Ge, Li, and Meng (2004) reported the reduction in strength behaviours was resulted from decreasing interfacial adhesion and homogeneity with increasing particles contents due to the presence of lignin and OH group in cellulose that influenced the agglomeration. With an increase of filler loading from 10% to 50%, the water absorption of composites increased by 750%, accountable to the increase in the filler surface area which was naturally hygroscopic (Kassim, 1999).

Apart from percentage of particles content in polymer composite, it is also understood that the effect of particles size on the properties of this type of composites was great (Kassim, 1999). Atuanya, Government, Okoye, and Onukwuli (2014) also confirmed the small effects of particles size on the strength performance of date palm wood-recycled low density PE composites. A reduction of the bamboo particles size used by Kassim (1999) from 0.25 to 0.12-mm showed that most of the properties such as modulus of elasticity increased by 17.8%, bending strength by 13.8%, tensile strength by 9.74%, and water absorption by 10.4%. The increases could be due to the small particles covered a larger surface area within composites than the same weight of large particles (Kassim, 1999). However, Atuanya et al. (2014) reported only a small increase of tensile strength when the date palm wood particles size was increased. Using date palm wood particles at 30% loading rate, the ultimate tensile strength of composites with 150-μm particles size was 9.48 MPa, 212-μm was 9.51 MPa, and 250-μm was 9.56 MPa. In a different occasion, particles size also influenced the mixing performance of thermoplastic composite (Atuanya et al., 2014). Dispersion uniformity in the matrix becomes poorer when using high aspect ratio particles (Bledzki and Gassan, cited in Kim, Shim, Kim, Lee, Min, Jang, Abas, and Kim, 2015). Satov (2008) added that the fine particles will cause higher melt viscosities which generally influence the processing performance. In a different report by Stark and Gardner (2008), increasing the wood particles size increased the equilibrium moisture content of the composites. Nevertheless, it was specified that the materials with larger aspect ratio are the best candidate for composite's reinforcement (Gardner, Han, & Wang, 2015). In accordance to Stark and Rowlands (2003), it was confirmed that increase in mechanical properties was found to correspond with increase in aspect ratio of particles. Filler with higher aspect ratio has improved the stiffness and stress transfer of composites (Kim et al., 2015). Although the aspect ratio of wood-based particles is only about 1 – 5, the properties of composite made out of these particles are sufficient for many uses (Clemons, 2008). However, the effect of particles size on the properties of WPC is not sufficiently clear due to many factors such as thermoplastic type, wood content, particles geometry, coupling agent type, processing methods, and any other technical considerations.

Despite the favourable features of bamboo particles for WPC products, the application of this type of particles was mostly limited to PP and PE oriented products as stated in the literature. Only few works have been done on polyvinyl chloride (PVC)-based composites. Kim, Peck, Hwang, Hong, Hong, Huh, and Lee (2008) and Wang, Sheng, Chen, Mao, and Qian (2010) focused on the surface modification of China bamboo particles for PVC composites using chemical treatments. Another study by Sheng, Qian, and Wang (2014) concerned the influence of potassium permanganate pre-treatment of China bamboo particles on the properties of PVC composites. Consequently, the use of PVC in thermoplastic composites production should be further increased due to its excellent chemical resilience, long term stability, good weatherability, and strength (Kim and Pal, 2010).

In this study, Malaysian bamboo particles were utilized as filler in PVC composites production, with the specific objectives were to analyse the particles size distribution, and to determine the effect of particles size, particles loading, and processing lubricants concentration level on the basic performance of PVC composites. This study is essential in order to increase the utilization of bamboo particles in PVC/WPC industry and to increase the understanding of its fundamental performance.

2. Experimental

2.1 Materials

Two Malaysian bamboo species, *Bambusa vulgaris* and *Schizostachyum brachycladum* were selected in this study. The matured bamboos were harvested from a natural bamboo stand in Raub, Malaysia. The bamboo culms were chipped into smaller pieces, at about 10 to 30-mm in length and 1 to 3-mm in thickness, before being air-dried for several weeks. The dried chips were milled using a hammer mill to produce two groups of small particles using two sieve sizes: 75-μm and 1-mm. These particles groups were selected in order to differentiate the effect of a very small and a very large particles size on the overall composite's performance. For composites production, PVC (Solvin, France) was used as a main matrix with other specific additives for PVC extrusion. The PVC's K-value was 63, powder size range was 100 to 150-μm, while T_g was ± 80°C. The additives used were stabilizer Mark CZ2000 (Chemtura, Philadelphia, USA), processing aid Paraloid K120 (Dow Chemical Co., Michigan, USA), internal lubricant Loxiol G60 (Emery Oleochemicals, Cincinnati, USA), external lubricant Loxiol G21 (Emery Oleochemicals, Cincinnati, USA), external lubricant Ligalub GT (Peter Greven Fettchemie GmbH, Bad Münstereifel, Germany), and external lubricant Licocene PE4201 (Clariant, Muttenz, Switzerland).

2.2 Analysis of Particles Size Distribution

In order to obtain the information on bamboo particles size distribution, the analysis of particles shape and geometry was carried out using the dynamic image analysis (DIA) optical system with QICPIC (Sympatec) sensor machine (Figure 1). A small amount of dried particles from two different sieve size groups (75-μm and 1-mm) were collected and air-conditioned at 22°C and 65% humidity for one week. The range of measurement area in this system was between 5 to 5120-μm (M6). The dispersion of the particles through the scanning optic was performed by a dry dispersion unit in DIA system with 1 bar air pressure. The particles were separated from each other by the transportation of air pressure from this dispersion unit. Typically, more than 1 million particles are needed for each measurement to reach the maximum error value below 1 % (International Organization for Standardization ISO/DIS 14488, 2007). The particles were oriented randomly and captured with the highest possible contrast in order to detect the precise images. The bamboo particles geometry was analysed based on their size (length) distribution and shape (aspect ratio and elongation). The evaluation was also conducted with x_{10}, x_{50} and x_{90} values of statistical interval (means of the x_y value is that y percent of the particles are smaller (shorter) than x) (Grüneberg, 2010). Some specific terms were used in the software analysis, such as length of fibre (LEFI) and diameter of fibre (DIFI). According to DIA system, LEFI is defined as the length of the shortest path between the two most distant end points of the fibres/particles, while DIFI as the projection area of the fibre/particle divided by the length of all fibres/particles sections.

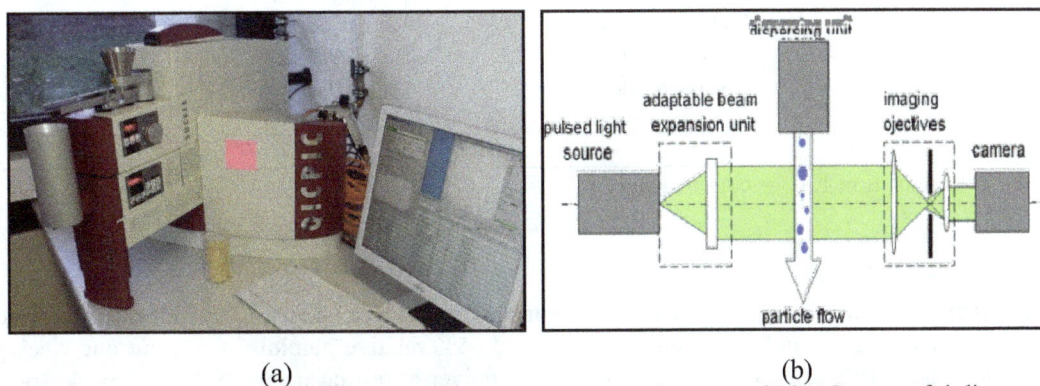

(a)   (b)

Figure 1. DIA system: (a) a combination of dynamic image analysis sensor QICPIC, powerful dispersers (dry disperser RODOS/L) and dry feeder, with image analysis software, (b) schematic diagram of optical set-up in combination with dispersion unit for image analysis process (Sympatec)

2.3 Composites Production

The production process of bamboo particles-PVC composites was separated into the following steps. The particles from both size groups and both species were dried in a drying oven at 103°C for several days to reduce the moisture contents to 2% – 3%, before being blended together with PVC and other additives in powder form. The compositions of PVC, additives and bamboo particles in parts per hundred (pph) for dry blend process are shown in Table 1. The content and function of additives such as Mark CZ2000, Paraloid K120, Loxiol G60, Loxiol G21, Ligalub GT and Licocene PE4201 are also listed in the table. Two different blending compositions (based on different processing lubricants concentration levels) were used: 1.2 and 3.6 pph for Loxiol G60 (dicarboxylic acid ester/internal lubricant), and, 0.15 and 0.45 pph for Licocene PE4201 (polyethylene wax/external lubricant). Composition 1 (C_1) indicated the ingredients with low concentration of these lubricants, while composition 2 (C_2) indicated high concentration. The different compositions of lubricants were considered in order to determine the influence of the usage of these lubricants on the overall composites' performance. In point of fact, an independent study on fusion behaviour of bamboo particles-PVC dry blend was conducted, which reported the compounding torque and temperature of dry blending process with different concentration of additives. Excellent mixing stability (low compounding torque and homogeneous temperature) was observed when the concentrations of Loxiol G60 and Licocene PE4201 were increased three times, reason of an increased lubricants concentration.

The bamboo particles, PVC, and additives were mixed together to dry blend powder in a hot-cool mixer (Reimelt Henschel, FM L 30 KM 85) until the blending temperature of 120°C (for hot section) and 40°C (for cool section) were reached. Bamboo particles were mixed at different percentage ratio: 25% and 50% w/w. These ratios were considered due to the intention in determining the influence of a very low and a very high content of bamboo particles on the composites performance. In this study, the maximum percentage of bamboo particles was 50%, attributable to high viscosity of the PVC (Müller, 2012) and numerous additives used in the blending process (Jiang and Kamdem, cited in Müller, 2012).

All dry-blend powders were compounded by counter-rotating screw extrusion (Leistritz MICRO 27 40D) to produce granules. The average compounding temperature in the extruder zone was 180°C with a screw rotation of 90 rpm. The granules were finally consolidated into compression molded boards using a tempered hydraulic press. The temperature, pressure and duration of hot press was 190°C, 60 bar and 5 minutes, respectively. Pure PVC (without bamboo particles) using C_1 were also processed under the same conditions for comparison purposes.

Table 1. Compositions of PVC, additives, and bamboo particles for dry-blend process

Raw materials	Contents	Functions	Parts per hundred (pph)	
			C_1	C_2
PVC (K value = 63)		Matrix	100	100
Mark CZ2000	Calcium/zinc	Stabilizer	2.5	2.5
Paraloid K120	Acrylic acid	Processing aid	1.0	1.0
Loxiol G60	Dicarboxylic acid ester	Internal lubricant	1.2	3.6
Loxiol G21	Fatty acid	External lubricant	0.2	0.2
Ligalub GT	Glycerol ester	External lubricant	1.2	1.2
Licocene PE4201	Polyethylene wax	External lubricant	0.15	0.45
Bamboo particles		Filler	106.25	108.95

Note: C_1 = composition 1, C_2 = composition 2

2.4 Performance Test of Composites

Performance of all composites were measured in a 3-point static bending, tensile, impact and water resistance tests. Samples for all tests were cut and conditioned at 22°C and 65% relative humidity for about one week prior to testing. Bending and tensile tests were conducted using a universal testing machine model Zwick/Roell (Z010 Allround Line) equipped with Test Expert II software, fitted with 10 kN load cell according to *Deutsches Institut für Normung* DIN EN ISO 178 (2003) and *Deutsches Institut für Normung* DIN EN ISO 527-1 (1993), respectively. The dimension of bending samples was 80 x 10 x 4 mm (length × width × thickness, respectively). The load was applied at a speed of 1 mm/min until failure occurred. Thickness and width of tensile samples at necked-down section was 4 and 10 mm respectively, while samples' length was 190 mm. The samples were held in small grips while testing, at a crosshead speed of 1 mm/min until fracture. The consequent bending modulus, bending strength, tensile modulus and tensile strength were determined in these tests. Samples with the dimension similar to bending were also prepared for impact and water uptake tests. Impact test was conducted according to *Deutsches Institut für Normung* DIN EN ISO 179 – 1 (2006) in an un-notched state. The test was performed in a Charpy impact machine with 1 Joule (J) hammer and 60 mm support span. Samples for water resistance test were oven-dried at 103°C to remove moisture prior to the immersion in distilled water at room temperature, according to International Organization for Standardization ISO 62 (2008). The samples were periodically taken out of water in between 1600 hours. The sample's surfaces were wiped before the measurement and weighed for water uptake determination using Equation 1:

$$\text{Weight changes} = (W_{submersed} - W_1/W_1) \times 100 \ (\%) \quad ... \text{Equation 1}$$

where:

W_1 is weight of dried sample (g), and

$W_{submersed}$ is change in weight (g).

3. Results and Discussion

3.1 Particles Size Distribution

Figure 2 shows the cumulative distribution of particles length from different sieve size (75-µm and 1-mm). Table 2 presents the particles length at x_{10}, x_{50} and x_{90} of statistical interval, while Table 3 displays the cumulative distribution and aspect ratio of particles. The elongation of particles is depicted in Figure 3.

A total number of 9.3 to 24.8 million particles were measured and analysed by DIA for each sieve size group (Table 2). Based on Figure 2, a wide distribution of particles length was recorded, varying from almost 0 to 1500-µm. As an example, analysis of 75-µm size group showed that 95% of *B. vulgaris* and 96% of *S. brachycladum* particles length were less than 250-µm, whereas for 1 mm, 80% and 70% of the respective *B. vulgaris* and *S. brachycladum* particles length were less than 250-µm (Figure 2). This implied that a wide range of particles size was recorded for each particles size group, suggested the insufficient amount of actual particles length distribution for each sieve size.

According to Table 2, the evaluation on x_{10}, x_{50} and x_{90} of statistical intervals confirmed the variation of particles length. X_{10} showed that both sieve size groups had a similar amount of fine particles (between 12.9 to 16.0-µm), whereas x_{50} recorded a wider distribution of particles (between 57.9 to 100.6-µm). Meanwhile, x_{90} showed the highest variation of particles length between different sieve size groups for both species, which ranged from 178.9 to 722.2-µm.

Based on Table 3, both bamboo species showed an increment of aspect ratio from small to large particles for both sieve size groups. A higher aspect ratio was recorded from the longest particles. For all groups, only 5.9% to 7.5% of particles had an aspect ratio of less than 1.7. About 99% of particles from *B. vulgaris*/75-µm group had an aspect ratio of less than 4.2. According to Gardner et al. (2015), the large-sized particles with high aspect ratio provided better mechanical properties than the small-sized particles. Transfer efficiency of load from matrix to wood fiber could be possibly increased with increasing fiber length/diameter ratio (Migneault, Koubaa, Erchiqui, Chaala, Englund, Krause, & Wolcott, 2008). Based on the small increase of aspect ratio in this study, reinforcement and properties of bamboo particles-PVC composite are possibly improved, although the reinforcing potential for this type of composite seems to be quite limited.

Based on the elongation analysis depicted in Figure 3, particles from 75-µm size group for both species were a little more elongated than 1-mm. However, its effect on the actual composite properties was not definitely ascertainable, as the amount of actual particles length in each sieve size group was inadequate. Many other factors influencing the composite properties (e.g. processing conditions and other raw material attributes such as particles loading and processing lubricants concentration) have to be considered. This phenomenon is explained through composite's performance results in the following discussion.

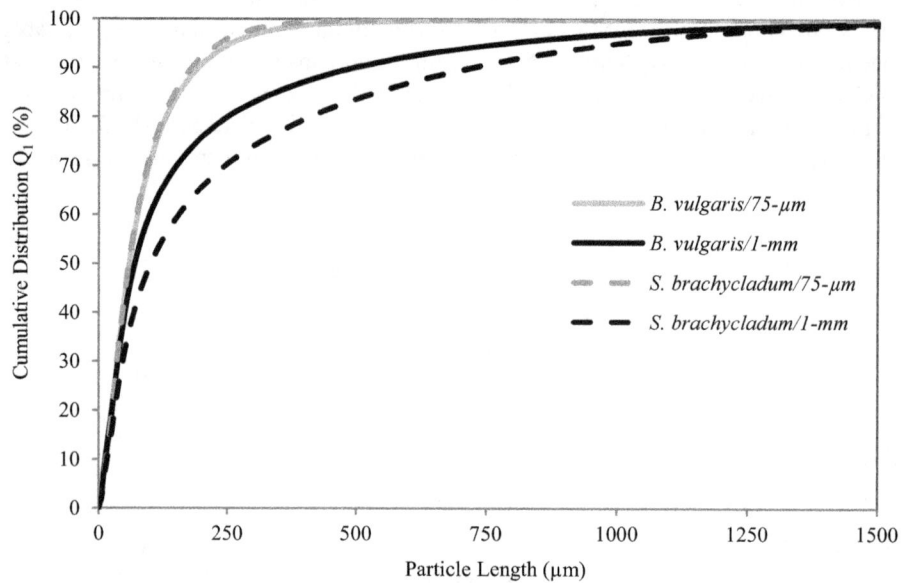

Figure 2. Particle length cumulative distribution of Malaysian bamboo species from different sieve size (75-μm and 1-mm) measured from DIA optical system

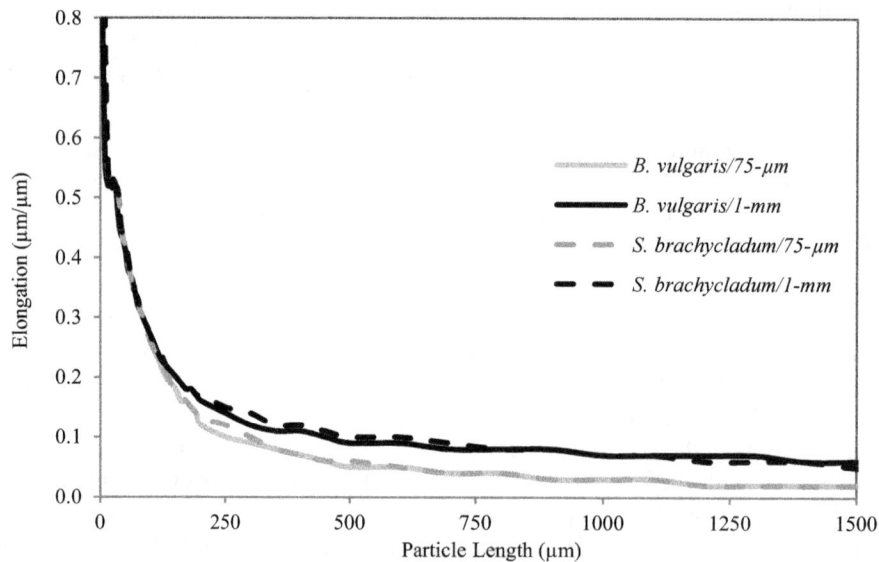

Figure 3. Particle elongation of Malaysian bamboo species from different sieve size (75-μm and 1-mm) measured from DIA optical system

Table 2. Evaluation on particle length of Malaysian bamboo species from different sieve size at x_{10}-, x_{50}- and x_{90}-value of statistical intervals

Species	Sieve size	No. of analysed particles (Million)	x_{10}-value (μm)	x_{50}-value (μm)	x_{90}-value (μm)
Bv	75-μm	24.1	14.1	59.2	192.5
	1-mm	13.3	12.9	69.8	495.8
Sb	75-μm	24.8	14.2	57.9	178.9
	1-mm	9.3	16.0	100.6	722.2

Note: Bv = *B. vulgaris*, Sb = *S. brachycladum*

Table 3. Cumulative distribution and aspect ratio of Malaysian bamboo particles

Species	Sieve size	Length (μm)	Cumulative distribution (%)	Aspect ratio
Bv	75-μm	10	6.7	1.7
		100	71.1	2.1
		500	99.4	4.2
	1-mm	10	7.5	1.7
		100	60.3	2.1
		500	90.1	4.0
Sb	75-μm	10	6.6	1.7
		100	72.2	2.1
		500	99.7	3.3
	1-mm	10	5.9	1.7
		100	49.9	2.2
		500	83.6	3.4

Note: Bv = *B. Vulgaris*, Sb = *S. brachycladum*

3.2 Composites Performance

3.2.1 Bending and Tensile Properties

Based on the observation, *B. vulgaris* and *S. brachycladum* were comparable in term of composites performance; established an understanding that both species behaved identically. Shown in Figures 4 and 5 are the bending properties of the composites with different particles size, particles loading, and processing lubricants concentration for *B. vulgaris* and *S. brachycladum*, respectively, whereas the tensile properties are shown in Figures 6 and 7 for the respective bamboo species. Impact property of the composites for both species is depicted in Figure 8. Water uptake property of the composites is displayed in Figures 9 and 10 for *B. vulgaris* and *S. brachycladum*, respectively.

According to Figures 4 (a) and 5 (a), Malaysian bamboo particles size had a very small effect on bending modulus properties of PVC composites, in which, a slight increase of bending modulus was recorded for composites with 1-mm particles size in comparison to 75-μm. The improvement of bending modulus of composites with large particles can be related to the increased aspect ratio of the longest particles as listed in Table 3 that influenced the better bending modulus. Particles or fibres with a higher aspect ratio may improve the stress transfer between matrix and particles/fibres and finally tend to strengthen the bending modulus properties of the composites (Gozdecki, Zajchowski, Kociszewski, Wilczyñski, & Mirowski, 2011). However, this trend is observed only for composites from C_1 formulation. Based on Figure 6 (a), in contrast, the *B. vulgaris* particles-PVC composites from C_1 showed slightly lower tensile modulus value when using 1-mm particles size. On the other hand, it was observed that the influence of particles size on bending modulus and tensile modulus of composites from C_2 formulation was uncertain (Figure 4 (a), 5 (a), 6 (a), and 7 (a)). In this condition, the variation of bending modulus and tensile modulus of composites from C_2 formulation does not depend on the different of particles sizes.

Furthermore, the incorporation of Malaysian bamboo particles has greatly improved the bending modulus and tensile modulus of PVC composites. According to Figures 4 (a), 5 (a), 6 (a), and 7 (a), composites with 50% bamboo particles loading had bending modulus of up to 34% higher and tensile modulus of up to 32% higher compared to composites with 25% bamboo particles loading. In relation to the pure PVC composites in Table 4, it was recorded that the bending modulus and tensile modulus of Malaysian bamboo particles-PVC composites were up to 81% higher and 59% higher for the respective bending modulus and tensile modulus as compared to pure PVC composites. The high bending modulus and tensile modulus of Malaysian bamboo particles-PVC composites in this study were influenced by the high ratio of particles that increased the stiffness of the composites. Based on the records, *S. brachycladum* particles-PVC composites with 1-mm particles size/50% particles loading/C_1 showed the highest bending modulus (5203 N/mm²), meanwhile *B. vulgaris* particles-PVC composites with 75-μm particles size/50% particles loading/C_2 showed the highest tensile modulus (5380 N/mm²) among others.

However, based on Table 4, Figures 4 (b), 5 (b), 6 (b), and 7 (b), the bending strength and tensile strength of composites with presence of bamboo particles and high particles loading were generally lower. The high ratio of bamboo particles in the composites did not help in supporting maximum bending rupture and maximum tensile rupture. Kassim, Rahman, and Ramlan (2007) in their study reported that reduced modulus of rupture of natural particles-filled thermoplastic composites was a consequence of decreased deformability of rigid interphase between particles and matrix. It is previously reported that micro voids were typically formed in thermoplastic composites based on natural filler, due to interfacial failure between natural fillers and matrix, which finally influenced the low tensile strength in general (Chen, Mao, Xue, Deng, & Lin, 2013). According to the current study, the maximum value of bending strength was recorded by *B. vulgaris* particles-PVC composites with 1-mm particles size/25% particles loading/C_1 (60 N/mm^2), whereas, *S. brachycladum* particles-PVC composites with 75-μm particles size/25% particles loading/C_1 showed the highest tensile strength value (31 N/mm^2).

Based on Figures 4 and 5, there were no significant effects of processing lubricants concentration level on bending and tensile properties of composites in this study. In this case, modifying the processing lubricants concentration has no impact on the bending and tensile properties of the Malaysian bamboo particles-PVC composites.

Overall results showed that the bending and tensile properties of Malaysian bamboo particles-PVC composites were superior when compared to the other study on bamboo-based thermoplastic composites. Chen, Guo, and Mi (1998) reported the values of tensile strength of bamboo fibre-PP composites (at 20% – 60% bamboo loading) were less than 20 MPa, tensile modulus values were less than 4000 MPa, whereas Mohanty and Nayak (2010) revealed the value of bending modulus and bending strength of untreated bamboo-high density PE composites (at 40% bamboo loading) were 2987.70 MPa and 25.35 MPa, respectively.

Table 4. Properties of pure PVC (without bamboo particles) composites using C_1

Properties	Mean values
Bending modulus (N/mm^2)	2866 (44.2)
Bending strength (N/mm^2)	73 (2.7)
Tensile modulus (N/mm^2)	3295 (546.1)
Tensile strength (N/mm^2)	52 (7.5)
Impact (kJ/m^2)	23 (0.9)

Note: Standard deviations in parentheses

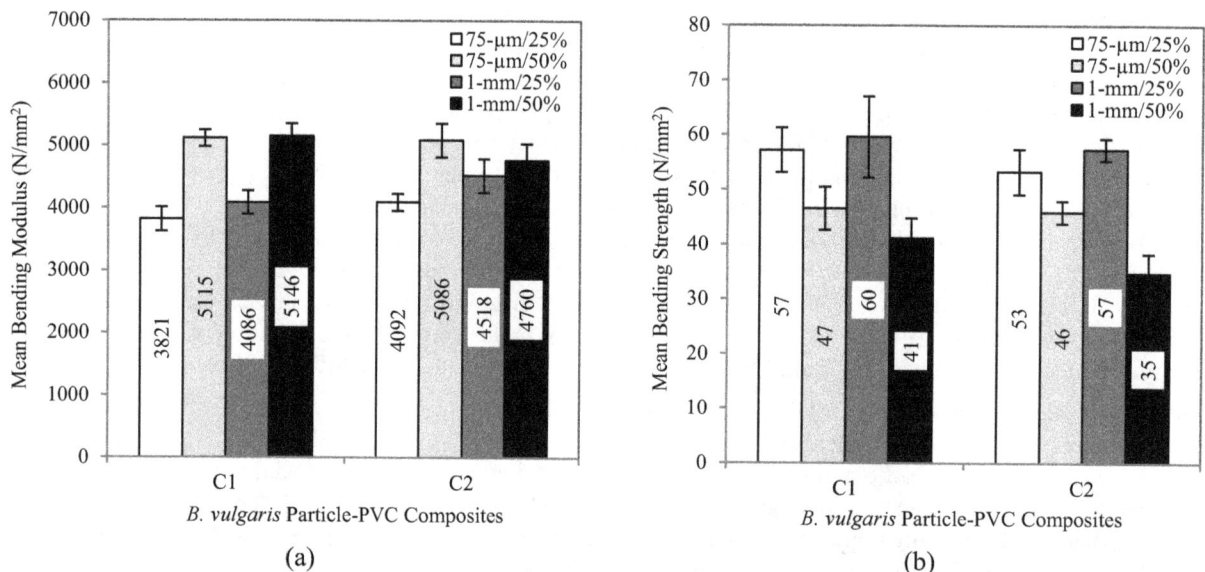

Figure 4. Bending property of *B. vulgaris* particles-PVC composites with different particles sieve size, particles mixing ratio and processing lubricants content level; (a) mean bending modulus, (b) mean bending strength

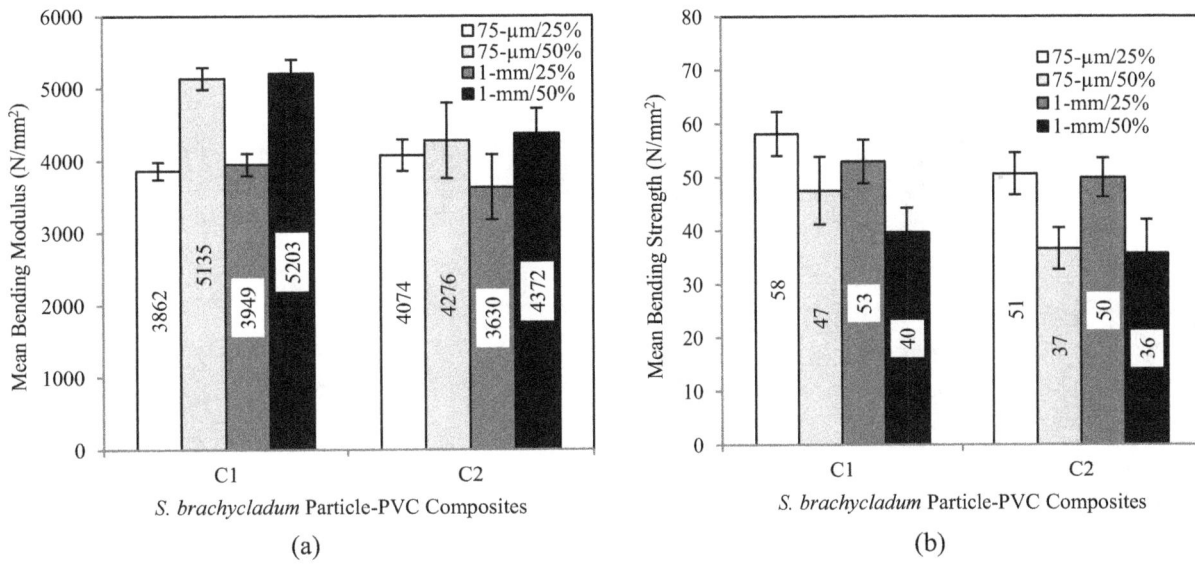

Figure 5. Bending property of *S. brachycladum* particles-PVC composites with different particles sieve size, particles mixing ratio, and processing lubricants content level; (a) mean bending modulus, (b) mean bending strength

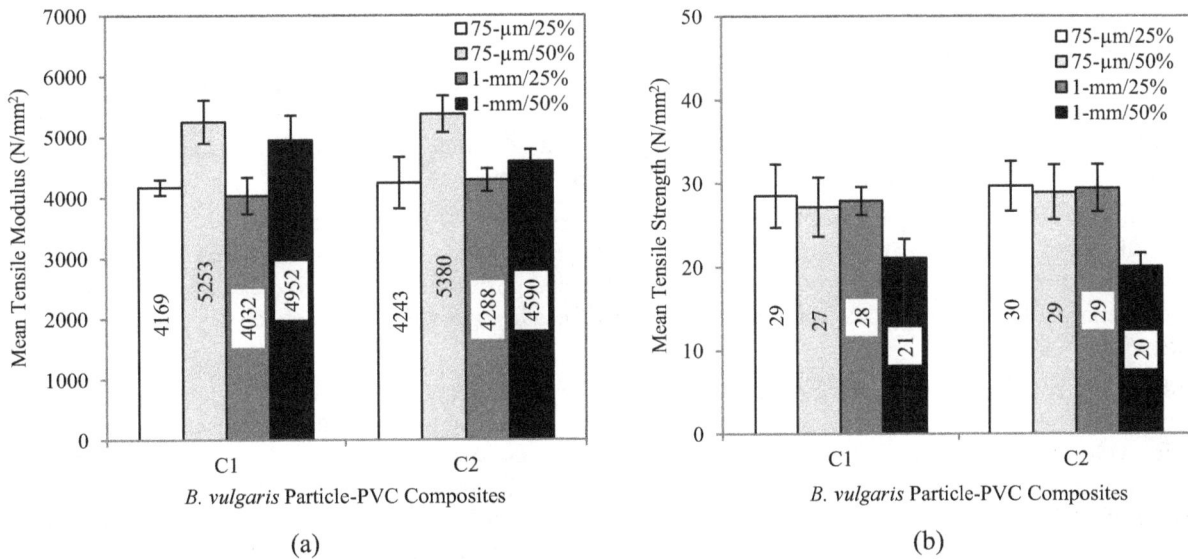

Figure 6. Tensile property of *B. vulgaris* particles-PVC composites with different particles sieve size, particles mixing ratio, and processing lubricants content level; (a) mean tensile modulus, (b) mean tensile strength

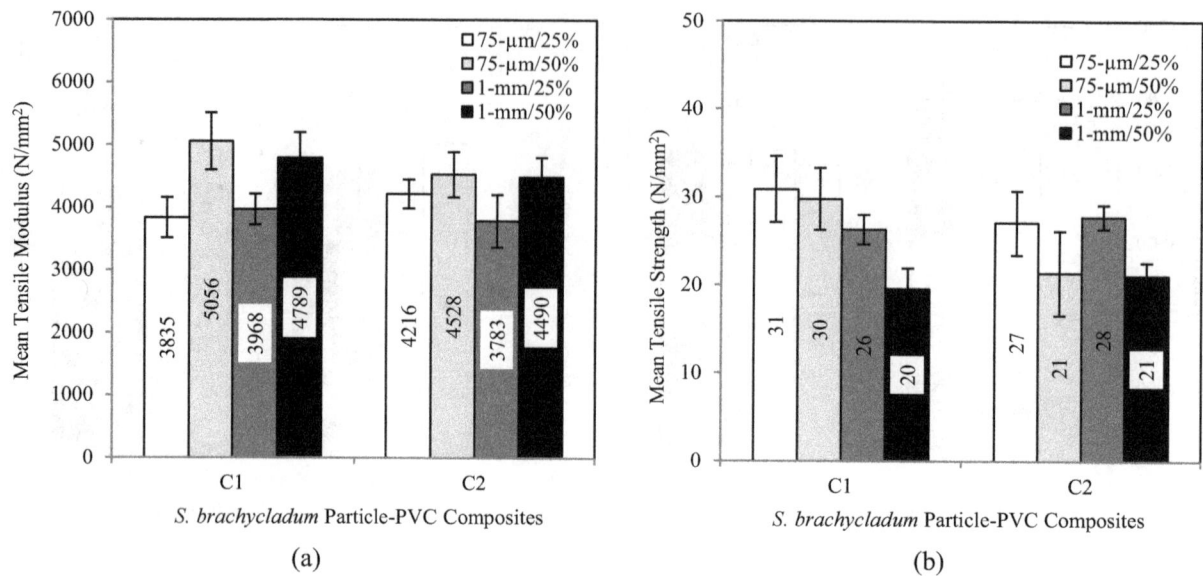

Figure 7. Tensile property of *S. brachycladum* particles-PVC composites with different particles sieve size, particles mixing ratio, and processing lubricants content level; (a) mean tensile modulus, (b) mean tensile strength

3.2.1 Impact Properties

According to Figure 8, the use of fine bamboo particles (75-μm) in the formulation was capable in influencing the increases of impact strength of the composites. This was probably due to the binding competency of particles with the matrix due to fine size. Fine natural fibres entangle less, have more resistance to breaking, and have a high surface area that make their distribution into the polymer matrix is more homogeneous (Zazyczny & Matuana, 2005), therefore, help to improve the impact resistance of the composites.

For particles loading aspect, the impact strength of composites with 25% bamboo particles was from 77 to 1150% greater than composites with 50% bamboo particles. However, in a general comparison to pure PVC composites in Table 4, the impact strength of bamboo particles-filled PVC composites is diminished by the presence of bamboo. The presence of bamboo particles in the formulation has possibly influenced the defieciency in supporting the sudden impact loads.

Furthermore, composites with high concentration of processing lubricants (C_2) exhibited higher impact strength compared to low concentration (C_1). The incorporation of high amount of these internal and external processing lubricants helped to improve the impact strength of the composites in this study. As the internal lubricant promote the fusion process and reduce the melt viscosity of PVC-filler mixing, the external lubricant tends to migrate to the surface of PVC composites mixture to reduce the friction between PVC melt and processing machine surfaces (Thacker, 2008), thus simultaneously help to improve the resistance toward impact load. According to Figure 8, the maximun mean impact strength value that the composite from C_2 can achieve was 4.9 kJ/m^2 (recorded from *S. brachycladum*/1-mm particles sieve size/25% particles mixing ratio).

While excellent bending modulus and tensile modulus have been recorded, impact result shows that the average impact strength of Malaysian bamboo particles-PVC composites are considered inferior in the context of high-rated sudden load applications. Under the best of circumstances, a balance of stiffness and impact resistance is desired for any kind of composites products to be used in load-bearing application (Robinson, Ferrigno, & Grossman, 2008). Due to the outstanding modulus of the composites regardless of particles size and particles loading, the formulation of 75-μm particles size/25% particles loading/C_2 for *B. vulgaris*, and 1-mm particles size/25% particles loading/C_2 for *S. brachycladum* may be used in order to achieve the optimum impact resistance stipulated in Figure 8.

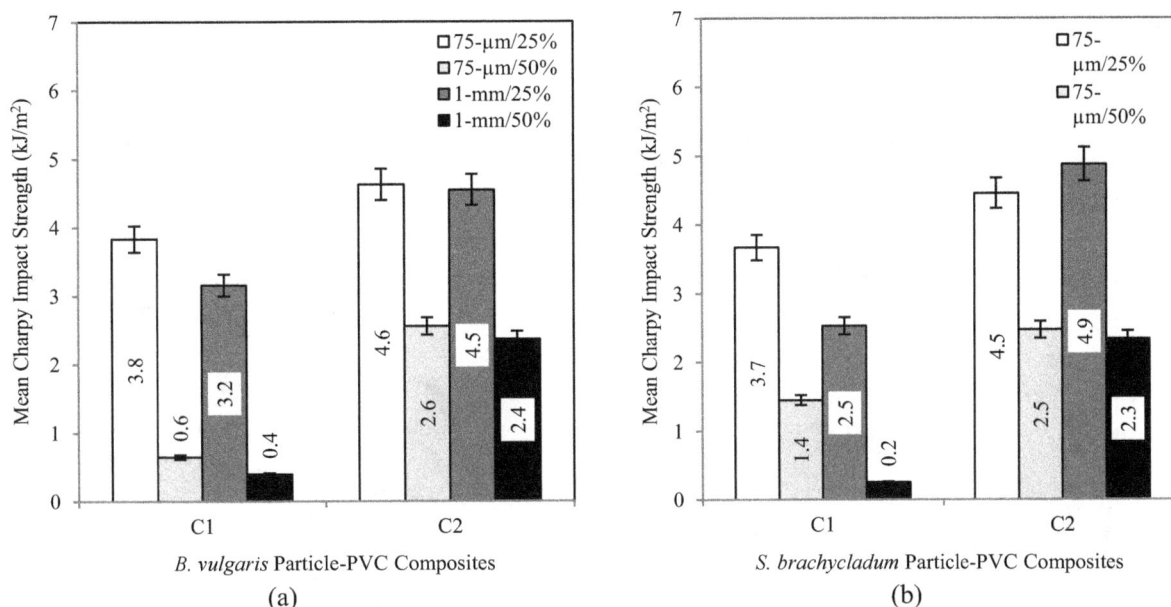

Figure 8. Impact of Malaysian bamboo particles-PVC composites with different particles size, particles mixing ratio, and processing lubricants content level: (a) *B. vulgaris*, (b) *S. brachycladum*

3.2.2 Water Resistance

An extensive increase of water uptake at the early stage of 1600 h water soaking was observed for composites from both bamboo species, especially for 50% particles loading (Figures 9 and 10). This was followed by a gradual raise thereafter, until the completion of water absorption measurement. Although increased rapidly, composites with 50% particles loading reached the equilibrium level faster than 25% loading. In relation to this observation, the water absorption of bamboo particles-PVC composites can be typically divided into two diffusion steps: the first one occurred over a rapid diffusion rate, while the second one was at a slower rate close to zero (Petchwattana, Covavisaruch, & Pitidhammabhorn, 2013).

A significant different of composites' water uptake property was recorded between different particles size. Composites with fine bamboo particles (75-μm) have less water uptake percentage compared to composites with large bamboo particles (1-mm). The low water uptake percentage of composites with fine particles was probably due to the compact mixture between fine particles and matrix which influenced the less water absorption. Compared with large filler in composition, the use of fine particles in the composition reduces the voids formation (Yu, Huang, & Yu, 2014). Consequently, fillers with a smaller particles size have a higher adhesion interaction with polymer than those with large size (Zazyczny & Matuana, 2005). This phenomenon reduces the tendency of composites to absorb water.

On the other hand, composites with 25% bamboo particles show less water uptake compared to composites with 50% particles. In a comparison to pure PVC (Figure 9 (a) and 10 (a)), water uptake of bamboo particles-filled PVC composites was extremely higher due to the high water uptake capability of bamboo particles as natural filler. The presence of hydrophilic –OH groups in the fibres/particles has influenced the high moisture uptake capacity of any types of natural filled-polymer composites (Samal et al., 2009).

Composites with high lubricants concentration (C_2) showed lower water uptake percentage compared to low lubricants concentration level (C_1). The presence of internal and external lubricants at high content level has reduced the water uptake ability. It is recorded in Figure 9 (b) and 10 (b) that composite from C_2 group using 75-μm particles sieve size with 25% particles mixing ratio exhibited the lowest water uptake percentage throughout the entire soaking period.

As a comparison, the water uptake property of Malaysian bamboo particles-PVC composites in this study is considered favourable when referring into the other study by Kassim (1999) (increases of 750% water absorption of composites from 10 to 50% bamboo fillers loading) and Kushawa and Kumar (2009) (water uptake values of bamboo fibre-epoxy composites (at 64% bamboo loading) of up to 40%).

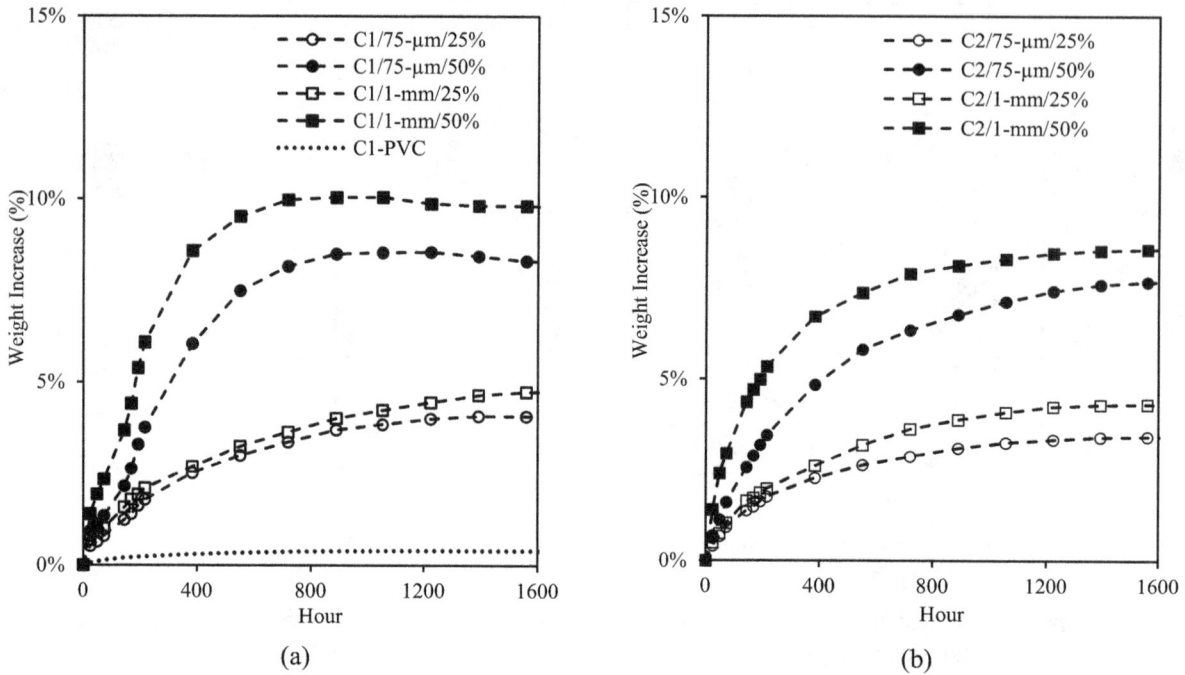

Figure 9. Water uptake of *B. vulgaris* particles-PVC composites for both C_1 and C_2 using different particles size, particles mixing ratio, and processing lubricants content level: (a) C_1, (b) C_2

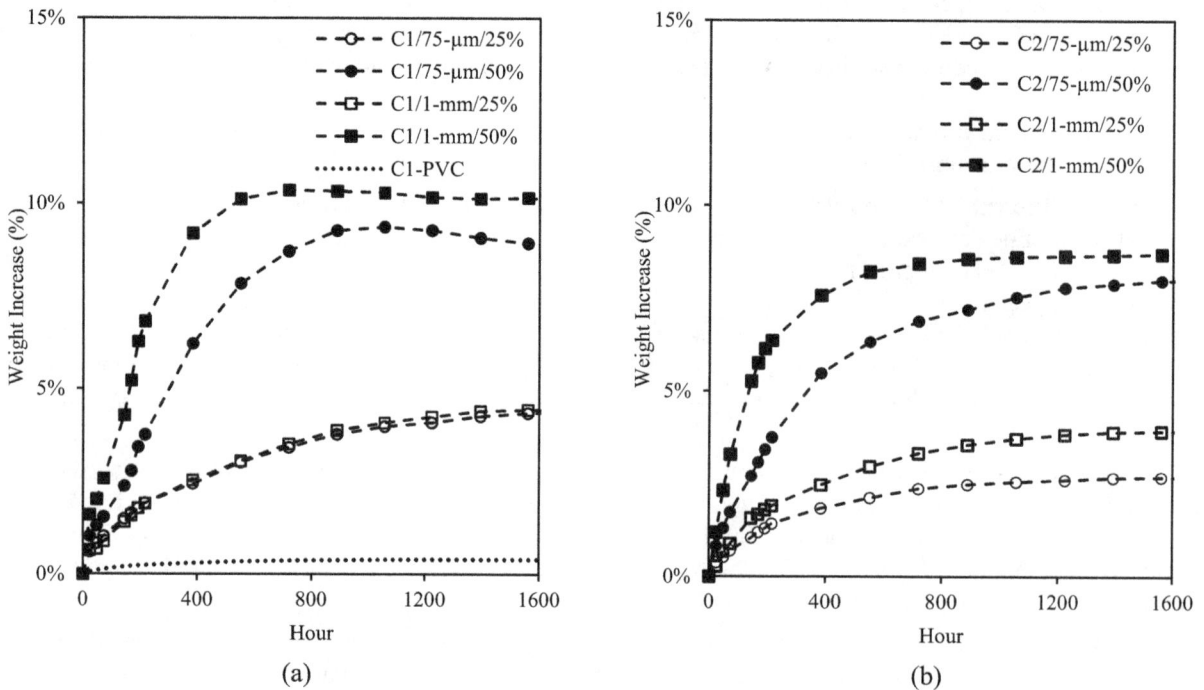

Figure 10. Water uptake of *S. brachycladum* particles-PVC composites for both C_1 and C_2 using different particles size, particles mixing ratio, and processing lubricants content level: (a) C_1, (b) C_2

4. Conclusion

DIA revealed a wide distribution of Malaysian bamboo particles (*B. vulgaris* and *S. brachycladum*) length ranged from 0 to 1500 μm. Insufficient amount of actual particles length distribution from each sieve size (75-μm and 1-mm) has been also recorded. Aspect ratio has been increased with particles length, although particles from small sieve size group exhibited a little more elongation than 1 mm. Due to these features, particles size had only a minor impact on the composites performance. Simultaneously, bamboo particles loading has tremendously increased the bending modulus and tensile modulus of PVC composites, however in contrast, decreased the resistance towards maximum bending load, maximum tensile load, impact and water uptake. The consumption of high processing lubricant content level in composites provided remarkably support with regard to impact load, and promoted better resistance towards water absorption. No significant different of composites performance has been recorded between different bamboo species. It is concluded that Malaysian bamboo particles are possible to be mixed with PVC to produce composites, with particles loading plays the most important role in the composite's properties.

Acknowledgements

This research has been supported by the German Research Foundation (DFG), grant GRK 1703/1 for the Research Group 'Resource Efficiency in Corporate Networks - Planning Methods to Utilize Renewable Resources'. The authors would like to thank to the Faculty of Applied Sciences, Universiti Teknologi MARA, Shah Alam, Malaysia, for providing machineries (cross-cut and hammer milling machines) during the initial preparation of bamboo particles in Malaysia.

References

Atuanya, C. U., Government, M. R., Nwobi-Okoye, C. C., & Onukwuli, O. D. (2014). Predicting the mechanical properties of date palm wood fibre-recycled low density polyethylene composite using artificial neural network. *International Journal of Mechanical and Materials Engineering, 7*(1), 20.

Bouza, R., Lasagabaster, A., Abad, M. J., & Barral L. (2008). Effects of vinyltrimethoxysilane on thermal properties and dynamic mechanical properties of polypropylene-wood flour composites. *Journal of Applied Polymer Sciences, 109,* 1197 – 1204.

Carus, M., & Eder, A. (2014). Biocomposites: 352,000 t of wood and natural fibre composites produced in the european union in 2012. Wood-Plastic Composites (WPC) and Natural Fibre Composites (NFC): European and Global Markets 2012 and Future Trends. Nova Institute GmbH.

Chen, X., Guo, Q., & Mi, Y. (1998). Bamboo fiber-reinforced polypropylene composites: a study of the mechanical properties. *Journal of Applied Polymer Science, 69,* 1891 – 1899.

Chen, Q., Mao, X., Xue, H., Deng, Y., & Lin, J. (2013). Preparation and characterization of bamboo fiber-graft-lauryl methacrylate and its composites with polypropylene. *Journal of Applied Polymer Science, 130*(4), 2377 – 2382. DOI: 10.1002/APP.39347.

Clemons, C. (2008). Raw materials for wood-polymer composites. In Wood-Polymer Composites, edited by Niska, K. O. & Sain, M. Woodhead Publishing, Ltd. Cambridge, UK.

Deutsches Institut für Normung (DIN). (2003). *Kunstoffe – bestimmung der biegeeigenschaften – Teil 1. Nicht instrumentierte Schlagzähigkeitsprüfung.* DIN EN ISO 178.

Deutsches Institut für Normung (DIN). (1993). *Kunstoffe – bestimmung der zugeigenschaften – Teil 1. Algemeine Grundsätze.* DIN EN ISO 527 – 1.

Deutsches Institut für Normung (DIN). (2006). *Kunstoffe – bestimmung der charpy-schlageigenschaften – Teil 1. Nicht instrumentierte Schlagzähigkeitsprüfung.* DIN EN ISO 179 – 1.

Gardner, D. J., Han, Y., & Wang, L. (2015). Wood-plastic composite technology. *Current Forestry Reports, 1,* 139 – 150.

Ge, X. C., Li, X. H., & Meng, Y. Z. (2004). Tensile properties, morphology, and thermal behavior of pvc composites containing pine flour and bamboo flour. *Journal of Applied Polymer Science, 93,* 1804 – 1811.

Gozdecki, C., Zajchowski, S., Kociszewski, M., Wilczyñski, A., & Mirowski, J. (2011). Effect of wood particle size on mechanical properties of industrial wood particle-polyethylene composites. *Polimery, 56*(5), 375 – 380.

Grüneberg, T. (2010). Improvement of selected properties of wood-polymer composites (wpc) – silane modification of wood particles. Doctoral dissertation, Georg-August-Universität Göttingen, Germany.

Gupta, A., & Kumar, A. (2008). Potential of bamboo is sustainable development. *Asia Pacific Business Review, IV*(3), 100 – 107.

Hua, L. Z., & Kobayashi, M. (2004). Plantation future of bamboo in china. *Journal of Forestry Research, 15*(3), 233 – 242.

International Organization for Standardization ISO 62. (2008). Plastics-determination of water absorption.

International Organization for Standardization ISO/DIS 14488. (2007). Sample preparation - sampling and sample splitting of particulate materials for the characterization of particulate properties.

Kassim, J. (1999). Properties of particleboard and particle-filled thermoplastic composite from bamboo (*GigantochloaScortechinii*). Doctoral dissertation, Universiti Putra Malaysia, Malaysia.

Kassim, J., Rahman, W. M. N. W. A., & Ramlan, M. N. B. (2007). Mechanical and physical properties of araucaria fibre-PP composite. *Journal Gading, 11*, 1 – 10.

Ke, C. H., & Jyh, H. W. (2010). Mechanical and interfacial properties of plastic composite panels made from esterified bamboo particles. *Journal of Wood Science, 56*, 216 – 221.

Kim, J. K., & Pal, K. (2010). Recent advances in the processing of wood-plastic composites. Springer-Verlag Berlin Heidelberg, Germany.

Kim, J. Y., Peck, J. H., Hwang, S. H., Hong, J., Hong, S. C., Huh, W., & Lee, S. W. (2008). Preparation and mechanical properties of poly(vinyl chloride)/bamboo flour composites with a novel block copolymer as a coupling agent. *Journal of Applied Polymer Science, 108*, 2654 – 2659.

Kim, J. H., Shim, B. S., Kim, H. S., Lee, Y. J., Min, S. K., Jang, D., Abas, Z., & Kim, J. (2015). Review of nanocellulose for sustainable future materials. *International Journal of Precision Engineering and Manufacturing – Green Technology, 2*(2), 197 – 213.

Kushwaha, P. K., & Kumar, R. (2010). Studies on water absorption of bamboo-epoxy composites: effect of silane treatment of mercerized bamboo. *Journal of Applied Polymer Science, 115*(3), 1846 – 1852.

Lee, S. Y., Chun, S. J., Doh, G. H., Kang, I. A., Lee, S., & Paik, K. H. (2009). Influence of chemical modification and filler loading on fundamental properties of bamboo fibers reinforced polypropylene composites. *Journal of Composite Materials, 43*(5), 1639 – 1657.

Lutz Jr., J. T., & Dunkelberger, D. L. (1992). Impact modifiers for pvc: the history and practice. Wiley, New York.

Migneault, S., Koubaa, A., Erchiqui, F., Chaala, A., Englund, K., Krause, C., & Wolcott, M. (2008). Effects of fibre length on processing and properties of extruded wood-fibre/HDPE composites. *Journal of Applied Polymer Science, 110*, 1085 – 1092.

Mohanty, S. & Nayak S. K. (2010). Short bamboo fiber-reinforced hdpe composites: influence of fiber content and modification on strength of the composite. *Journal of Reinforced Plastics and Composites, 29*(14), 2199 – 2210.

Müller, M. (2012). Influence of wood modification on the properties of polyvinyl chloride based wood polymer composites (wpc). Doctoral dissertation, Cuvillier Verlag, Göttingen, Germany.

Petchwattana, N., Covavisaruch, S., & Pitidhammabhorn, D. (2013). Influence of water absorption on the properties of foamed poly(vinyl chloride)/rice hull composites. *Journal of Polymer Research, 20*, 172 – 178. DOI: 10.1007/s10965-013-0172-y.

Robinson, S., Ferrigno, T. H., & Grossman, R. F. (2008). In R. F. Grossman (Ed.), *Handbook of vinyl formulating*. John Wiley and Sons Inc., USA.

Samal, S. K., Mohanty, S. & Nayak, S. K. (2009). Polypropylene-bamboo/glass fiber hybrid composites: fabrication and analysis of mechanical, morphological, thermal, and dynamic mechanical behavior. *Journal of Reinforced Plastic and Composites, 28*(22), 2729 – 2747.

Satov, D. V. (2008). Additives for wood-polymer composites. In Wood-Polymer Composites, edited by Niska, K. O. & Sain, M. Woodhead Publishing, Ltd. Cambridge, UK.

Sheng, K. C., Qian, S. P., & Wang, H. (2014). Influence of potassium permanganate pretreatment on mechanical properties and thermal behavior of moso bamboo particles reinforced pvc composites. *Polymer Composites, 35*, 1460 – 1465.

Stark, N. M., & Gardner, D. J. (2008). Outdoor durability of wood-polymer composites. In Wood-Polymer Composites, edited by Niska, K. O. & Sain, M. Woodhead Publishing, Ltd. Cambridge, UK.

Stark, N. M., & Rowlands, R. E. (2003). Effects of wood fiber characteristics on mechanical properties of wood/polypropylene composites. *Journal of Wood and Fiber Science, 35*(2), 167 – 174.

Tewari, D. N. (1993). Monograph on bamboo. International Book Distributors, India.

Thacker, G. A. (2008). Formulating rigid pvc for extrusion. In Grossman, R. F. (Ed.), *Handbook of vinyl formulating*. John Wiley and Sons Inc., USA.

Wahab, R., Samsi, H. W., Ariffin, W. T. W., & Mustafa, M. T. (1997). Industri pembuatan pepapan laminasi buluh. *FRIM Technical Information Handbook* No. 11, Forest Research Institute Malaysia (FRIM), Malaysia.

Wang, H., Sheng, K. C., Lan, T., Adl, M., Qian, X. Q., & Zhu, S. M. (2010). Role of surface treatment on water absorption of poly (vinyl chloride) composites reinforced by *phyllostachys pubescens* particles. *Composites Science and Technology, 70*, 847 – 853.

Wong, K. M. (1995). The bamboo of peninsular Malaysia. *Forest Research Institute Malaysia (FRIM)*, Malaysia.

Yu, Y. L., Huang, X. A., & Yu, W. J. (2014). High performance of bamboo-based fiber composites from long bamboo fiber bundles and phenolic resins. *Journal of Applied Polymer Science, 131*(12), 1 – 8. http://dx.doi.org/10-1002/APP.40371.

Zazyczny, J. M., & Matuana, L. M. (2005).Fillers and reinforcing agents. In C. E. Wilkes, J. W. Summers, & C. A. Daniels (Eds.), *PVC Handbook* (pp. 235 – 275).

8

Deactivation Mechanism of Titania Catalyst

Mansour A. Al-Shafei[1], Ahmed K. Al-Asseel[1], Abdulhadi M. Adab[2], Hasan A. Al-Jama[1], Amer A. Al-Tuwailib[1] & Shouwen X. Shen[1]

[1] Research & Development Center, Saudi Aramco, Dhahran, Saudi Arabia

[2] Shedgum Gas Plant Department, Saudi Aramco, Shedgum, Saudi Arabia

Correspondence: Mansour Al-Shafei Research & Development Center, Saudi Aramco, Dhahran, Saudi Arabia, E-mail: mansour.shafei@aramco.com

Abstract

Catalyst deactivation is a well-recognized phenomenon in the petroleum and chemical processing industries. Identifying the root causes of this phenomenon is an important factor for enhancing catalyst efficiency and preventing undesirable failures. In this study, state-of-the-art instruments were utilized to investigate the causes of catalyst deactivation that led to the replacement of the catalyst bed in one of the sulfur recovery units at a Saudi Aramco gas plant. Titania catalysts have been examined to determine the inherent deactivation mechanism and also to find out the possibilities of its curement. Understanding the root cause of the deactivation is mandatory for field engineers to minimize future catalyst deactivation. The collected analysis data revealed that the deactivation mechanism occurred for the Ti catalyst due to irreversible chemical phase transformation of the catalyst caused by a temperature runway in the catalytic converter.

Keywords: gas processing, catalyst deactivation, XRD, ESEM, TGA, Claus, titania

1. Introduction

Hydrogen sulfide is a naturally occurring component of crude oil and natural gas.

Petroleum oil and natural gas are the products of thermal conversion of decayed organic matter (kerogen) that is trapped in sedimentary rocks. High-sulfur kerogens release hydrogen sulfide during decomposition, and this H_2S stays trapped in the oil and gas deposits (1). Significant quantities of natural gas resources around the world known to contain H_2S. These have been difficult to produce in the past because of the tendency of sour gas to cause factual damage to the health, processes and environment (1). In the processes, H_2S causes corrosion problems especially in pipelines. However, the advent of corrosion resistance materials and advanced manufacturing techniques make it possible to handle sour gas as a by-product to manufacture highly valuable products (2). Therefore, several processes are used to handle natural gas in order to remove hydrogen sulfide (2). In natural gas processing, Claus process is the most predominant desulfurization technique that convert hydrogen sulfide into elemental sulfur (3, 4). Typically, the fundamental behind this process relies on two conversion stages which are thermal and catalytic stages (5).

A typical Claus unit consists of a thermal stage followed by two or three catalytic stages (three is more common). Before each catalytic stage, reheating process take place to maintain the reaction conditions (5). A typical process flow diagram for Claus unit is shown http://hengyeinc.com/claus-catalyst/The thermal stage consists of a Reaction Furnace (RF), Waste Heat Boiler (WHB), and a condenser. In this stage, 1/3 of the H_2S feed is burned inside the RF to SO_2 and elemental sulfur. The resulting Sulfur is condensed by the 1st condenser and collected in the sulfur pit. Some of the generated SO_2 further reacts with the H_2S to produce elemental sulfur and H_2O while the remaining SO_2 and H_2S are processed to the catalytic converters. It is important to note that the heat generated in this stage is recovered in the WHB. The main chemical reactions are shown below:

(Thermal Stage)

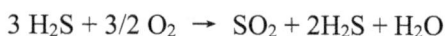

$$3\ H_2S + 3/2\ O_2 \rightarrow SO_2 + 2H_2S + H_2O$$

$$3H_2S + 3/2\ O_2 \rightarrow 3/2S_2 + 3H_2O$$

$$2\ H_2S + SO_2 \rightarrow 3/2\ S_2 + 2H_2O$$

The generated SO_2 in the Reaction Furnace then reacts over the catalyst bed in the catalytic converters with the remaining H_2S to form elemental sulfur, which is removed from the process gas in the condenser by cooling and condensation process. The catalyst in the catalytic reactors serves to boost the sulfur yield according to the following chemical reaction:

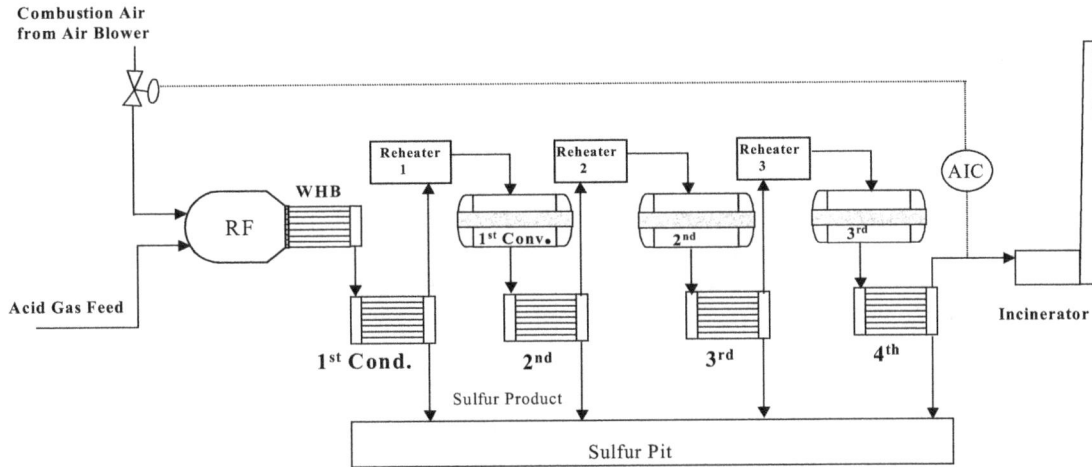

$$2\,H_2S + SO_2 \quad \rightarrow \quad 3/xS_x + 2H_2O \ \text{(Catalytic Stage)}$$

Figure 1. Claus Process Flow Diagram

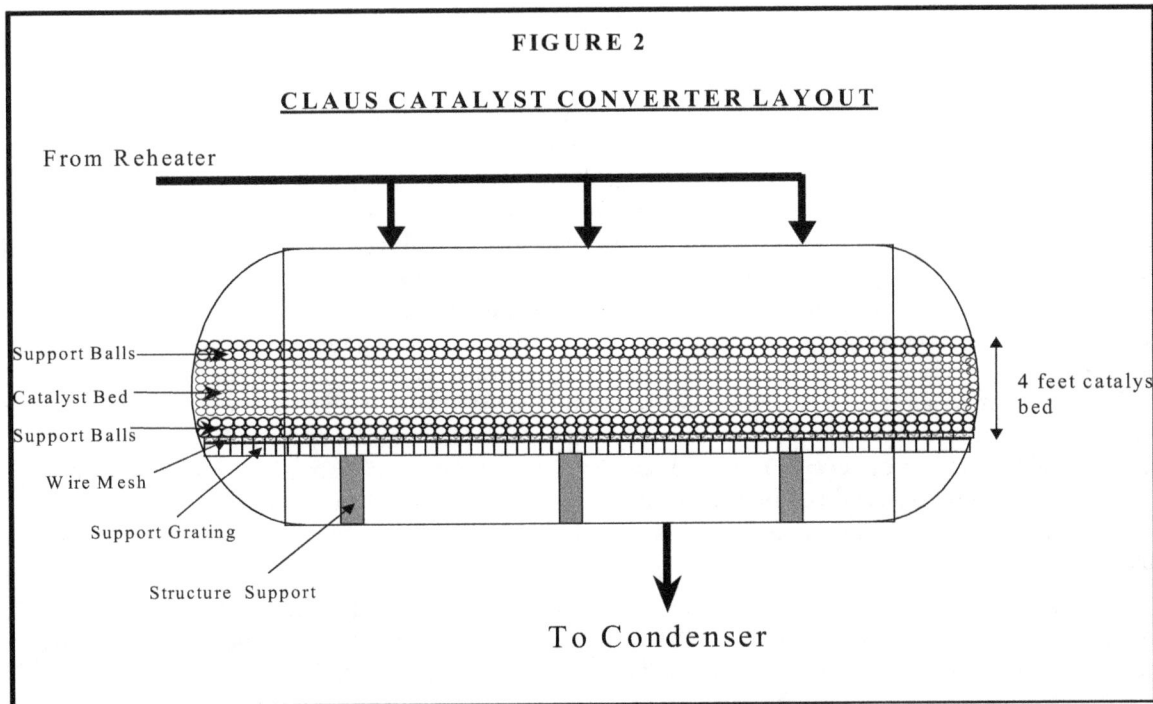

Figure 2. Claus Catalyst Converter Layout

The catalytic stage, which consists of a reheater, catalytic reactor, and condenser, is repeated two or three times to achieve the maximum overall sulfur recovery. Depending on the acid gas feed H_2S % and the reaction furnace configuration, the sulfur conversion in the Reaction Furnace can reach up to 75% (6). The remaining is recovered in the subsequent catalytic stages (6). Figure 2 represents a typical layout of a catalyst converter. It consists mainly of 3-4 feet of rectangular activated alumina and titanium catalyst bed. The catalyst bed is typically covered by a

3-6 inches layer of a bigger support alumina balls to hold the catalyst and also serves as gas distributor. Another support ceramic balls layer is used at the bottom of the catalyst bed to support the catalyst bed. This layer is placed on top of a wire mesh to prevent catalyst migration to condensers. The whole bed is supported by a support grating.

1.1 Claus Catalyst

Catalysts are the workhorses of chemical transformations in most industries. Approximately 85–90% of the products of chemical industry are made in catalytic processes (6). Claus process, where most of the applied catalysts are highly porous and with high surface areas such as aluminum oxide (Al_2O_3), and titanium dioxide (TiO_2). (7, 8) It is well known that catalysts differ from each other based on materials involved, prepration methods, and conditions (7). TiO_2 based materials play a key role due to high stability in alkaline and acid media (7). Titanium dioxide (also known as Titina) is used as heterogeneous catalyst in many applications. For example, it is used as a photo-catalyst in solar cells for the production of hydrogen and electric energy. Alumina Catalyst is a hard working product that offers a high surface area for the conversion of H_2S and O_2 to form elemental sulfur and water (7). The addition of a promoter to an activated alumina catalyst increases the hydrolysis of CS_2 and COS into H_2S, adds resistance to thermal aging, lower operating temperatures in the first Claus reactor, and an increase in working capacity during life span (7). Moreover, TiO_2 is used in the Claus process to aid in the hydrolysis of carbonyl sulfide (COS) and carbon disulfide (CS_2) into hydrogen sulfide (H_2S), which can then be converted in elemental sulfur; this allows much higher quantities of sulfur to be recovered and lowers the quantity of COS and CS_2 from going to flare (7).

TiO_2 exists in three crystalline forms; anatase and rutile are the most common types, and the crystalline size of the rutile is always larger than the anatase phase (Bagheri, Muhd Julkapli, & Bee Abd Hamid, 2014). Brookite is the third structure form, an orthorhombic structure, which is rarely utilized, and is of no interest for most applications (Bagheri, Muhd Julkapli, & Bee Abd Hamid, 2014). Rutile phase is the most thermal stable among the three phases. Brookite and anatase crystalline, above 600C, experience a phase transition and convert into the rutile phase (Bagheri, Muhd Julkapli, & Bee Abd Hamid, 2014). The anatase phase contains zigzag chains of octahedral molecules linked to each other, while the rutile consists of linear chains of opposite edg-shared octahedral structure (Bagheri, Muhd Julkapli, & Bee Abd Hamid, 2014). Generally, the anatase-to-rutile phase transformation occurs between 600-700°C, but for certain applications, it is required that TiO_2 anatase be stable at 900°C (Bagheri, Muhd Julkapli, & Bee Abd Hamid, 2014). Generally, the anatase TiO_2 nanoparticles are stabilized by the addition of cations (Bagheri, Muhd Julkapli, & Bee Abd Hamid, 2014). The synthesis techniques of TiO_2 usually require high temperatures to crystallize the amorphous material into one of the phases of TiO_2, such Brookite, anatase, and rutile, consequently leading to large particles and typically nonporous materials (Bagheri, Muhd Julkapli, & Bee Abd Hamid, 2014). Recently, low temperature synthesis methods resulted in crystalline TiO_2 with a higher degree of control over the formed polymorph and its intra- or interparticle porosities (Bagheri, Muhd Julkapli, & Bee Abd Hamid, 2014). There are reports on the formation of crystalline nanoscale TiO_2 particle via solution based approach without thermal treatment with special focus on the resulting polymorphs, surface area, particle dimensions, and crystal morphology (Bagheri, Muhd Julkapli, & Bee Abd Hamid, 2014). There are exceptional emphases on the sol-gel method via glycosylated precursor and also the miniemulsion method (Bagheri, Muhd Julkapli, & Bee Abd Hamid, 2014). TiO_2 possesses good mechanical resistance and stabilities in acidic and oxidative environments. In addition, TiO_2 based catalyst has outstanding resistance towards corrosion in different electrolytic media and can be regarded as a support for heterogeneous catalysts which guarantees stability in electrochemical environment and commercial availability (Bagheri, Muhd Julkapli, & Bee Abd Hamid, 2014). These properties make TiO_2 a prime candidate for heterogeneous catalyst (Bagheri, Muhd Julkapli, & Bee Abd Hamid, 2014).

There are only a few studies reporting a rutile catalyst support which resulted in higher catalytic activity compared to anatase, such as the oxidation of toluene, xylene, and benzene over rutile-supported Cu catalyst. In comparison, rutile is preferred as a model support for particles of metals in surface science studies (Bagheri, Muhd Julkapli, & Bee Abd Hamid, 2014), due to its high crystal phase's thermodynamic stability. Furthermore, it is indicated that rutile and anatase differ noticeably in their ability of fixing particles of metals onto their respective surface (Bagheri, Muhd Julkapli, & Bee Abd Hamid, 2014); whereas the strong metal support interaction is normally shown on anatase, this effect is not as significant on rutile. Inopportunely, the thermodynamic stability of TiO_2 is comparatively low, and calcination would usually lead to the collapse of the porous structures (Bagheri, Muhd Julkapli, & Bee Abd Hamid, 2014). Additionally, it is reported that calcination above 465°C has always resulted in the phase transition from anatase to rutile (Bagheri, Muhd Julkapli, & Bee Abd Hamid, 2014). The phase transition could be connected to the growth of crystal size, which results in a severe reduction in specific surface area (Bagheri, Muhd Julkapli, & Bee Abd Hamid, 2014). Consequently, this should also influence the overall catalytic performance of metal heterogeneous catalysts.

However, there are drawbacks in using TiO_2 in Claus process which will be discuss in more details in this paper. Catalyst deactivation is very challenging task. In general, deactivation is inevitable, but it can be slowed or prevented and some of its consequences can be avoided. Several deactivation mechanisms have been identified and discussed extensively in the literature. These intrinsic mechanisms can be classified into the following distinct types which are poisoning, coking, sintering, solid-state transformation, and others. In general, they can be categorized as chemical, mechanical, or thermal mechanism (Coulsen & Richardson, 1989). Understanding catalyst deactivation mechanisms is essential in order to better support field engineers to elongate the lifetime of processes catalysts.

1.2 Deactivation Mechanisms

Deactivation of catalysts is a problem that causes loss of catalytic rate with time. It classifies by type such as chemical, thermal, and mechanical. However, it is also classified by mechanisms such as poisoning, fouling, thermal degradation, vapor formation, vapor-solid and solid-solid reactions, and attrition/crushing. (10)

1.2.1 Poisoning

Poisoning is the loss of a catalyst activity due to the strong chemisorption on the active sites of impurities present in the feed stream. A poison may act simply by blocking an active site or may alter the adsorptivity of other species essentially by an electronic effect. Poisons can also modify the chemical nature of the active sites or result in the formation of new compounds so that the catalyst performance is altered (Gamson & Elkins, 1953).

1.2.2 Coking

Coking for catalytic reactions involving hydrocarbons (or even carbon oxides) side reactions occur on the catalyst leading to the formation of carbonaceous residue (coke or carbon) which tend to physically cover the active surface. Coke deposits may deactivate the catalyst either by covering of the active sites, and by pore blocking. Mechanisms of carbon deposition and coke formation on catalysts have been detailed in literature (Gamson & Elkins, 1953). The amounts of coke deposited into the catalyst pores may be estimated by burning the coke with air and recording the weight changes via TGA techniques and by monitoring the evolution of the combustion products CO_2 and H_2O.

1.2.3 Sintering

Sintering refers to the loss of active surface due to structural modification of the catalyst. This is a thermal activated process. Sintering occurs both in supported metal and unsupported catalysts where the key variable is temperature, so operation at low temperatures greatly reduces the sintering rate (Gamson & Elkins, 1953).

1.2.4 Solid-State Transformation

It is a process of deactivation that occurs at high temperatures and lead to the transformation of one crystalline phase into a different one. These processes may involve both metal-supported catalysts and metal oxide catalysts as well. For example, transformation of γ- into δ-Al_2O_3 with-wise decrease in the internal surface area from about 150 m2/g to less than 50 m2/g (Goar, 1968). Several of transformations are limited by the rate of nucleation. This process may occur due to the presence of foreign compounds in the lattice or on the surface. For example, V_2O_5 and Fe_2O_3 have been reported to favor the transformation of anatase-to-rutile (Gamson & Elkins, 1953; Coulsen & Richardson, 1989).

1.2.5 Other Mechanisms of Deactivation

Other deactivation mechanisms include masking or pore blockage, caused by physical deposit of substances on the outer surface of the catalyst which hinder the reactants from the active sites. In additional to the coke deposition, masking may occur during the processes where metals such as corrosion products deposit on the catalyst external surface (Gamson & Elkins, 1953).

This paper presents a failure analysis that was conducted on Ti-based catalyst utilizing several analytical techniques to investigate the root causes resulted in the catalyst deactivation. This catalyst was deactivated in less than a year although the manufacturer specification stated that the lifetime of this catalyst is expected to last for five years. The preliminary investigation indicated that the deactivation was a result of temperature run-away in the catalytic converter. This study will investigate the changes occurred to the physical and chemical characteristic of titania catalyst that resulted in its deactivation.

2. Experimental Section

In this study, fresh and plant spent catalyst samples were physically and chemically investigated using several analytical techniques. The spent Claus catalyst samples were collected from different rector layers:

2.1 Surface Area and Pore Volume Analysis

This analysis was performed to determine the surface areas before and after deactivation of the catalysts. The fresh and spent catalyst samples were ground to fine powder using a mortar and a pestle. The principle of the measurement is to use nitrogen molecule adsorption on the surface at 77 Kelvin to induce multi-layer physical adsorption. By measuring adsorbed amount of nitrogen at specific pressure of nitrogen, surface area and pore size distribution are estimated with B.E.T and B.J.H equations, respectively.

2.2 Crush Strength Analysis

This analysis was used to determine the strength of fresh and spent catalyst samples. It also determines the resistance of formed catalysts and catalyst carriers to compressive force. Such information will reflect the severity of the process to the catalyst.

2.3 Total Carbon Analysis

This analysis was carried out to determine carbon deposition in both the surface soot and the pore soot. 2-3 mg of the samples were weighted into disposable tin capsules and injected into a high temperature combustion reactor at 900°C in pure oxygen under static conditions. This method is intended to determine the amount of carbon deposited in/on catalysts using EA 1112 Elemental Analyzer.

2.4 Total Sulfur Analysis

This analysis was used to determine sulfur deposition in both the surface soot and the pore soot. The samples were analyzed using high-temperature tube furnace combustion method to determine the amount of sulfur deposited in/on the surface of the deactivated catalyst using ASTM D-4239-08 method.

2.4 X-Ray Powder Diffraction and X-Ray Florescence Analysis

X-ray powder diffraction (XRD) was utilized to determine the catalyst crystalline inorganic phases/compounds for fresh and spent plant catalyst samples. In addition, XRD was also used to carry out an experiment known as **In-situ XRD** analysis for fresh catalyst sample. The experiment is designed to determine phase transformation during heating the sample from ambient temperature up to 1200° C.

2.5 X-Ray Florescence

(XRF) was also used to determine the elemental chemical composition of the fresh and spent catalysts. For XRD analysis, the samples were ground to fine powder using a mortar and a pestle, and the fine powder was mounted in the XRD sample holder by back pressing.

The identification of the inorganic crystalline compounds present in the titania catalyst samples was achieved by using X-Pert High Score Plus program. The quantifications of the data were completed using either Rietveld refinement methods or RIR methods. For XRF, approximate 4 g of the powder was mixed and homogenized with 0.9g of binder (Licowax C micropowder PM (Hoechstwax)). Then, the powder was pressed with 20 tons to pellet with 31 mm diameter prior to testing.

2.6 Thermogravimetric Analysis (TGA)

This technique was used to provide information on the chemical transformation of catalytic phases to non-catalytic phases. The samples were analyzed using TGA analyzer from ambient to 900°C at a rate of 10°C per minute under air. This was done to determine the material thermal stability and the loss on ignition (LOI) using SALAM 530-06 method.

2.7 Environmental Scanning Electron Microscopy (ESEM)/Energy Dispersive X-ray Analysis (EDS)

ESEM was utilized to study topographic features of the catalyst samples to better understand the mechanisms of catalyst failures. A beam of electrons are focused on the catalyst surface to ionize the atoms of the catalyst in order to produce characteristic X-rays. These X-rays were processed to generate an EDS spectrum that represented the elemental composition and their intensity that are proportional to the concentration of the elements in the catalyst sample.

The electron beam penetrates 2-3μm into the material so the composition is an average of the sample. The samples were analyzed using FEI Quanta 400 ESEM applying the procedures outlined in the Saudi Aramco Laboratory Analytical Method "SALAM" 024. The ESEM was operated at 15-30 kV, and 10 mm working distance.

Backscattered electron images together with EDS X-ray spectra were acquired from different parts of the catalyst samples.

3. Results and Discussion

Deactivation is well-recognized phenomenon in the petroleum and chemical processing industries. Identifying the root causes of this phenomenon is an important factor to enhance catalyst efficiency and prevent undesirable failures.

The fresh and spent titanium dioxide TiO_2 catalyst samples were physically and chemically characterized using different techniques such as surface area, pore volume, crush strength, thermal stability, chemical composition, and surface morphology. The collected data are presented in Table 1. Since it is very important to determine the level of CS_2 and COS in/on the surface of the catalyst. Because the activity of titania catalyst for pure Claus reaction at industrial operation conditions is so high and it always performs at equilibrium level, instead of the direct determination of the Claus activity the measurement of CS_2 conversion over the catalyst becomes almost a unique and standard procedure (11). In addition, C-S compound, COS, is also formed in the same stage by a multitude of reactions. The presence of both CS_2 and COS in the process gas flowing to the catalytic converters is very important as, if not destroyed in the 1st converter, these substances place a limit on the total sulfur capture ability in the plant. (11). It is very important to determine the level of CS_2 and COS in/on the surface of the catalyst.

Table 1 shows the properties of Ti fresh catalyst: The XRD data (Figures 4) showed the presence of anatase peaks which is the major phase of fresh titania catalyst with a minor phase of calcium sulphate. The patterns clearly exhibited strong diffraction peaks at 25 and 48 indicating TiO_2 in the anatase phase. On the other hand, the XRD difractogram of the spent catalyst exhibited strong diffraction peaks at 27, 36 and 55 indicating TiO_2 in the rutile phase with a minor phase of perovskite that was detected at 34 and 48 peak position (Figure 5). Nevertheless, titania normally undergoes anatase-rutile phase transformation in the temperature range 600-700°C (Gamson & Elkins, 1953; Goar, 1968; Clark, Dowling, & Huan, n.d.). For high temperature conversion of titania, a stable anatase phase is necessary. Both anatase and rutile are tetragonal in structure. Both structures are consisted of TiO_6 octahedral, sharing four edges in anatase and two in rutile as illustrated in Figure 3. The anatase to rutile transformation is reconstructive transformation, which means that the transformation involves the breaking and forming of bonds (Goar, 1968). This is in contrast to a displacing transformation, in which the original bonds are distorted but retained. According to the XRD data the crystallite sizes are 11.5 nm and 408.9 nm of anatase in the fresh and rutile in the spent catalyst samples respectively. Changes in the titania phase is proportional to the significant changes in the surface area. The total surface area lost is about 98% which attributed to the formation of rutile phase (Table 1). These changes in the arrangements of TiO_6 have effects chemical properties of titania catalysts.

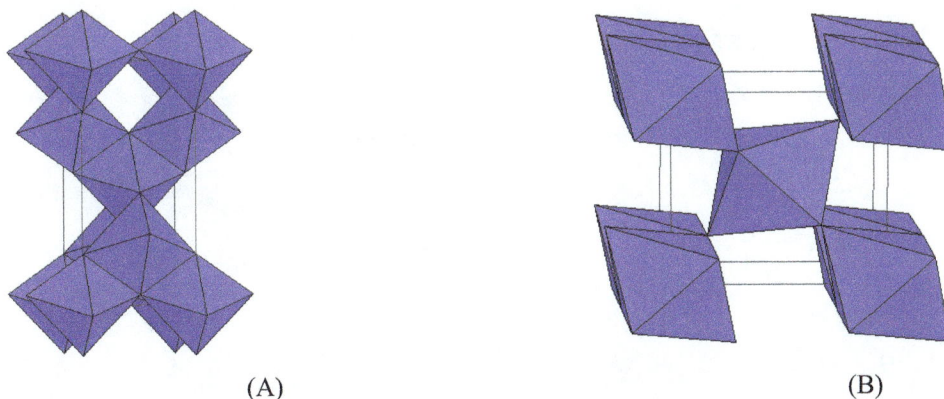

(A)   (B)

Figure 3. Three-dimensional representation of the (A) rutile and (B) anatase

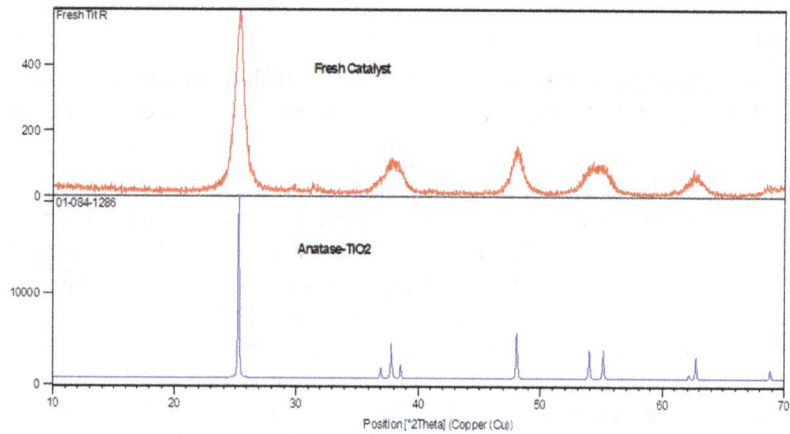

Figure 4. XRD Pattern of the fresh Ti Catalyst Sample

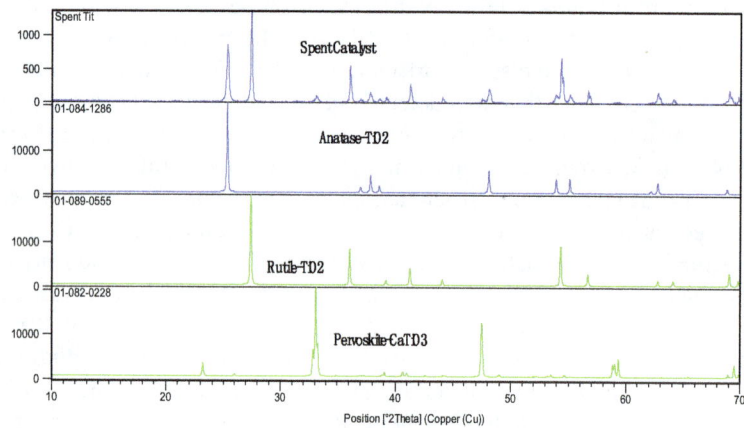

Figure 5. XRD Pattern of the spent Ti catalyst Sample

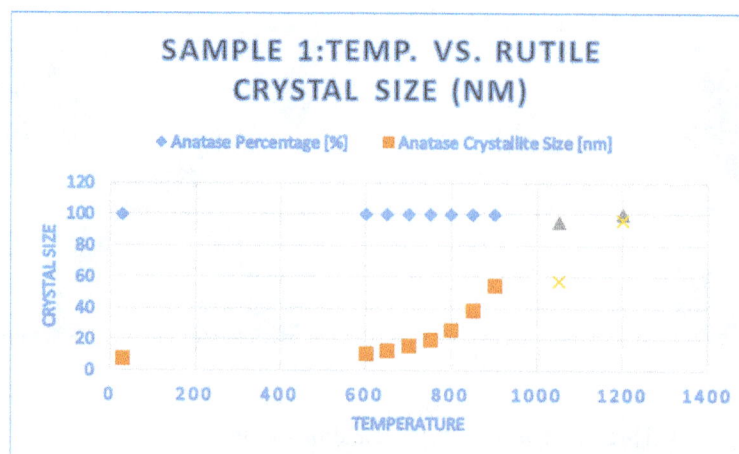

Figure 6. Temperature vs. anatase crystal size (nm) formation

Another significant change was observed in the pore volume analysis. It revealed substantial decrease in the porosity for the deactivated titania catalyst compared to the fresh catalyst samples. The change in the pore volume of the deactivated titania catalyst can be correlated with the increased crystallite size which lowered the porosity due to phase transformation as demonstrated by the X-ray diffraction patterns. Furthermore, a change was also observed in the crush strength where the fresh Ti catalyst showed superior mechanical strength compared to spent Ti catalyst (Crush strength 35.5 N for fresh Ti catalyst vs. 11.7 for spent Ti catalyst).

In literature, morphology of the catalyst plays a major role that influences its activity and selectivity (10). With the aid of ESEM images, it is possible to visualize the effects caused by temperature runway in the catalytic converter by comparing the micro-images of fresh and spent catalyst samples. The deactivation changed the morphological properties of the fresh catalyst. Fresh and deactivated catalyst samples were analyzed by SEM and ESEM. In this study it was found that the catalyst deactivation is not directly dependent on CS_2 and COS content but it is highly influence by the catalyst shapes and pores structure. However, hydrocarbons deposition were detected on the surface of the deactivated catalyst surfaces such deposition contributes to decrease the catalyst porosity and ultimately affect the performance of the catalyst activity. Figure 6 shows homogenous structure with well-defined open pore with an average dimension of 10-100 nm. Such structure maximizes the contact between the reactants and the active parts of the catalyst and hence increases the activity. Maintaining these structural properties will certainly maximize the life time of the catalyst. In comparison, Figure 7 shows non-homogeneous structure with much closed pores. The micrographs also show collapsed pores full of hydrocarbons. It seems that runway temperature caused damaged to all types of pores which illustrates the mechanisms of deactivation. Nevertheless, the images in Figure 7 (C, D, and E) show that the catalyst cannot be regenerated or rejuvenated due to sever collapse of the catalyst pores. The EDS analysis data shows hydrocarbons laying down on the surface of the deactivated catalyst as shown in Figure 7 (E, F). However, it was difficult to look inside the pores since the SEM images shows that all the catalyst pores are fully damaged due to overheating temperature that caused to deactivation.

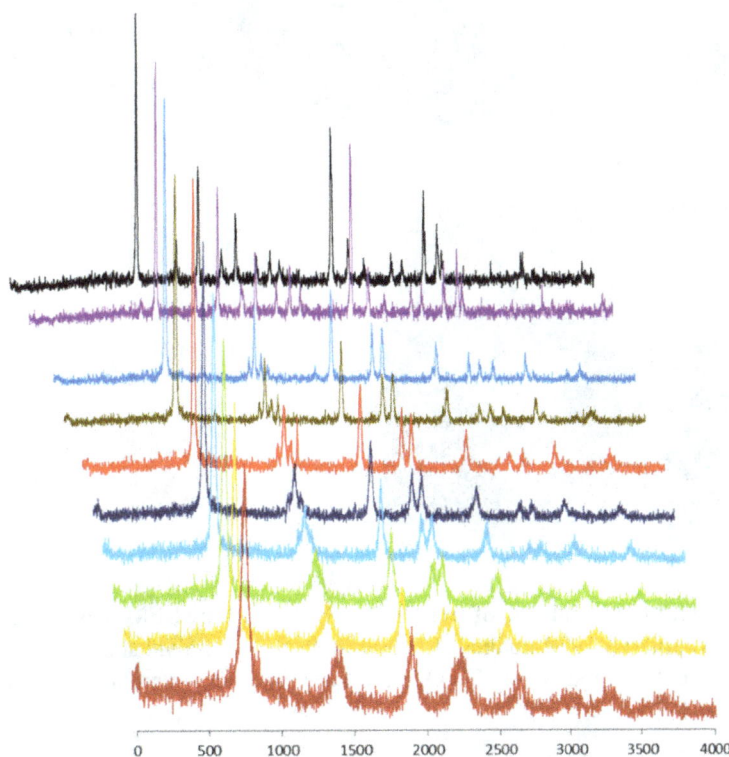

Figure 7. In-situ XRD Pattern: In-house calcination experiment of anatase from ambient to 1200°C

In theory, catalyst mechanical properties are highly effected by structure of the catalyst. Due to deactivation, the spent Ti catalyst lost 67% of its strength (Table 1). In Addition, the lack of strength generates large amounts of dust which blocks catalyst pores which results in loss of catalytic active sites.

The TGA analysis in Figure 8 shows the difference between fresh and spent Ti catalyst samples. The spent Ti catalyst shows straight line from ambient temperature up to 900°C which illustrates that all the catalyst components had been decomposed. Where the fresh catalyst lost 2.5 (% weight) at 130°C, 4.3 (% weight) at 600°C, and 6.2 (%weight) at 900°C.

All results are in line with the phase transformation which is in this case cannot be regenerated or rejuvenated to the original phase due to breaking and formation of new bonds.

According to literature, anasate has a certain structure that effectively increases the sulfur conversion by reacting the generated SO_2 with H_2S to produce elemental sulfur in the Claus process (6). Moreover, the catalyst performance is significantly affected by lay down of hydrocarbons that blocked the active sites and also phase transformation that caused breaking and formation of new bonds.

In order to better understand the Ti catalyst deactivation mechanism, in-house calcination experiments were carried out in 7 fresh catalyst samples calcined for 2 hours from room temperature up to 500°C, 600°C, 700°C, 800°C, 900°C, 1000°C, and 1200°C. The resulted catalyst samples were analyzed using XRD, and ESEM techniques.

The XRD phase identification and quantification data (Table 2) shows that anatase TiO_2 phase started to transform to rutile TiO_2 phase at 800°C. It shows that only 3 wt. % of anatase was converted to rutile phase, where 10 wt. % of anatase phase remained at 900°C. This clearly shows that the temperature during the failure was beyond 900°C, because the spent Ti catalyst contains only rutile phase as shown in Figure 5.

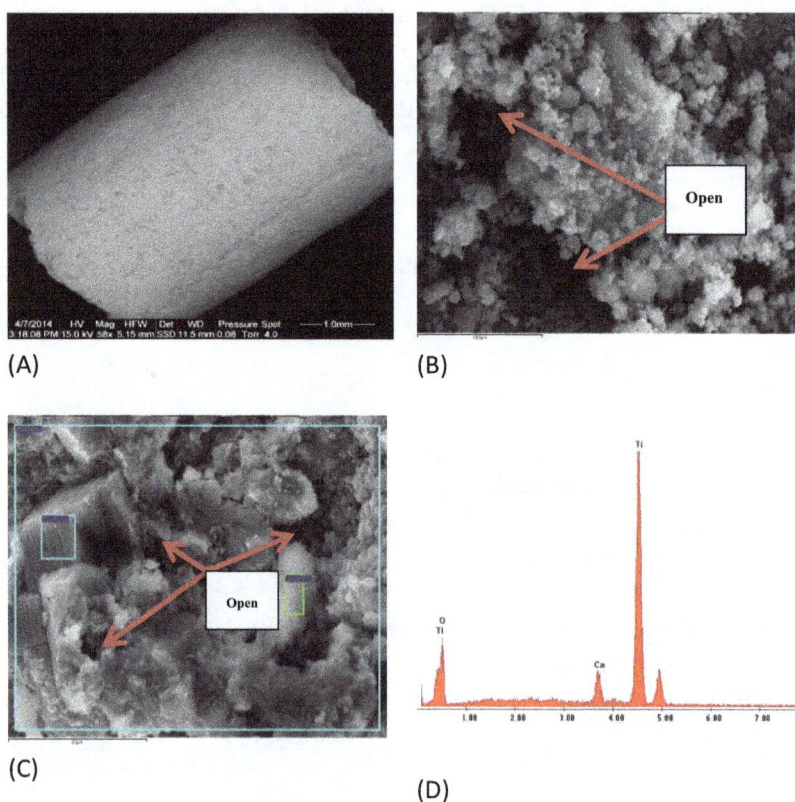

Figure 8. SEM images and EDS analylsis of fresh Ti catalyst: (A) Low SEM magnification of fresh Ti catalyst sample as received, (B) and (C) SEM images at 5000x of fresh Ti Catalyst showing open pores. And (D) EDS spectrum of fresh Ti catalyst that shows Ti, O and Ca are the preeminent elements in the catalyst sample

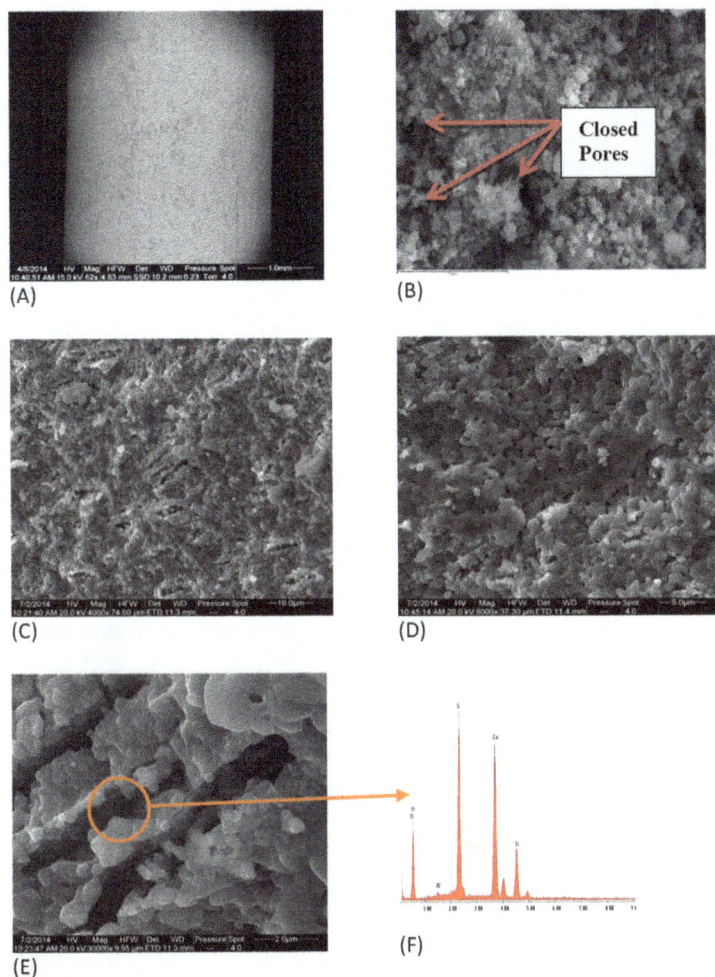

(A) (B) (C) (D) (E) (F)

Figure 9: SEM images and EDS analysis of spent Ti catalyst: (A) Low SEM magnification of spent Ti catalyst sample as received. (B) SEM image of spent catalyst showing closed pores due to deactivation process. (C) (D) SEM images of damaged pores caused by temperature runway in the catalytic converter. (E) (F) SEM image of damaged Ti catalyst surface showing deposition of sulfur content (EDS) caused by deactivation

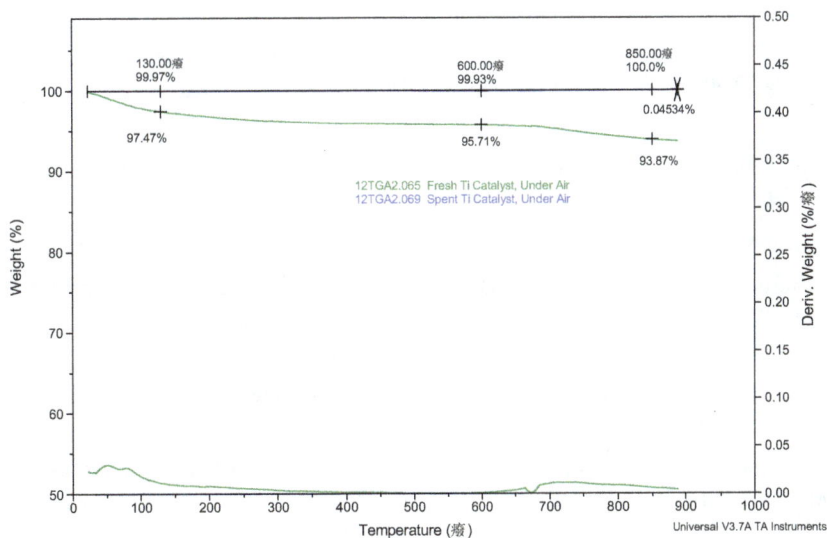

Figure 10. TGA analysis of fresh and spent Ti catalyst

Table 1. Physical properties and XRD composition of the fresh and spent Titania catalyst

Property	Fresh Ti Catalyst	Spent Ti Catalyst	Remark
Surface Area, m^2/g	134.2	2.1	Lost 98% of surface area
Pore Volume, cc/g	0.27	0.47×10^{-2}	Lost 98% of pore volume
Crush Strength, N (Ibs)	35.5	11.7	Lost 67% of strength
Major Crystalline Phase (XRD)	Anatase-TiO$_2$	Rutile-TiO$_2$,	Inactive catalyst
Minor Crystalline Phase (XRD)	CaSO$_4$	Perovskite-CaTiO$_3$	Reactions due to temperature runaway
Colour	White	Grey	Change in color

Table 2. XRD Phase Identification and Quantification (in wt.%) of Catalyst Samples

Compounds	Catalyst Treated Temperature				
	500 °C	600 °C	700 °C	800 °C	900 °C
Anatase-TiO$_2$	100%	100%	100%	97%	16%
Rutile-TiO$_2$	-	-	-	3%	84%

4. Conclusion

- Different structures lead to different physical properties.

- Hydrocarbon compounds were detected on the surface of the deactivated catalyst.

- Loss of catalyst activity was due to chemical conversion of anatase to rutile caused by temperature runaway inside the Catalytic Converter.

- Increasing the temperature caused significant changes in the textural properties of fresh catalyst.

- Changes in the titania phase was paralleled by significant changes in the surface area at temperature above 650°C. The spent catalyst lost 98% of the total surface area which attributed to the formation of rutile phase.

- The Nitrogen adsorption analysis showed that the deactivated titania catalyst suffered significant loss of micro-pores.

- The chemical characterization data showed that the deactivation was not due to surface contamination. However, the detected hydrocarbons on the surface of the deactivated catalyst samples was due inability of the catalyst to covert the reactants to products.

- The presented ESEM images illustrated a wide application of this method in the field of heterogeneous catalysis. ESEM investigations combined with spectroscopic diffraction and other spectroscopic methods were very successful in understanding the cause of catalyst deactivation.

- It is found that the anatase to rutile phase transformation was accelerated by increasing temperature.

Acknowledgment

Authors would like to thank Sayed R. Zaidi, and Akram A. Alfliow for their involvements in the research project. Ki-Hyouk X. Choi, and Abed T. Al-Harthi for their contributions to this study.

References

Al-Haji, M. N., & A-Adab, A. M. (2001). Claus Catalyst Sulfur Wash: A Successful Technique to Restore Normal Pressure Drop. *Oil & Gas Journal, 99*(24).

Al-Shafei, M. (n.d.). Internal technical report.

Argyle, M. D., & Bartholomew, C. H. (2015). Heterogeneous catalyst deactivation and regeneration: A review. *Catalysts, 5*(1), 145-269.

Arrouvel, C., Digne, M., Breysse, M., Toulhoat, H., & Raybaud, P. (2004). Effects of morphology on surface hydroxyl concentration: a DFT comparison of anatase–TiO 2 and γ-alumina catalytic supports. *Journal of Catalysis, 222*(1), 152-166.

Bagheri, S., Muhd Julkapli, N., & Bee Abd Hamid, S. (2014). Titanium dioxide as a catalyst support in heterogeneous catalysis. *The Scientific World Journal, 2014*.

Bartholomew, C. H. (2001). Mechanisms of catalyst deactivation. *Applied Catalysis A: General, 212*(1), 17-60.

Batzill, M., Morales, E. H., & Diebold, U. (2006). Influence of nitrogen doping on the defect formation and surface properties of TiO 2 rutile and anatase. *Physical review letters*, *96*(2), 026103.

Beauchamp, R. O., Bus, J. S., Popp, J. A., Boreiko, C. J., Andjelkovich, D. A., & Leber, P. (1984). A critical review of the literature on hydrogen sulfide toxicity. *CRC Critical Reviews in Toxicology*, *13*(1), 25-97.

Clark, P. D., Dowling, N. I., & Huan, M. (n.d.). Fundamental and Practical Aspects of the Claus Sulfur Recovery Process, Alberta Sulfur Research Ltd., Calgary, Alberta, CanadaP.D. Clark, N.I. Dowling and M. Huan., Fundamental and Practical Aspects of the Claus Sulfur Recovery Process, Alberta Sulfur Research Ltd., Calgary, Alberta, Canada.

Clark, P. D., Dowling, N. I., & Huang, M. (2008). Reversible deactivation of TiO 2 in CS 2 conversion associated with Claus reaction. *Applied Catalysis A: General*, *343*(1), 104-108.

Clark, P. D., Dowling, N. I., & Huang, M. (2008). Reversible deactivation of TiO 2 in CS 2 conversion associated with Claus reaction. *Applied Catalysis A: General*, *343*(1), 104-108.

Claus Catalyst. (n.d.). In *Hengye*. Retrieved from http://hengyeinc.com/claus-catalyst

Coulsen, J. M., Richardson, J. F. (1989). *Chemical Engineering* (Vol 2, 3rd Ed.). Pergamon Press.

Farrauto, R. J., Dorazio, L., & Bartholomew, C. H. (2016). *Introduction to Catalysis and Industrial Catalytic Processes*. Retrieved from http://www.wileyvch.de/books/sample/3527316728_c01.pdf

Forzatti, P., & Lietti, L. (1999). Catalysis Today 52 (1999). 165±181.

Gamson, B. W., & Elkins, R. H. (1953). Sulfur from Hydrogen Sulfide. *Chemical Engineering Progress, 49*(4), 203-215.

Goar, B. G. (1968). Today's Sulfur Recovery Processes. *Hydrocarbon Processing*, 248-252.

Larraz, R. (1999). *Hydrocarbon Processing*.

Mohammad, N. A. H., & Abdulhadi, M. A. A. (2001). Catalyst Sulfur Wash. *51st Annual Laurance Reid Gas Conditioning Conference*, Norman, Oklahoma, February 25 – 28, 2001.

Shafei, M. A., & Dutta, T. K. (2000). *Proceding of the world petroleum Congress*. Calgary, Canada, 2000

9

Heat Treatment of Duplex Stainless Steel 2205 by Inserting Nano Nd2FeB14 in HIP Manifolds Under the Scope of Category Theory

Mohamed Atef Mohamed Gebril[1]

[1] Department of material engineering, Faculty of petroleum and minning engineering, Suez University, Egypt

Correspondence: Mohamed Atef Mohamed Gebril, Department of material engineering, Faculty of petroleum and minning engineering, Suez University, Egypt. E-mail: atefmohamed840@yahoo.com

Abstract

In this paper, improving mechanical properties of duplex stainless steel by spinodal reaction inhibitors has been discussed. Spinodal gaps can be minimized when the host particles are minimized to the nanoscale. The Cahn-Hilliard Equation and the Allen-Cahn Equation on Manifolds with Conical Singularities can be related to manifolds produced by hot isostatic pressing (PM HIP) are used to improve mechanical strength. Spinodal reactions will be Inhibited by magnetic treatment and Nd-Fe-B Magnets. Grain refiners will be used to retard spinodal reactions. Mathematically, category theory is used to establish links between these different concepts.

Keywords: Nd2Fe B14 magnetic thin films, Duplex stainless steel 2205, Impact energy, spinodal reactions, cahn hillard equations, Ising model, Ohta–Kawasaki energy, Fractals - solobev space, Finsler manifold, The Lusternik-Schnirelman category

1. Introduction

Duplex stainless steels are called "duplex" because they have a two-phase microstructure consisting of grains of ferritic and austeniticstainlesssteel.Duplexstainlesssteelshavebeenincreasinglyusedfor a variety of applications in marine construction, chemical industries and power plants due to their excellent combinational of mechanical properties and corrosion resistance The superior properties of duplex stainless steels come primarily from approximately equivalent amounts of austenite (γ) and δ. It is well known that in Fe–Cralloys, there is a miscibility gap, where the ferrite phase may decompose into an iron-rich BCC phase(α)and a chromium-enriched BCC phase (α') by spinodal decomposition the effect of isothermal treatment temperatures ranging between (400and500C) on the embrittlement of a2205duplex stainless steel (with 45% ferrite-55% austenite, vol.%) has been investigated. The impact toughness and hardness of the aged specimens were measured. The results show that the steel is susceptible to severe embrittlement when exposed at 475 C. High-resolution transmission electron microscopy has revealed that anisotropic spinodal decomposition occurred in the steels during aging at 475C;theδ-ferrite decomposed into a nanometer-scaled modulated structure with a complex interconnected network, which contained a Fe-rich BCC phase(α)and Cr-enriched BCC phase (α').It is deduced that the locking of dislocations in the modulated structure leads to the severe embrittlement. Spinodal reactions are retarded by inserting nanoparticles in certain places. Also, we can reduce these spinodal reactions by grain refiners as they will be used to modify gibbs free energy for the total systems. By applying magnetic field and making magnetic bubbles, we can use their properties to control in the thermal charachteristics of the entire system (Alvarez-Armas & Degallaix-Moreuil, 2009; Adams, Olubambi, Potgieter, & Van Der Merwe, 2010; Kangas & Newman, 1998).

Permanent magnet is a ferromagnetic material that has a magnetization even if there is no external applied field. Using Ising model we can explain their properties, and the Ising model has special charchteristic in manifolds. Nd-Fe-B magnets has special properties, Fractal Magnetic Domains on Multiple Length Scales in Nd2Fe14BThe industrial strength ferromagnet, Nd2Fe14B has become a prototypical system for the study of magneticdomain structures. Below the Curie temperature, Tc ¼ 565 K, the Nd and Fe moments order ferromagnetically (Byrne, Spence, Olsen, Houghton, & MCMAHON, 1994).

Nowadays, many companies use HIP thechniques to create manifolds to enhance mixed (mechanical and electrochemical) properties. Also, we will study the effect of nano particles in the spinodal gap and their strong relation to cahn -callihard spinodal reactions.

Many theories in the group theory are used to describe the actions of group, in addition, functors and classes in the category theory are used to establish many beautiful relations between different mathematical structures (Bucur, Deleanu, & Hilton, 1968; May, 1999).

The paper is organized as follow. In section 2, properties of duplex stainless steel were briefly reviewed. In section 3, Concepts in the theory of categories were briefly discussed. In section 4, many experimental data are collected to achieve progress in the development of the properties of duplex steel. In section 5, we have mentioned the results and the discussion how to develop duplex stainless steel using mathematical theories. The work closes with some concluding remarks in section 6.

2. Duplex Stainless Steel

Duplex stainless steels are becoming more common. They are being offered by all the major stainless steel mills for a number of reasons:

· Higher strength leading to weight saving

· Greater corrosion resistance particularly stress corrosion cracking

· Better price stability

· Lower price

There is a conference on the subject of duplex every 2-3 years where dozens of highly technical papers are presented. There is a lot of marketing activity surrounding these grades. New grades are being announced frequently.

Yet, even with all this interest, the best estimates for global market share for duplex are between 1 and 3%. The purpose of this article is to provide a straightforward guide to this steel type. The advantages and disadvantages will be described.

Ferritic – low strength (a bit higher than austenitic, 250 MPa 0.2% PS), poor weldability in thick sections, poor low temperature toughness. In addition, the high nickel content of the austenitic types leads to price volatility which is unwelcome to many end users. The basic idea of duplex is to produce a chemical composition that leads to an approximately equal mixture of ferrite and austenite. This balance of phases provides the following:

· Higher strength – The range of 0.2% PS for the current duplex grades is from 400 – 550 MPa. This can lead to reduced section thicknesses and therefore to reduced weight. This advantage is particularly significant for applications such as:
o Pressure Vessels and Storage Tanks
o Structural Applications e.g. bridges

· Good weldability in thick sections – Not as straightforward as austenitics but much better than ferritics.

· Good toughness – Much better than ferritics particularly at low temperature, typically down to minus 50 C, stretching to minus 80 C

Duplex 2205 is a nitrogen enhanced duplex stainless steel that was developed to combat common corrosion problems encountered with the 300 series stainless steels. "Duplex" describes a family of stainless steels that are neither fully austenitic, like 304 stainless, nor purely ferritic, like 430 stainless. The structure of 2205 duplex stainless steel consists of austenite pools surrounded by a continuous ferrite phase. In the annealed condition, 2205 contains approximately 40-50% ferrite. Often referred to as the work horse grade, 2205 is the most widely used grade in the duplex family of stainless steels.

It is well known that in Fe–Cr alloys, there is a miscibility gap, where the ferrite phase may decompose into an iron-rich BCC phase(α)and a chromium-enriched BCC phase (α') by spinodal decomposition the effect of isothermal treatment temperatures ranging between (400 and 500C) on the embrittlement of a2205duplex stainless steel (with 45% ferrite-55% austenite, vol.%) has been investigated. The impact toughness and hardness of the aged specimens were measured. The results show that the steel is susceptible to severe embrittlement when exposed at 475 C. It is deduced that the locking of dislocations in the modulated structure leads to the severe embrittlement. Spinodal reactions are retarded by inserting nanoparticles in certain places (Alvarez-Armas & Degallaix-Moreuil, 2009; Hsieh, Tsai, Chang, & Yang, n.d.; Huang & Shih, 2005).

3. Concepts in the Theory of Categories

Category theory formalizes mathematical structure and its concepts in terms of a collection of *objects* and of *arrows* (also called morphisms). A category has two basic properties: the ability to compose the arrows associatively and the existence of an identity arrow for each object. The language of category theory has been used to formalize concepts of other high-level abstractions such as sets, rings, and groups.

Several terms used in category theory, including the term "morphism", are used differently from their uses in the rest of mathematics. In category theory, morphisms obey conditions specific to category theory itself.

Samuel Eilenberg and Saunders Mac Lane introduced the concepts of categories, functors, and natural transformations in 1942–45 in their study of algebraic topology, with the goal of understanding the processes that preserve mathematical structure.

Category theory has practical applications in programming language theory, in particular for the study of monads in functional programming.

A *category C* consists of the following three mathematical entities:

-   A <u>class</u> ob(C), whose elements are called *objects*;

-   A class hom(C), whose elements are called <u>morphisms</u> or <u>maps</u> or *arrows*. Each morphism f has a *source object a* and *target object b*.
    The expression $f: a \rightarrow b$, would be verbally stated as "f is a morphism from a to b".
    The expression **hom(a, b)** — alternatively expressed as **hom$_C$(a, b)**, **mor(a, b)**, or **C(a, b)** — denotes the *hom-class* of all morphisms from a to b.

-   A <u>binary operation</u> ∘, called *composition of morphisms*, such that for any three objects a, b, and c, we have hom(b, c) × hom(a, b) → hom(a, c). The composition of $f: a \rightarrow b$ and $g: b \rightarrow c$ is written as $g \circ f$ or gf, governed by two axioms (Kim et al., 2005):

    -   <u>Associativity</u>: If $f: a \rightarrow b$, $g: b \rightarrow c$ and $h: c \rightarrow d$ then $h \circ (g \circ f) = (h \circ g) \circ f$, and

    -   <u>Identity</u>: For every object x, there exists a morphism $1_x: x \rightarrow x$ called the <u>*identity morphism for x*</u>, such that for every morphism $f: a \rightarrow b$, we have $1_b \circ f = f = f \circ 1_a$.

From the axioms, it can be proved that there is exactly one <u>identity morphism</u> for every object. Some authors deviate from the definition just given by identifying each object with its identity morphism (Kreyssig et al., 2009; Borceux, 1994; May, 1999).

4. Experimental Procedure and Methods

1. The chemical composition of wrought 2205 duplex steel studied was Fe-22.62Cr-5.12Ni-3.24Mo-1.47Mn -0.38Si-0.02C (wt%). The as-received 2205 duplexstainless steel rods for this research was produced by Gloria Material Technology Corporation through the 4-folded forging of a cast slab at 1160 - 1180°C and annealing at 1050°C for 30 min, followed by water quenching. The annealed steel gave a dual-phase structure (with 45 δ-ferrite – 55 austenite, vol%) without other secondary phases. Experiments in this report contained two parts. The first part was to investigate the microstructure of the unaged and aged specimens without deformation, including dislocation characterization and dislocation density calculation. The second part focused on the evolution of microstructures and dislocations in the unaged and aged specimens after Charpy impact test. In the first experiment, specimens 6mm in length and 3mm in diameter were machined from the steel bar and aged at 475°C for 64 h. Optical microscopy specimens sliced from the aged specimens were mechanically polished and then electrically etched in 10N NaOH solution at 9V. Transmission electron microscopy specimens were also sliced from the aged specimens, thinned to 0.06 mm by abrasion on SiC papers and twin-jet electropolished using a mixture of 5% perchloric acid, 25% glycerol, and 70% ethanol in the temperature range of -5 to -10°C and 40V etching potential. The microstructures and microanalyses of the specimens were investigated using a FEI Tecnai G2 T20 transmission mission electron microscopy. In the second experiment, Charpy specimens in which the elongated austenite grains were perpendicular to V-notch were machined in the standard 10mm × 10mm × 55mm dimensions and aged at 475°C for 64h.

After the aging treatment, Charpy impact tests were carried out at room temperature and the specimens were sliced longitudinally for optical microscopy and scanning electron microscopy observation.

To avoid the interference from deformation, the unaged and aged specimens were observed without impact test first. Fig. 1a and Fig. 1b show the OM of the unaged and aged specimens in transverse sections, from which and no difference between their microstructure can be found. Elongated austenite grains showed in Figure 1c. shows the macrographs of the specimens after the Charpy impact tests. The unaged specimen in Figure 3a was heavily deformed with rough failure surface and the impact energy was 278J (Hsieh, Tsai, Chang, & Yang, n.d.).

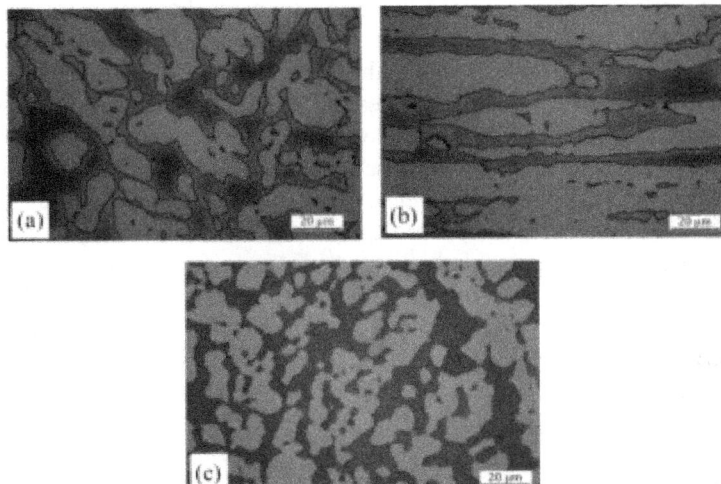

Figure 1. Optical micrographs of the 2205 duplex stainless steel bar: (a) transverse section of the unaged specimen; (b) longitudinal section of the unaged specimen; (c) transverse section of the aged specimen

Table 1. Dislocation densities and average lengths of the unaged and aged δ-ferrite grains

	Dislocation density (m^{-2})	Dislocation length (nm)
Unaged δ-ferrite	$8.3 \pm 2.1 \times 10^{14}$	234 ± 23
Aged δ-ferrite	$8.2 \pm 2.5 \times 10^{14}$	262 ± 36

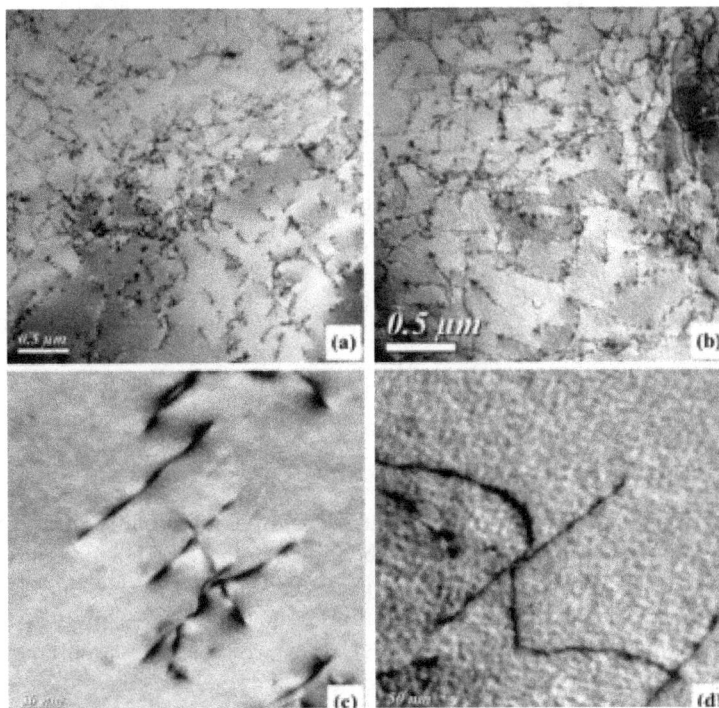

Figure 2. Transmission electron micrographs of the: (a) unaged δ-ferrite; (b) aged δ-ferrite; (c) unaged δ-ferrite with high magnification; (d) aged δ-ferrite with higher magnification

2. Probing Fractal Magnetic Domains on Multiple Length Scales in Nd2Fe14BThe industrial strength ferromagnet, Nd2Fe14B has become a prototypical system for the study of magneticdomain structures. Below the Curie temperature, Tc ¼ 565 K the Nd and Fe moments order ferromagnetically. The crystal-electric field produces a strong magnetic anisotropy with the easy axis along the tetragonal c direction. Below the spin-reorientation temperature, TSR ¼ 135 K, the magnetic structure (and easy axis direction) changes via a second order transition and becomes conelike in which the moments are canted away from the c direction by an angle that increases from 0 at T SR to 28 at 4 k. The moments lie in one of the four symmetry-equivalent f1 1 0 g planes in agreement with calculations of crystal-electric field effects. A multitude of techniques have been used to image magnetic domains at exposed surfaces of Nd2Fe14B, such as Bitter decoration, Kerr microscopy, Lorentz [and holographic transmission electron microscopy, scanning electron, and magnetic force microscopy. For imaged surfaces perpendicular to the c direction, domains have been observed with dimensions between 2–5 m at 4 K (Kim et al., 2005) and about 0:1–0:6 m at room temperature, respectively. These values are very close to the single domain size of about 0:2–0:4 m determined by magnetization measurements on polycrystalline samples manifest smooth, mirrored surfaces and can have volumes as large as 1 cm^3 (Condette, Melcher, & Süli, 2011; Kreyssig et al., 2009).

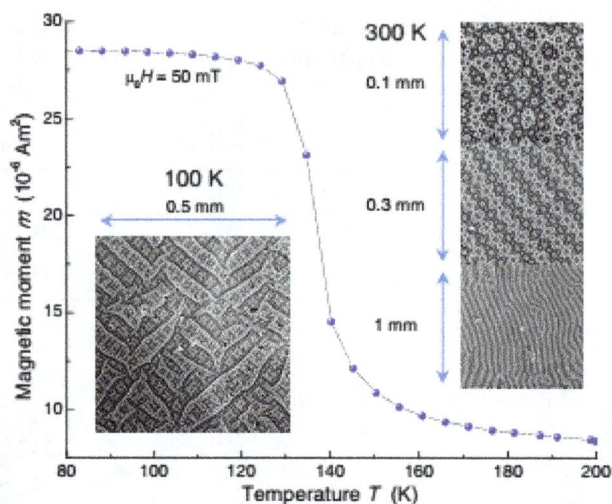

Figure 3. Temperature dependence of magnetization and the magnetic domain patterns Nd2Fe14B single crystal. The magnetization was measured at $\mu_0 H$ =50mT applied along C axis

3. conventional methods of manufacture of high pressure process manifolds in duplex stainless steelshaveemployed3basic approaches.

a) pipe and butt weld outlets

b)extruded branches(including non standardtees)

c)swept outlets.

These methods are all labour intensive manufacturing routes requiring the manufacture of usually, nonstandard seam less or welded pipe of thick section. This then has to be forged or machined prepared to accept the out let sort manifold fabrication. As such, these routes can be associated with high costs and long lead times. Also when duplex and super duplex stainless steel involved considerations of quality of fabrication, distortion due to residual stresses and ease of inspection and interpretation of ND Edatacan become problematical. Many of these difficulties can be avoided if the technique of Hot Isostatic Pressing (HIP) isused for the manufacture of near net shape manifold sections with integral branches and branch connections. Generally, HIPDISDSS can achieve 70 Joules average Charpy "V" notch impact test results at – l0°C which is commensurate with a conventional forging. However, the form of the full impact energy vs temperature transition curve is quite different, since HIP items do not exhibitan impact transition temperature like conventional wrought products, Figure 5. In this case HIP products mirror the behaviour of DISDSS weld metals, where the oxygen content lowers the upper shelf impact energy and extends its range to lower temperatures. This impact transition behavior of HIPD/SDSS is not currently recognised by design 1 fabrication codes which are real based up on the behavior of wrought ferritic or austenitic steels. Hence codes call for high (uppershelf) levels of charpy toughness at minimum design temperature or below dependent upon the thickness compensation factor (Byrne, Spence, Olsen, Houghton, & MCMAHON, 1994; Park & De Cooman, 2014; Udarne & Nerjavnega, 2015).

4. By imserting nano Nd2FeB14 in the HIP manifolds of duplex stainless steel and then making analysis for dislocation denisity we find the following:

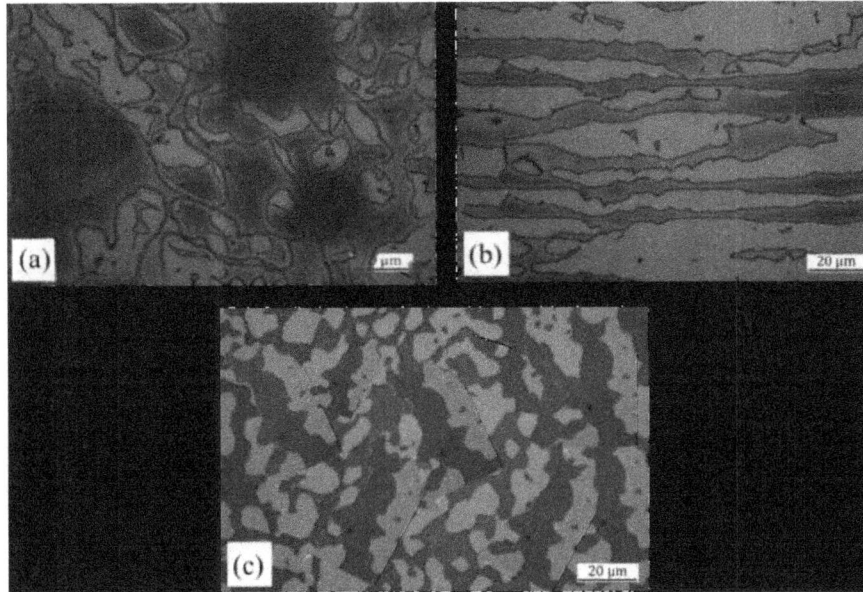

Figure 4. Optical micrographs of the 2205 duplex stainless steel bar after inserting Nd2Fe B14 particles in HIP manifold.(dislocation densities vary considerably) : (a) transverse section of the unaged specimen; (b) longitudinal section of the unaged specimen; (c) transverse section of the aged specimen

And also by applying high magnetic field we have enhanced impact energyat aging from 22-25 J to 45 to 100J depends on the magnetic field as showed in the figure (Udarne & Nerjavnega, 2015; Maehara, Koike, Fujino, & Kunitake, 1983; Hsieh, Tsai, Chang, & Yang, n.d.).

You can see high degree of brightness as we have applied high and intensive magnetic field (Maehara, Koike, Fujino, & Kunitake, 1983; Park & De Cooman, 2014; Frost & Ashby, 1982; Hsieh, Tsai, Chang, & Yang, n.d.).

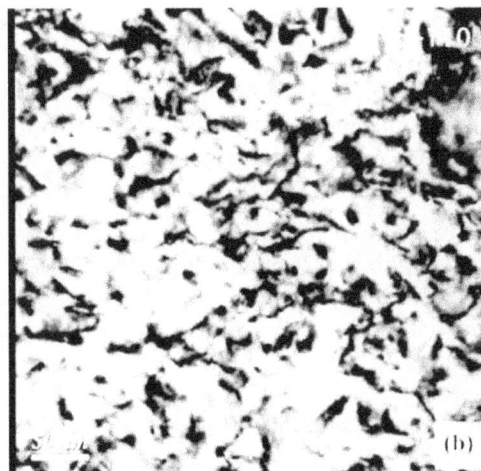

Figure 5. Aged δ-ferrite grain after impact test after inserting Nd2 Fe B14 particles in HIP manifold and the application of high magnetic field

5. Results and Discussions

In Duplex stainless steel 2205, spinodal reactions have bad effects in the mechanical properties of duplex stainless steel. The spinodal reactions have strong relations to the gibbs free energy (Alvarez-Armas & Degallaix-Moreuil, 2009; Park & De Cooman, 2014; Hsieh, Tsai, Chang, & Yang, n.d.).

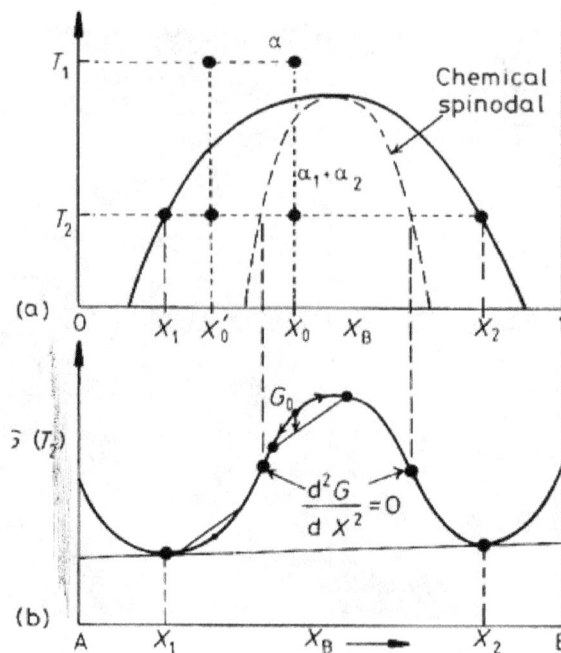

Figure 6. (a)Boldlinedesignatesthe (incoherent)miscibility gap and dashed line the (chemical) spinodal regions (b) Gibbs energy-composition curve at T2

To enhance the mechanical properties of duplex stainless steel and to hinder the separation of gibbs free energy, we must raise the energy. The best treatment to hinder this separation is to apply magnetic field that can raise the thermal content of duplex stainless steel. We can use ferromagnetic material that has special movement in its magnetic domain like Nd2Fe14B. By its fractal movement, we can equalize and raise the gibbs free energy as Nd2Fe14B is permenant magnet in it self (Kreyssig et al., 2009).

Also raising the gibbs free energy, give the austenitic stabalizers their chance to hinder the spinodal separation and hinder the chromium diffusion as we will discuss soonly.

Recently, many companies tend to manufacture manifolds in duplex stainless steel by HIP to enhance the mechanical properties of them.

The key word between the spinodal reactions, Nd2Fe14B and applied magnetic field is the manifold as the cahn hillard equations have special solutions in the manifold and Ising model has special holonomicity charachteristics on the manifold with the canonical singularities. We will show by category theory to establish mathematical relations between them.

By 1982 the latest version of the Fe-Cr phase diagram was that due to Kubaschewski, see figure 2.8a. Later on, Andersson and Sundman (Carpineti & Giglio, 1992) optimizedall the available experimental data up to 1987 and constructed another version of the Fe-Cr phase diagram. This phase diagram was rather differentfrom Kubaschewski's. However, these two did not issue any commentsonthespinodal line; this may be due to the fact that the spinodal is notreallypartofthephase diagram. Figures 2.8a and b also contain results of a study by Dubieland Inden (Roidos & Schrohe, 2013) which was aimed at further characterization of the miscibilitygaplineaswellas eutectoid temperature of the reaction σ→α + α'. They performedlongterm annealing (2 to 11 years at 460, 500 and 510oC) on Fe-Cr binaryalloysof15,20,48 and 70 at.% Cr and determined the chemical compositionofαorα'phases applying Mössbauer Spectroscopy (MS). Starting by a fully ferritic structure they observed that at 500oC no σ phase was formed even after 4 years.

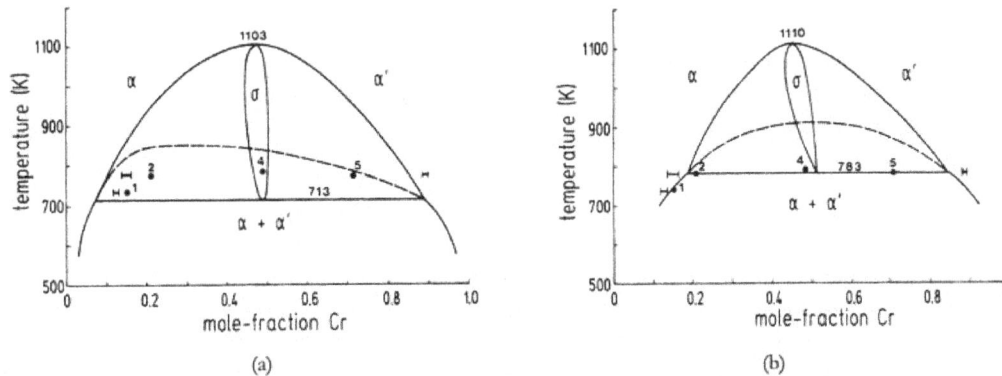

Figure 7. Fe-Cr phase diagram due to (a) Kubaschewski and (b) Andersson and Sundman. Numbers in the figures denote the initial composition of aged alloys by Dubiel and Inden. Horizontal bars correspond to the compositionrange of α and α' phases measured by means of Mössbauer spectroscopy. These bars denote the miscibility gap boundary (Adams, Olubambi, Potgieter, & Van Der Merwe, 2010; Kim et al., 2005; Udarne & Nerjavnega, 2015; Maehara, Koike, Fujino, & Kunitake, 1983; Park & De Cooman, 2014; Frost & Ashby, 1982)

Pattern formation processes due to phase separation of binary mixtures can conveniently be modelled by means of nonlocal Cahn–Hilliard equations. By this we mean H₋₁ gradient flows associated with functionals of Ginzburg–Landau type, which may include a dipolar interaction term. In a spatially periodic setting, i.e., using the three-dimensional torus $T = 2\pi R^3 / Z^3$ as spatial domain, such functionals typically have the form

During the aging treatment, the spinodal nanostructure was formed in the δ-ferrite. The spinodal structure would limit the lengths and movements of the dislocations during deformation. The function ghom(c) is the free energy per molecule of a homogeneous system of uniform concentration c, which is non-convex in systems exhibiting phase separation. The gradient penalty tensor **K** is assumed to be a constant independent of **x** and c. Then the diffusional chemical potential(in energy per molecule) is the variational derivative of Gmix (Burch & Bazant, 2009)

$$G_{mix}[c] = \int_V [g_{hom}(c) + .5(\nabla c).(K \nabla c)]\rho dV$$

The function ghom(c) is the free energy per molecule of a homogeneous system of uniform concentration c, which is non-convex in systems exhibiting phase separation. The gradient penalty\tensor **K** is assumed to be a constant independent of **x** and c. Then the diffusional chemical potential(in energy per molecule) is the variational derivative of Gmix,

$$\mu(X,T) = \frac{\partial g}{\partial c} - \nabla.(K \nabla c)$$

The Cahn-Hilliard equation is a phase-field or diffuse interface equation which is mainly used to model phase separation of a binary mixture, e.g. a two-component alloy, but many other applications are encountered., one finds the equation stated in various forms. We shall consider here the version As usual, we model a manifold with conical singularities by a manifold with boundary B of dimension n + 1, n ≥ 1, endowed with a conically degenerate Riemannian metric. On one hand, working on a manifold with boundary simplifies the analysis; on the other hand, the degeneracy of the Riemannian metric entails that geometric operators such as the Laplacian show the typical degeneracy they have on spaces with conic points in Euclidean space (Roidos & Schrohe, 2013; Vertman, 2016; Bahuaud & Helliwell, 2011). Permanent magnet is a ferromagnetic material that has a magnetization even if there is no external applied field. Using Ising model we can explain their properties, and the Ising model has special charchteristic in manifolds. Nd-Fe-B magnets has special properties, Fractal Magnetic Domains on Multiple Length Scales in Nd2Fe14BThe industrial strength ferromagnet, Nd2Fe14B has become a prototypical system for the study of magneticdomain structures. The Fractal movement is described by fameous Ginzburg Landau equations (Byrne, Spence, Olsen, Houghton, & MCMAHON, 1994).

$$G(z(x)) = G_0 + \int_W G(z(x), \nabla z(\mathrm{x})) \, \mathrm{dV} \tag{1}$$

where F(Z(x), ∇Z(x)) is the free energy density; $\nabla Z = \partial Z / \partial x$.. For the Ginzburg-Landau potential, this density is F (Z(x), ∇Z(x)) = 1. Here D is a fractal mass dimension of the fractal medium, and dV_D is an element of the D-dimensional volume:

$$dV_D = C_3(D, x) \, dV_3 \tag{2}$$

The location of the transition is a function of the temperature, field amplitude and frequency. A finite-size scaling analysis of large-scale Monte Carlo simulations of the kinetic Ising model in an oscillatory field has shown that the dynamic phase transition is in the same universality class as the equilibrium Ising model. A result confirmed in a recent study of a time-dependent Ginzburg-Landau model in an oscillatory field. Pattern formation processes due to phase separation of binary mixtures can conveniently be modeled by means of nonlocal Cahn–Hilliard equations. By this we mean H^{-1} gradient flows associated with functionals of Ginzburg Landau type, which may include a dipolar interaction term. Ohta-Kawasaki model, including the limiting case of $\sigma = 0$ which corresponds to the classical Cahn–Hilliard model. However, the numerical method proposed here can be adapted to more general functions ˆσ in (1.1), including, e.g., dipolar stray-field interaction in magnetic garnet films (Condette, Melcher, & Süli, 2011)

$$E(u) = \frac{1}{2} \int_{T^3} \varepsilon^2 (\nabla u)^2 + \frac{1}{2}(1-u^2)^2 \, dx + \frac{1}{2} \sum_{k \in \mathbb{Z}^3 / \{0\}} \sigma(k) u^2(k)$$

Second, we shall consider $\frac{1}{2} \sum_{k \in \mathbb{Z}^3 / \{0\}} \sigma(k) u^2(k) = \sigma \int_{T^3} (\nabla \varphi)^2 \, dx$ and E(u) becomes the Ohta–Kawasaki energy. One

of the most remarkable and possibly the most accurate experimental verification for the universal glassy dynamics theoretically predicted for disordered elastic systems has been the measurement of the ultra-slow creep motion of magnetic domain walls in ultrathin. Nd-Fe-B magnets ferromagnetic films with perpendicular anisotropy driven

by a very small (well below the depinning threshold defined below) applied magnetic fields.

$V \sim \exp(-\frac{U_c}{k_B T}(\frac{H_c}{H})^\mu)$, where Uc gives a characteristic energy scale in the creep regime and the applied field is

$H \ll H_c$ with Hc the so called depinning threshold. oth Uc and Hc are material dependent parameters and increase with the strength of the disorder. elastic manifold and its equilibrium roughness exponent ζeq through the relation.

$\mu = \dfrac{d - 2 + 2\zeta eq}{2 - \zeta eq}$ (Frost & Ashby, 1982). The roughness exponent ζeq measures the rate at which the interface

width grows with its linear size w ~ Lζeqat equilibrium and is in turn universal: it depends only on d, on the nature of the disorder correlations and on the short-ranged character of elastic interactions. This makes this experimental system a paradigmatic exampleof the universal physics predicted for elastic manifolds weakly pinned by random impurities. The theory for driven disordered elastic systems applied to magnetic domain walls also predicts two additional dynamical regimes as a function of H.26 For largefields H $\gg$ Hc (Vertman, 2016; Bahuaud & Helliwell, 2011; Elliott & Songmu, 1986).

In general permanent magnets are deeply explained by Ising model which exhibits high holonomicity behaviour under the scope of solobev spaces. Y. Yang generalized this definition to Finsler manifolds using Busemann volume form and Osculating Riemannian metric. We use this definition of Sobolev spaces in Finsler manifolds to prove two density theorems. Historically, one the significant density theorems is proved by S. B. Myers in 1954 for compact Riemannian manifolds and then in 1959 by M. Nakai for finite-dimensional Riemannian manifolds. Next in 1976 T. Aubin has investigated density theorems on Riemannian manifolds, Recently, in 2009 an extension of Myers-Nakai theorem to infinite-dimensional, complete Riemannian manifolds]. A similar result for the so-called finite dimensional Riemann-Finsler manifolds is given in in 2010. Next in 2011 the Myers-Nakai theorem is extended to the Finsler manifolds of class C^K. As usual, we model a manifold with conical singularities by a manifold with boundary B of dimension n + 1, n ≥ 1, endowed with a conically degenerate Riemannian metric. On one hand, working on a manifold with boundary simplifies the analysis; on the other hand, the degeneracy of the Riemannian metric entails that geometric operators such as the Laplacian show the typical degeneracy they have on spaces with conic points in Euclidean space. We measure smoothness in terms of weighted Mellin-Sobolev spaces $H_p^{s,\gamma}(B)$. Here s is a smoothness index, γ a weight, and 1 < p < ∞. They coincide with the usual L^p (Roidos & Schrohe, 2013; Vertman, 2016) -Sobolev spaces away from the singularities. As long as the integrands one considers are of a basically quadratic nature, hilbert manifolds of maps belonging to some Sobolev space L^p are the natural manifolds in which to formulate the problems. However for more general integrands it becomes necessary to use Banach manifolds, for example manifolds of maps belonging to one of the more general Sobolev spaces L^p. Because of the volume-constraint, we shall be mainly concerned with situations where test functions have zero integralIn this case the following dual estimate will be crucial: if uN ∈ XN and vN ∈ XN, their characterizing the homogeneous Sobolev–Slobodetskiĭ space H° s(T3) as the closure, with respect to the induced homogeneous $H(T^3)$ we use the Yang's method to define certain Sobolev spaces onFinsler manifolds. Let dvF be the Busemann volume form, f ∈ C^∞ (M) and denote $L^p(M) = \{f : M \to R\}$ where R is measurable. Let us consider the vector space C(kp) of C∞ functions φ, such that $\nabla\phi \in L^p(M)$

Let J be a nonnegative, real-valued function, on the space of C^∞ functions with compact support on IRn, denoted by C^∞ (Rn) and having properties :

$$J(x) = 0, x \geq 1$$

$$\int_{R^n} J_c(x)dx = 1$$

Chrome percentage varies with time of heating and also the magnetic domains have many fractal shapes. To emphaise conditions in all directions. $\lim D(b) = 1$, for K<Kc(T>Tc), $D(b) =1.875$, for K=Kc(T= Tc), $\lim D(b) = 2$ K ≥Kc(T>Tc) is $J / k_B T$, $^{h}k_B^\infty$ is boltzman constant, Tis the temperature, where **J** defined by the following Hamilton (Falconer, 2004; Ito & Suzuki, 1987). $H = -l / 2; SiSj \left(Si = \pm l\right)$. Also, heat kernels of dislocations on finsler manifolds details.

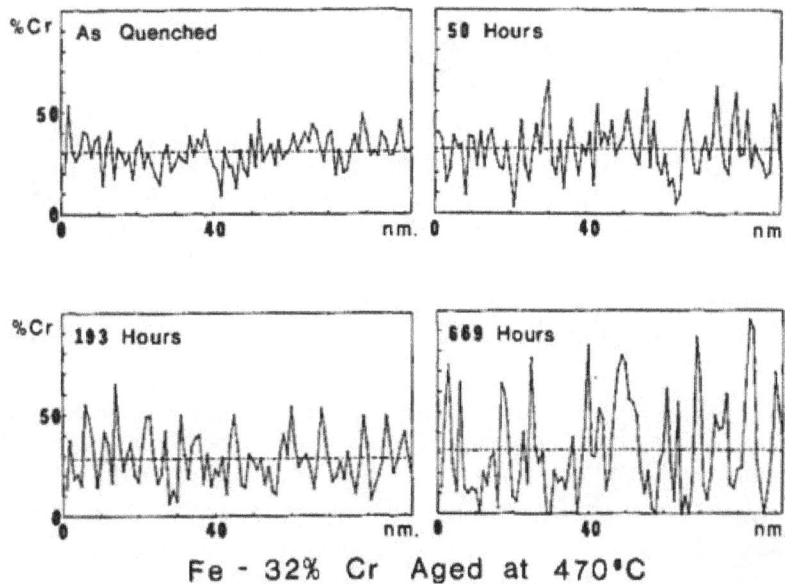

Figure 8. Concentration profiles of Fe-32at.%Cr samples in as-quenched and aged states. Numbers on the upper left of diagrams show the aging time in hours. Each data point in the diagrams corresponds to mean chemical composition of a volume element of the material 0.8nm thick and 1nm in diameter consisting of 30-50 atoms

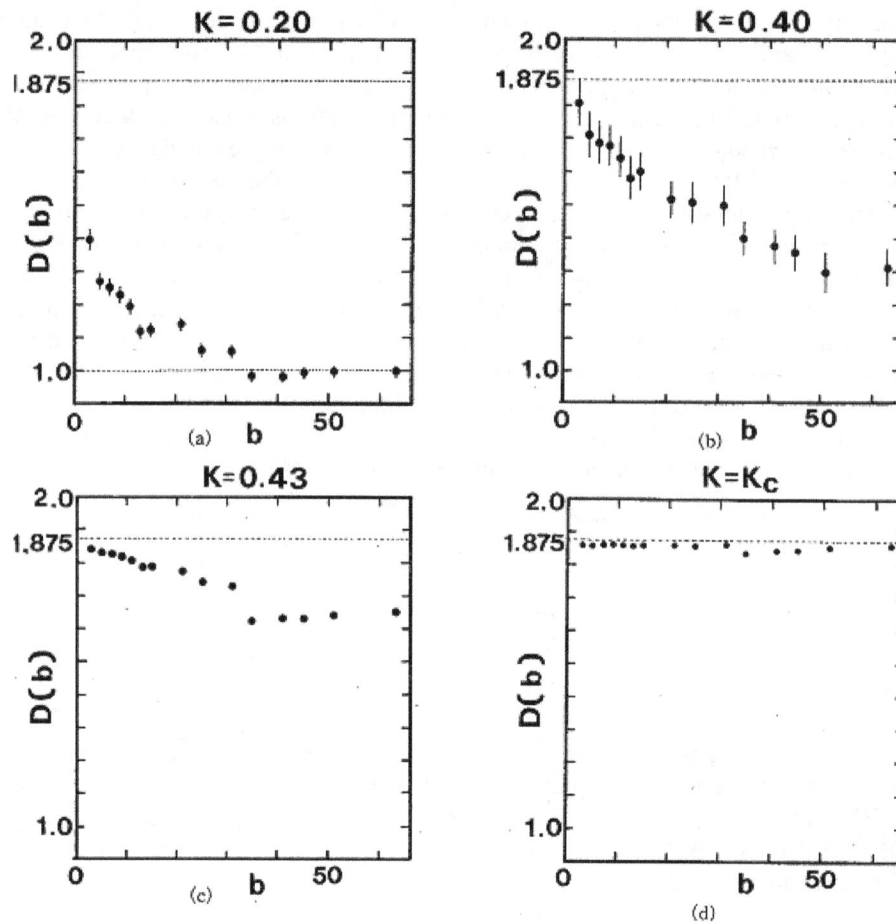

Figure 9. Values D(b) defined by equation(3) are shown as function of scale factor b for several temperatures. The system size 64×64 (a)K=.2 when the sacle b is large the values D(b) approaches to unity, which corresponds to random percolation case.(b)K=.4(c)K=.43(d)K=kc=.4407 D(b) converges to the expected value 1.875 irrespective of the used scale factor (Ito & Suzuki, 1987)

So, we can use the category theory to express about these finite objects as we have mentioned 4 shapes of fractals and four cr percentage with heating times. But the best way to regulate these parameters is to use the Lusternik-Schnirelman category as cahn hillard and magnetic fields are strongly related to solobev spaces as mentioned above.

DEFINITION. The Lusternik-Schnirelman category of A in X, cat(A: X), is the least integer n such that A can be covered by n closed subsets of X each of which is contractible in X. If no such integer n exists we put cat (A; X) = ∞ . We define cat(X) = cat(X; X)

The Lusternik-Schnirelman category is used for finslers spaces especially. As we mentioned above, solobev space where Ising model and cahn hillard equations are highly related to it. Solobev space is suppoed to be defined on finsler manifold (Antonelli, Ingarden, & Matsumoto, 2013; Bao, Chern, & Shen, 1996).

MAIN THEOREM OF LUSTERNIK-SCHNIRBLMANTHEORY: Let M *be a complete* C^2 *Finsler manifold without boundary,* $f : M \rightarrow \mathbb{R}$ *as* C^2 *map. Let K denote the set of critical points of f and for a E* R *let K, = K of* -'(a) and M= $f(-\infty, a)$ .Assume for each a E R that a is not a limit point of $\mathrm{int}\, f(k - k_\alpha)$

$$c_m(f) = Inf\left\{a \sim R \,\|\, cat\left(M,; M_\alpha\right) \geq m\right\} \text{ then } c_m(f) \leq c_{m+1}(f), \ 0 \leq m \leq n \leq cat(M), \ c_m(f) \leq \infty \ \text{ hence if}$$

M is connected then $\dim K_c \geq n - m$ (Bejancu & Blair, 1990; Stanley, 2002).

Gibbs free energy is hindered as it can not be used to express about all these parameters, so we the best solution is to use category theory.

So we can write the entire thermal content of the system as $C_T(M) = D(c(m))E(c(n))$

$D(c(m))$ is fractal measurement depends on c(m) as defined in main theorem of Lusternik Schnirbl main theory

$E(c(n))$ is dipolar field interaction depends on c(n) as defined in main theory of lusternik schinirby main theorem

$C_T(M)$ is a thermal term of the entire system depends on fractal movement of magnetic domain as the morophology of the system depends on the applied magnetic field showed by figure (red spots) and also $C_T(M)$ depends on dipolar field interaction, it is a function in M finsler manifold as cahn hillard spinodal is related to manifolds with canonical singularities.

It has been showed the morophology of duplex stainless steel at aging in the case of application of high magnetic field figure.

6. Conclusion

By inserting Nd2FeB14 magnetic thin films in the HIP manifolds of duplex stainless steel, there will be enhancing in the mechanical properties of duplex stainless steel 2205 and the the impact energy will be raised to high levels. Impact energy will be enhanced as the spinodal reaction will be hindered. Pattern formation processes due to phase separation of binary mixtures can conveniently be modeled by means of nonlocal Cahn–Hilliard equations. By this we mean H^{-1} gradient flows associated with functionals of Ginzburg Landau type, which may include a dipolar interaction term. Ohta-Kawasaki model, including the limiting case of σ = 0 which corresponds to the classical Cahn–Hilliard model. The Lusternik-Schnirelman category has been used to express about all the prameters and the objects which are used to enhance the mechanical properties of duplex stainless steel 2205. The Lusternik-Schnirelman category is used to show how categories behave on finsler manifold. Finsler manifolds are strongly related to solobev spaces in which cahn hillard and Ising model show great charachterisics.

References

Adams, F. V., Olubambi, P. A., Potgieter, J. H., & Van Der Merwe, J. (2010). Corrosion resistance of duplex stainless steels in selected organic acids and organic acid/chloride environments. *Anti-Corrosion Methods and Materials, 57*(3), 107-117. http://dx.doi.org/10.1108/00035591011040065

Akbar-Zadeh, H. (1988). Sur les espaces de Finsler à courbures sectionnelles constantes. *Acad. Roy. Belg. Bull. Cl. Sci.(5), 74*(10), 281-322.

Alvarez-Armas, I., & Degallaix-Moreuil, S. (Eds.). (2009). *Duplex stainless steels.* John Wiley & Sons.

Antonelli, P. L., Ingarden, R. S., & Matsumoto, M. (2013). *The theory of sprays and Finsler spaces with applications in physics and biology* (Vol. 58). Springer Science & Business Media.

Bahuaud, E., & Helliwell, D. (2011). Short-time existence for some higher-order geometric flows. *Communications in Partial Differential Equations, 36*(12), 2189-2207.

Bao, D., Chern, S. S., & Shen, Z. (1996). Preface for "Finsler geometry over the reals" (pp. 3–13).

Basyigit, A. B., & Kurt, A. (2014). Corrosion Properties and Impact Toughness of 2205 Duplex Stainless Steel after TIG Welding*. *Materials Testing, 56*(10), 786-794. doi: 10.3139/120.110628

Bejancu, A., & Blair, D. E. (1990). *Finsler geometry and applications.* New York: Ellis Horwood.

Bidabad, B., & Shahi, A. (2013). On Sobolev spaces and density theorems on Finsler manifolds. *arXiv preprint arXiv:1310.8027.*

Borceux, F. (1994). *Handbook of Categorical Algebra I: Basic Category Theory (Encyclopedia of Mathematics and Its Applications).* Cambridge University Press.

Bucur, I., Deleanu, A., & Hilton, P. J. (1968). *Introduction to the Theory of Categories and Functors* (Vol. 13, p. 13). London: Wiley.

Burch, D., & Bazant, M. Z. (2009). Size-dependent spinodal and miscibility gaps for intercalation in nanoparticles. *Nano letters, 9*(11), 3795-3800. DOI: 10.1021/nl9019787 ·

Byrne, G., Spence, M. A., Olsen, B., Houghton, P. J., & MCMAHON, J. (1994). Advantages of hot isostatic pressing (HIP) production routes for process manifolds, Duplex Stainless Steels' 94. In *Proc. of Int. Conf. on Duplex Stainless Steels, Glasgow, Scotland, Cambridge TWI, paper* (Vol. 19).

Campbell, P. (1996). *Permanent magnet materials and their application.* Cambridge University Press. DOI: http://dx.doi.org/10.1017/CBO9780511623073

Carpineti, M., & Giglio, M. (1992). Spinodal-type dynamics in fractal aggregation of colloidal clusters. *Physical review letters, 68*(22), 3327. DOI:10.1103/PhysRevLett.68.3327

Chiu, L. H., Hsieh, W. C., & Wu, C. H. (2003). Cooling rate effect on vacuum brazed joint properties for 2205 duplex stainless steels. *Materials Science and Engineering: A, 354*(1), 82-91. http://dx.doi.org/10.1016/S0921-5093(02)00911-5

Coey, J. M. (2010). *Magnetism and magnetic materials* (pp. 128-174). Cambridge University Press. http://dx.doi.org/10.1017/CBO9780511845000

Condette, N., Melcher, C., & Süli, E. (2011). Spectral approximation of pattern-forming nonlinear evolution equations with double-well potentials of quadratic growth. *Mathematics of Computation, 80*(273), 205-223.

Dos Santos, D. C., & Magnabosco, R. (2016). Kinetic Study to Predict Sigma Phase Formation in Duplex Stainless Steels. *Metallurgical and Materials Transactions A, 47*(4), 1554-1565. http://dx.doi.org/10.1007/s11661-016-3323-z

Elliott, C. M., & Songmu, Z. (1986). On the Cahn-Hilliard equation. *Archive for Rational Mechanics and Analysis, 96*(4), 339-357. DOI: 10.1007/BF00251803

Falconer, K. (2004). *Fractal geometry: mathematical foundations and applications* (pp. 44-66). John Wiley & Sons. doi : 10.1002/0470013850

Frost, H. J., & Ashby, M. F. (1982). Deformation mechanism maps: the plasticity and creep of metals and ceramics.

Hsieh, Y. C., Tsai, Y. T., Chang, Y. L., & Yang, J. R. (n.d.). The interaction between dislocations and the spinodal nanostructure in a 2205 duplex stainless steel.

Huang, C. S., & Shih, C. C. (2005). Effects of nitrogen and high temperature aging on σ phase precipitation of duplex stainless steel. *Materials Science and Engineering: A, 402*(1), 66-75. http://dx.doi.org/10.1016/j.msea.2005.03.111

Ito, N., & Suzuki, M. (1987). Fractal configurations of the two-and three-dimensional Ising models at the critical point. *Progress of theoretical physics, 77*(6), 1391-1401.

Kangas, P., & Newman, M. (1998). Performance of duplex stainless steels in organic acids. *Anti-Corrosion Methods and Materials, 45*(4), 233-242. http://dx.doi.org/10.1108/00035599810223698

Kim, S. C., Zhang, Z., Furuya, Y., Kang, C. Y., Sung, J. H., Ni, Q. Q., ... & Kim, I. S. (2005). Effect of Precipitation of. SIGMA.-Phase and N Addition on the Mechanical Properties in 25Cr-7Ni-4Mo-2W Super Duplex Stainless Steel. *Materials transactions, 46*(7), 1656-1662.

Kreyssig, A., Prozorov, R., Dewhurst, C. D., Canfield, P. C., McCallum, R. W., & Goldman, A. I. (2009). Probing fractal magnetic domains on multiple length scales in Nd 2 Fe 14 B. *Physical review letters, 102*(4), 047204.

Le Calvez, P. (2006). From Brouwer theory to the study of homeomorphisms of surfaces. In *International Congress of Mathematicians* (Vol. 3, pp. 77-98). doi:10.4171/022-3/4

Liu, X. Y., Xia, K. D., Niu, J. C., Xiang, Z., Yan, B., & Lu, W. (2015). Effects of Heat Treatment on Microstructure and Pitting Corrosion Resistance of 2205 Duplex Stainless Steel. *International Journal of Electrochemical Science, 10*, 9359-9369.

Maehara, Y., Koike, M., Fujino, N., & Kunitake, T. (1983). Precipitation of σ phase in a 25Cr-7Ni-3Mo Duplex Phase Stainless Steel. *Transactions of The Iron and Steel Institute of Japan, 23*(3), 240-246. DOI: 10.2355/isijinternational1966.23.240

May, J. P. (1999). *A concise course in algebraic topology.* University of Chicago press.

Palais, R. S. (1966). Lusternik-Schnirelman theory on Banach manifolds. *Topology, 5*(2), 115-132. DOI: 10.1016/0040-9383(66)90013-9

Park, H., & De Cooman, B. C. (2014). Creep Deformation of Type 2205 Duplex Stainless Steel and its Constituent Phases. *ISIJ international, 54*(4), 945-954. http://doi.org/10.2355/isijinternational.54.945

Phillips, N. S., Chumbley, L. S., & Gleeson, B. (2009). Phase transformations in cast superaustenitic stainless steels. *Journal of Materials Engineering and Performance, 18*(9), 1285-1293.

Pickering, F. B. (1976). Physical metallurgy of stainless steel developments. *Int. Met. Rev., Dec. 1976, 21,* 227-268.

Roidos, N., & Schrohe, E. (2013). The Cahn-Hilliard equation and the Allen-Cahn equation on manifolds with conical singularities. *Communications in Partial Differential Equations, 38*(5), 925-943. DOI: 10.1080/03605302.2012.736913.

Rudraraju, S., Van der Ven, A., & Garikipati, K. (2015). Mechano-chemical spinodal decomposition: A phenomenological theory of phase transformations in multi-component, crystalline solids. *arXiv preprint arXiv:1508.05930.*

Stanley, D. (2002). On the Lusternik-Schnirelmann category of maps. *Canadian Journal of Mathematics, 54*(3), 608-633. http://dx.doi.org/10.4153/CJM-2002-022-6

Udarne, P., & Nerjavnega, I. D. (2015). Impact-Toughness Investigations of Duplex Stainless Steels. *Materiali in tehnologije, 49*(4), 481-486. doi:10.17222/mit.2014.133

Vertman, B. (2016). The biharmonic heat operator on edge manifolds and non-linear fourth order equations. *Manuscripta Mathematica, 149*(1-2), 179-203. 10.1007/s00229-015-0768-0

Warren, A. D., Harniman, R. L., Guo, Z., Younes, C. M., Flewitt, P. E. J., & Scott, T. B. (2016). Quantification of sigma-phase evolution in thermally aged 2205 duplex stainless steel. *Journal of Materials Science, 51*(2), 694-707. http://dx.doi.org/10.1007/s10853-015-9131-9

10

Development of Electrically Assisted Rapid Heating for Metal Forming of Hot-Stamping Process

Mahmudun Nabi Chowdhury[1] & Dinh Thi Kieu Anh[1]

[1] School of Mechanical Engineering Department, University of Ulsan, Ulsan, South Korea

Correspondence: Mahmudun Nabi Chowdhury, School of Mechanical Engineering Department, University of Ulsan, Ulsan 44610, South Korea. E-mail: noyoncuet@gmail.com

Abstract

Two different concepts of electrically assisted (EA) rapid heating of Al–Si coated hot-stamping steels are compared. In "along the surface" EA heating (or simply EA surface heating), the electric current is simply applied to a specimen by clamping the each end of the specimen length with a set of flat rectangular electrodes. In "through the thickness" EA heating (or simply EA thickness heating), the electric current is applied to a specimen by attaching a set of electrodes with multiple contact points on upper and lower surfaces of the specimen. While the EA surface heating generally requires a shorter heating time due to a higher electrical resistance in the length direction, the EA thickness heating also may provide a technical advantage that the heating area can be more easily configured in a case of partial austenization.

Keywords: electrically assisted, hot-stamping, partial austenization, rapid heating, intermetallic

1. Introduction

Due to the demand for weight reduction of vehicle, improved safety and crashworthiness qualities, increasing hot stamping parts are used for automobile structural components from ultra-high strength steel (Chao, Yisheng, XiaoWei, Bin, & Jian, 2012). As a higher weights leads to increased fuel consumption the body in where must lighter to compensate the weight of additional components. Therefore local heating is applied on those components of automotive body where a high strength or stiffness is necessary. In the conventional hot stamping forming process, the ultra-high strength steel is heated up to the austenization temperature on a heating furnace, then formed and quenched after being transferred into a die during dwell stage (Karbasian & Tekkaya, 2010).

It has been reported that the full martensitic parts are in nature of high strength (1000 – 1500 MPa) and poor ductility generally less than 5% (Zhu, Zhang, Li, Wang, & Ye, 2011). Therefore, such types of high strength steel but low ductile parts are not suitable especially for energy absorbing parts of automotive vehicles (Wang, & Liu, 2014). However, the safety improvement of vehicles and the compatibility of mechanical properties have been reported that various mechanical properties are needed in different regions or joints of the same parts. Such properties are similar as tailored properties (Liang, Wang, & Liu, 2014). This can be done by applying different temperatures on the blank. The automotive part as A-pillar and B-pillar which can be formed by such tempering and meet the crashworthiness requirement of roof and side.

Many scholars have carried out a series of studies on how to realize the tailored mechanical properties of hot stamping parts. Recently, efforts have been made to implement the partially heating and forming of the blank (referred as tailored tempering). To balance the mechanical properties of metal many scholars have given their own different ways of efforts. Some scholars focused on controlling the cooling rate during local heating (Mori, Maeno, & Mongkolkaji, 2013). According to Hein and Wilsius (2008) have partially prevented quenching by reducing the cooling rate with partially heated tools. On the other hand, Erturk et al. (2011) have predicted hardness distribution of tailored tempered blanks by finite element simulation using a thermo-mechanical metallurgical model. However, it is difficult to heat the target area of blank and in this case tailored tempering is really helpful. Mori, Maki, and Tanaka (2005) have developed warm and hot stamping process of ultra-high strength steel using resistance heating. Mori, Maeno, and Fukui (2011) have applied resistance heating to heat the side wall of a cup between two electrodes to decrease flow stress. Mori, Maeno, and Fuzisaka (2012) have improved the quality of sheared edges punching process of ultra-high strength steel sheets by means of electrical resistance heating. The rapid resistance heating is applied to the heating of a local zone by passing current through the local zone.

In this present study, electrically assisted (EA) partial thickness heating process by using multi contact points of electrodes in hot stamping was developed to produce steel parts having strength distribution and intermetallic behaviour of Al-Si coated steel. In this process blank was partially heated to the austenite and was subsequently hot stamped. Only the heated region of the blank was hardened.

2. Experimental Setup

2.1 Materials

In this paper, the studied high strength steel belongs to specialized for hot stamping process, named BP4120B (HYSCO Steel, Korea). The thickness of the specimen was used in experiment was 1.0 mm. 1 mm thickness hot stamping material which was hot dip coated with a type 1 aluminized coating (87% Al - 15% Si) on both side was used. This type of coating is widely used for hot stamping material in order to prevent oxidation and decarburization during heating process.

2.2 Experimental Process

The experiments of ultra-high strength hot-stamping steel of rectangular blanks were implemented on the local resistance heating system by multiple points of electrode. Specimens were prepared as the size of $65 \times 95 \times 1$ mm^3. Custom made jig used for this experiment. Jig's upper plate was automatically moves by using hydraulic press. Figure 1 (a) and (b) showed the full system and close view of specimen set up with electrodes respectively. Two bus-bars were used as high electric current needs to flow over the specimen from the transformer to reach the high temperature about $900 \sim 1000°C$. Also two rectangular shapes ($50 \times 80 \times 10$mm^3) custom made electrodes with multi contact points designed/made to heat the blank locally for high temperature (Figure 2). The distance among the tip points of the electrode was 15 mm each. The tip point was not sharp actually rather it has flat shape at tip of each point to observe the flow of thermal effect over specimen. The schematic of experimental electrical connection is shown in Figure 3. For the electrically assisted rapid heating experiment under a non-continuous electrical electric current was generated by Vadal SP-1000U welder (Hyosung, South Korea) with a programmable controller.

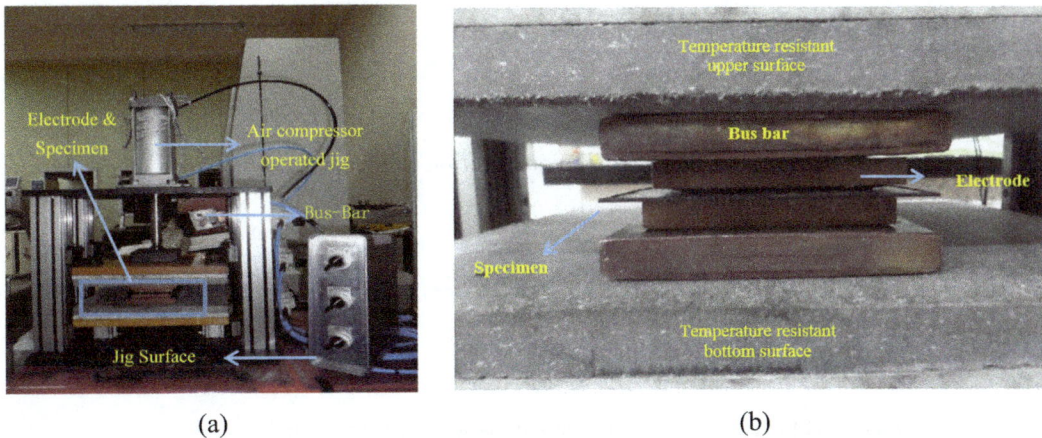

| (a) | (b) |

Figure 1. (a) Experimental setup (b) close view of experimental setup (electrode and specimen)

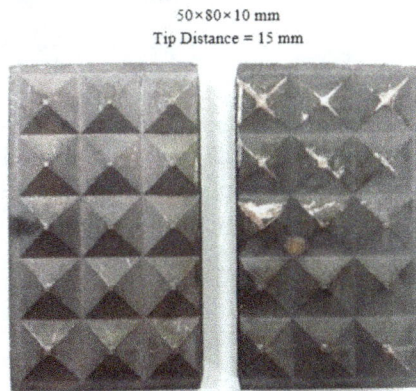

Figure 2. Multi contact points of electrode

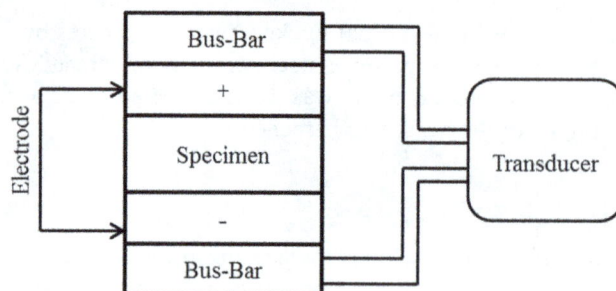

Figure 3. Schematic of electrical current connection

During experiment electrode's set up of both side of specimen was not parallel. The setup is schematically shown in Figure 4. By setting up electrodes of both side of specimen according to the Figure 3 we observed thermal flow through the point contacts. Temperature profile provides uniform temperature over specimen. A Multi pulse of electric current was used to heat the specimen locally. For electrically assisted rapid heating the temperature of the specimen through thickness was measured by a FLIR-T621 infra-red thermal imaging camera (FLIR, Sweden). Special insulation was used in jig so that there was no heat energy loss during experiment. At the starting of the experiment the room temperature was 12°C.

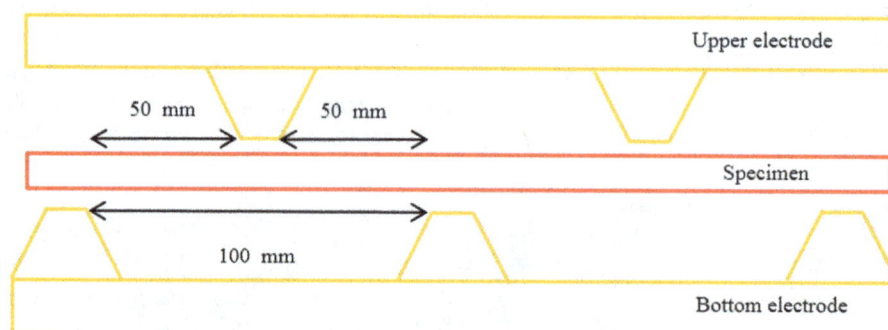

Figure 4. Electrode points position over specimen

2.3 Temperature Calculation

Due to passage of electric current through a specimen, the temperature is generated. The amount of heat released is proportional to the square of the current such that:

$$Q = I^2 R t$$

The temperature increment (ΔT) is calculated from the adiabatic approximation:

$$Q = I^2 R t = mc\Delta T$$

$$\Delta T = \frac{I^2 R t}{mc}$$

Where, ΔT (^{0}C) is the temperature increment of specimen, I (A) is the current traveling through the specimen, R is resistance of specimen, t (sec) is the duration of applying current into specimen, m (kg) is the quality of specimen, c (J/Kg.K) is the specific heat of specimen. The concept of electrically assisted local rapid heating was examined by employing different electrical pulse with a fixed amount of current and duration, as listed in Table 1.

Table 1. Experimental Parameters

Size of electrode (Tip distance)	Current (A)	Number of Pulses
15 mm	3000	17

3. Results

3.1 Temperature

The temperature rising rate of the blank by electrically assisted local heating is shown in Figure 5. Heating the blank to austenite temperature about 900°C required very short period by using multiple contact points of electrode heating. During heating multi pulsed electric current was applied. At certain temperature around 620°C holding time was used for cooling. Holding time was around 10 seconds. By using the holding time temperature was spread over the specimen evenly as well as Al-Si intermetallic was increased. Intermetallic is useful to protect from the oxidation and decarburization of Al-Si coated steel.

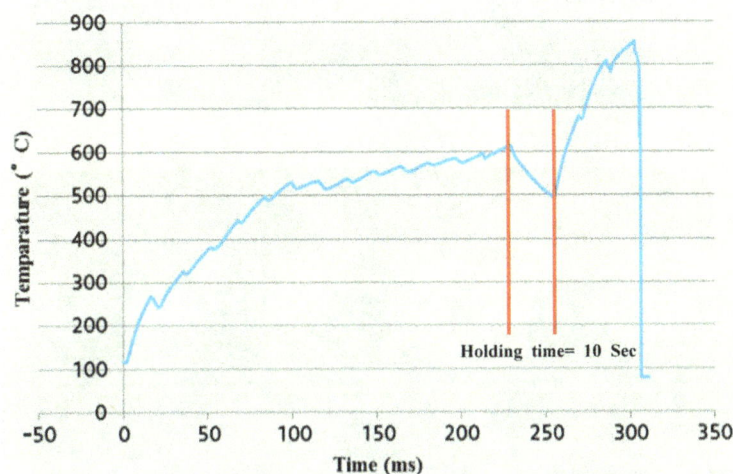

Figure 5. Temperature profile by multi pulsed electric heating

3.2 Intermetallic Developement

By using holding time during rapid heating of Al-Si coated steel at room temperature, the Al matrix and the Si distribution cannot be distinguished because of their equivalent atomic weights. Comparing with base materials, the intermetallic layers of heated specimens are thicker due to the reaction of the coating with the substrate steel. As can be summarized in Figure 6 that electrically assisted (EA) heating specimen achieved the highest result of 5.04 ± 0.66 μm of intermetallic layer while it was 3.75 ± 0.13 μm in case CH and 3.24 ± 0.58 in case of induction heating. Figure 6 (a) and (b) show the OM view of development of intermetallic after heating the blank with respect to no heat treatment. Use of electric current for heat treatment obtained result in intermetallic evolution which includes thin layer of Fe_2Al_5 phase constituent close to the Fe substrate and thick layer of Fe_4Al_{13} & $FeAl_{13}$ phase constituents.

3.3 Temperature Distribution

The steady temperature variation in the blank is shown in Figure 7. The temperature of the blank fluctuated due to heat generation in contact points of electrodes by multiple pulses of electric current. As we see in Figure 7, temperature generated at the initial stage of heating was not uniform over the blank. We marked two points (A, B) on thermal image of different stages of heating and analyzed the heat distribution. At initial stage Figure 7 (a) (300°C) between two points we can see unsteady temperature distribution of heating that is marked with red box. Furthermore, in Figure 7 (b) higher temperature (470°C) the unsteady temperature region was reduced due to heat generation and spreading over the blank. Finally, at 900°C in Figure 7(c) we see the uniform temperature distribution of blank.

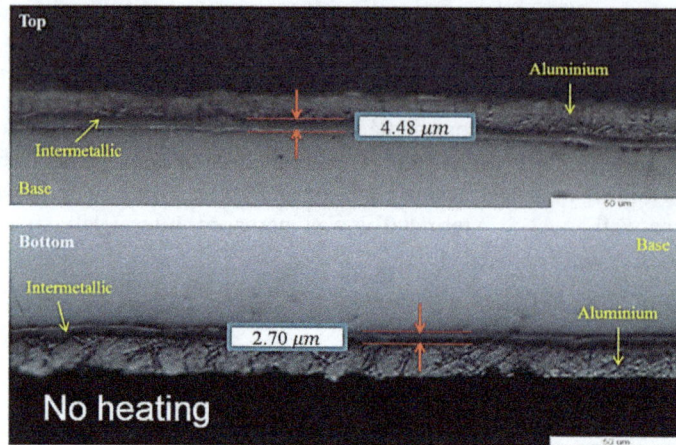

(a)

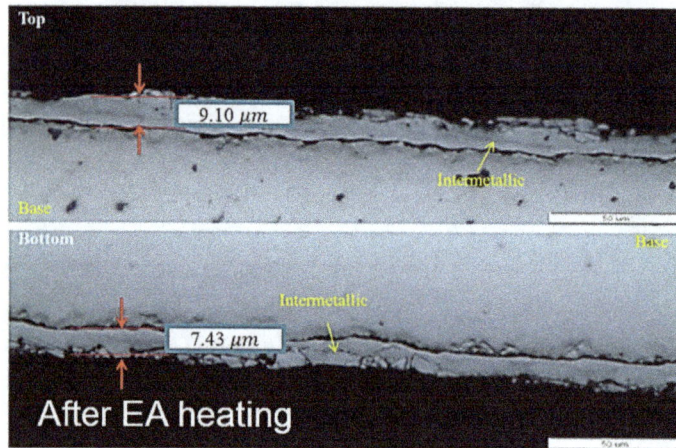

(b)

Figure 6. Development of intermetallic due to the effect of EA rapid heating (a) & (b)

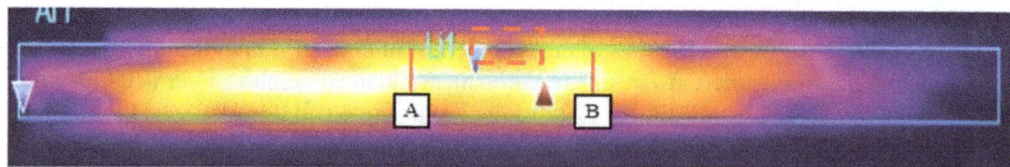

(a)

(b)

(c)

Figure 7. (a) Temperature distribution at 300°C over specimen; (b) Temperature distribution at 450°C over specimen; (c) Temperature distribution at 850°C over specimen

3.4 Tensile Test

After rapid heating, specimen was observed carefully. Centre part of the blank was locally heated shown in Figure 8 (a). Discolored parts inside the heated are is being considered as result of cooling the specimen consisting of spraying through the copper electrodes. Tensile test was carried out by tensile specimen which produced in a heated area, as shown in Figure 8 (a). Tensile speed was 2mm/min. The average value of tensile strength from locally heated blanks is about more than 1 GPa. Tensile test result is shown in Figure 8 (b).

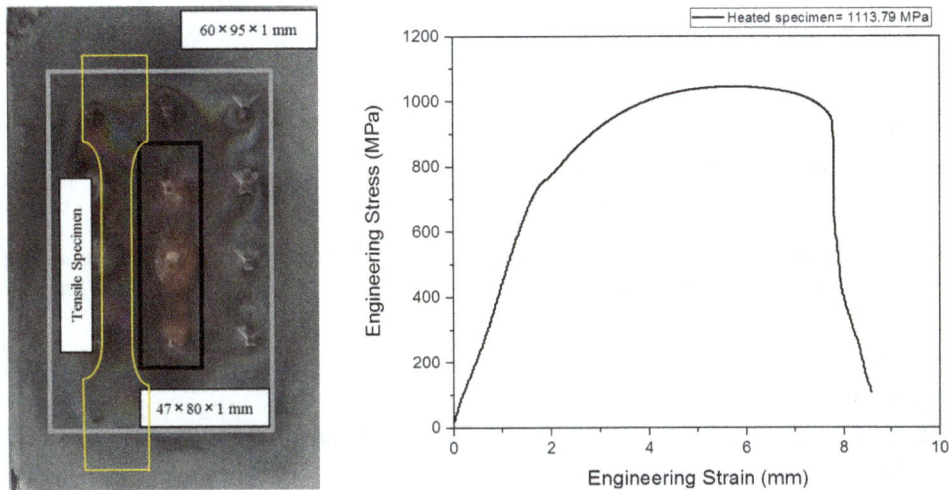

Figure 8. (a) Heated specimen and schematic of tensile specimen for the heated zone;
(b) Tensile test after rapid heating

3.5 Hardness Distribution

The hardness was tested under 430 SVD Vickers Instrument. The hardness data of the blanks represented an average measurement of five different points considering heated zone, transition zone and unheated zone. The heated area's hardness varies from 620 HV to 680 HV, which indicated full martensite obtained in this area. After electrically assisted rapid heating the area under heated zone became full martensite that represents the hardness value in graph. While the hardness of transition area varies from 453 HV20 to 643 HV20. Actually in transition region mixsturcutre was found and that's why fluctuation was observed in hardness value calculation. Finally the unheated area hardness was measured and the value is 149.5 HV. Unheated are had no effect of heat conduction even and the values are checked even after heating experiment. Every hardness value was measured by double checking with its repeatability value. These results demonstrated the hot stamping blank with tailored properties was realized by means of local resistance heating. Hardness value is shown in Table 2/Figure 9.

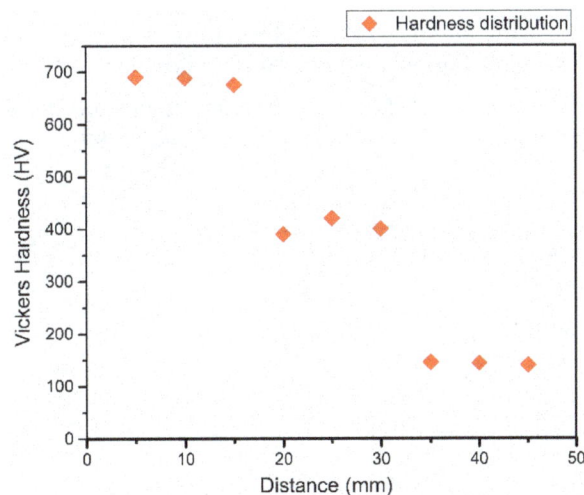

Figure 9. Hardness distribution of heated blank of different regions

4. Conclusions

The effect of hot stamping processes by using point contacts electrode local resistance heating of steel was studied. The processes included hot stamping and cooling by quenching. It was concluded that:

- The temperature rising profile was even at highest temperature.

- Temperature profile can be achieved more smooth and uniform without or less interval during electrical pulse.

- Partial local heating was properly done.

- Hardness of different region of heated specimen shows better result to get better mechanical properties.

- Tensile tests satisfied for better tailored properties.

- Intermetallic was observed by the effect of electric pulsed current.

Acknowledgments

This work is supported by Ulsan University fund.

References

Chao, W., Yisheng, Z., XiaoWei, T., Bin, Z., & Jian, L. (2012). Thermal contact conductance estimation and experimental validation in hot stamping process. *Scientia Sinica Tech., 55*(7), 1852. http://dx.doi.org/10.1007/s11431-012-4871-0

Erturk, S. (2011). Simulation of Tailored Tembering with a thermo-mechanical metallurgical model in autiform. *AIP-Proceedings, 1383*, 610. http://dx.doi.org/10.1063/1.3623664

Hein, P., & Wilsius, J. B. (2008). Status and innovation trends in hot stamping of USIBOR 1500 P. *Steel Research Int., 79*, 85-91.

Karbasian, H., & Tekkaya, A. E. (2010). A review on hot stamping. *J. of Mat. Process. Techn., 210*(15), 2103-2118.

Liang, W., Wang, L., & Liu, Y. (2014). Hot stamping parts with tailored properties by local resistance heating. *J. of Tech. of Plasticity., 81*, 1731-1736. http://dx.doi.org/10.1016/j.proeng.2014.10.222

Mori, K., Maeno, T., & Fukui, Y. (2011). Spline forming of ultra-high strength gear drum using resistance heating of side wall of cup. *CIRP Annals - Manuf. Tech., 60*, 299-302. http://dx.doi.org/10.1016/j.cirp.2011.03.144

Mori, K., Maeno, T., & Fuzisaka, S. (2012). Punching of ultra-high strength steel sheets using local resistance heating of shearing zone. *J. of Mat. Process. Techn., 212*, 534-540. http://dx.doi.org/10.1016/j.jmatprotec.2011.10.021

Mori, K., Maeno, T., & Mongkolkaji, K. (2013). Tailored die quenching of steel parts having strength distribution using bypass resistance heating in hot stamping. *J. of Mat. Process. Techn., 213*, 508-514. http://dx.doi.org/10.1016/j.jmatprotec.2012.10.005

Mori, K., Maki, S., & Tanaka, Y. (2005). Warm and hot stamping of ultra-high tensile strength steel sheets using resistance heating. *CIRP Annals - Manuf. Tech., 54*, 209-212. http://dx.doi.org/10.1016/S0007-8506(07)60085-7

Wang, Z., & Liu, P. (2014). Hot stamping of high strength steel with tailored properties by two methods. *J. of Tech. of Plasticity., 81*, 1725-1730. http://dx.doi.org/10.1016/j.proeng.2014.10.221

Zhu, B., Zhang, Y., Li, J., Wang, H., & Ye, Z. (2011). Simulation research of hot stamping and phase transition of automotive high strength steel. *Mat. Researc. Innov., 15*, 426-430. http://dx.doi.org/10.1179/143307511X12858957675435

11

Quartz Crystal Microbalance Monitoring of Poly(Vinyl Alcohol) Sol during the Freeze–Thaw Process

Yoshimi Seida[1] & Mitsuteru Ogawa[2]

[1] Dept. Policy Study, Natural Science Laboratory, Toyo University, Japan

[2] Fuji Silysia Chemical Ltd., Japan

Correspondence: Yoshimi Seida, Dept. Policy Study, Natural Science Laboratory, Toyo University, Japan.
E-mail: seida@toyo.jp

Abstract

A quartz crystal microbalance (QCM) working under cryo-conditions was applied to analyzing the gelation and/or phase behavior of poly(vinyl alcohol) (PVA) sol during repeated freeze–thaw processes. The development of a porous structure with the gelation of PVA sol during the freeze–thaw cycle was examined in terms of the thermal behavior of the water in the sol and the viscoelastic behavior of the sol through thermal and QCM analyses. Water was liberated from the hydrophilic PVA during the freeze–thaw process through the aggregation of PVA. The water decreased the freezing temperature and increased the melting temperature because of the development of the porous structure with gelation by the thermal treatment. The state of the water during the gelation was estimated from the phase transition temperature and enthalpy change of the water during the thermal scan by using water-saturated silica gels with a series of pore size distributions. The viscoelasticity of the PVA sol during the freeze–thaw process was measured by cryo-QCM using admittance analysis (QCM-A). The free water and/or porous structure in the PVA sol was found to increase from the viscoelastic point of view by QCM measurements showing a shift in the resonance parameters (f_s, R_1). A hard gel was confirmed to form by the decrease in f_s and the increase in R_1 with the thermal scan treatment. The cryo-QCM was found to be an effective probe for clarifying the gelation and phase behavior of PVA sol in detail with high resolution.

Keywords: hydrogel, quartz crystal microbalance, poly(vinyl alcohol), freeze–thaw, gelation, pore water

1. Introduction

The quartz crystal microbalance (QCM) is used in biological, medical and industrial fields as a molecular probe because of its fine spatiotemporal resolution of several tens of nanometers to sub-microns and its economic advantages (Janshoff, Galla, & Steinem, 2000; Marx, 2007; Iniewski, 2012; Diethelm, 2015). Modern QCMs realize a stable resonator by using a network analyzer with a sufficient power supply. This development has inspired the application of QCMs to large loading system associated with solid phases of samples and soft materials. The physical properties of soft/wet materials in a cryo-environment and/or at low temperature are also of interest. The shear-mode megahertz oscillation of the quartz in a QCM works as a probe for the rheological measurement of viscous samples. Thus, a QCM can be applied to observing the structural changes in soft materials from a viscoelastic point of view. QCMs have been applied in measuring the phase behavior of crosslinked thermo-responsive hydrogels, which are a typical soft material (Nakano, Kawabe, & Seida, 1998; Nakano, Seida, & Nakano, 2007; Seida, 2007; Seida, 2013). Viscoelasticity changes significantly depending on hydration behavior in hydrogel and hydration behavior is temperature-responsive in general. Thus we have investigated a temperature controllable QCM as a tool observing a change in hydration with high resolution based on the viscoelastic sensing function of QCM in the previous study. The use of QCM for the analysis of soft materials has several advantages. The shear-mode dynamic viscoelastic behavior of a micro-volume sample can be observed. Because a QCM has a high resonant frequency of several megahertz, the viscoelasticity of a small sample can be observed with high sensitivity. The QCM allows the hydration behavior of a thermo-responsive hydrogel to be observed in its collapse phase at higher temperature than the volume phase transition temperature (i.e., 306 K). The QCM is effective at identifying the phase behavior of a hydrogel and/or the temperature dependence of the hydration behavior at polar sites of the hydrogel, which is difficult with other more conventional techniques during the collapse phase of the gel.

The present study examined the potential of cryo-QCM for observing the gelation process of soft materials, where poly(vinyl alcohol) (PVA) sol was used as a representative soft material. In PVA, gelation occurs with a change in viscoelasticity, so it can be observed by using QCM with high-resolution. A QCM with admittance analysis (QCM-A) was applied to observing the viscoelasticity of PVA sol during its freeze–thaw process. An aqueous solution of PVA with a large degree of saponification is known to develop a strong and physically crosslinked hydrogel through repeated freeze–thaw thermal scan cycle treatments (Peppas & Stauffer, 1991) due to the formation of dense polymer aggregates and/or bundled polymers (micellar crystalline). A porous network structure is created with repeated freeze–thaw treatments of PVA sol. The morphology and viscoelastic properties of the hydrogel depend on the molecular weight, saponification, concentration of PVA, freeze–thaw temperature, time, and number of the thermal scan cycles (Peppas & Stauffer, 1991). The micellar crystallization and/or bundle formation of PVA via hydrogen bonding plays an important role in the formation of hydrogel through freeze–thaw cycles. There have been no reports of in situ mechanical measurements of PVA sol during its freeze–thaw process. The gelation process of PVA sol includes a phase change and is performed under cryo-conditions so that the mechanical measurement of the gelation itself is difficult to perform with conventional methods. QCM-A was applied to the mechanical measurement of the gelation of PVA sol through repeated freeze–thaw thermal scan cycles. The QCM response in the repeated thermal cycle process was examined from a viewpoint of hydration behavior of PVA obtained through thermal analysis of water in the PVA, referring to a pore structure-hydration behavior relationship obtained from thermal analysis of a series of water saturated porous silica gels with controlled pore size distributions. The silica gel had a hydroxyl group on its surface that was similar to that of porous PVA gel. Thus, silica gels with a series of pore size distributions were employed in the thermal analysis to investigate the influence of the pore structure on the state of water and to identify the gelation mechanism observed by the QCM in terms of the phase behavior of water in the sol.

2. Experimental

2.1 Samples

2.1.1 PVA Sample

PVA (molecular weight: 2000, saponification: 98%) purchased from Wako Pure Chemicals Co. Ltd. (Japan) was used as received. A 15 wt% aqueous PVA solution was prepared by dissolving the PVA in pure water at 371 K under vigorous stirring. The prepared viscous solution was gently stirred for 1 day to remove bubbles that were produced during the preparation stage. The sol showed gelation when it was kept at 253 K for 1 h in a freezer followed by thawing at room temperature.

2.1.2 Silica Gel

Silica gels with a series of pore size distributions d_p were supplied from Fuji Silysia Chemical Ltd. The N_2 adsorption/desorption method was used to characterize the porous structure of the silica gels at 77 K. The Brunauer–Emmett–Teller (BET) method was used to determine the specific surface area S_{BET} from the N_2 desorption isotherm of each silica gel. The Barrett–Joyner–Halenda (BJH) method was used to determine the pore size distribution and pore volume V_p by using the N_2 desorption isotherms (Gregg & Sing, 1982). Figure 1(a) shows the pore size distribution of each silica gel. The silica gels used in this study had relatively narrow pore size distributions: sharp with a single peak. Figure 1(b) shows the specific surface area as a function of the pore diameter. The water absorption capacity of each silica gel was determined based on the mass balance between water-saturated and dry samples. The water-saturated silica gels were prepared by immersing the series of silica gels in a large amount of distilled water for 48 h with ultrasonication for 10 min in the initial stage of immersion. The water content of the series of the silica gels was a simple linear function of the pore volume of the samples as shown in Figure 1(c). Table 1 summarizes the structural characteristics of the series of silica gels.

Table 1. Characterized structural parameters of the silica gels

Sample	A	B	C	D	E	F
d_p (nm)	3.3	6.2	10.6	16.0	32.6	51.1
V_p (ml/g)	0.27	0.66	1.01	1.02	0.97	0.88
S_{BET} (m^2/g)	511.0	370.0	309.7	198.0	101.0	69.8
Water content (%)	35.9	55.0	63.3	64.9	60.7	61.1

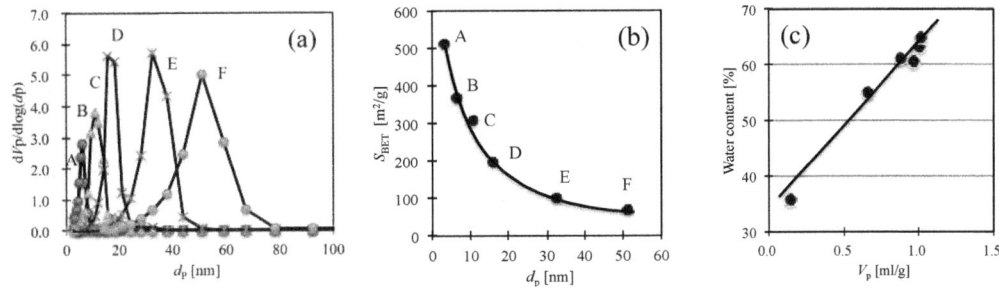

Figure 1. (a) The pore size distribution, (b) the relationship between the pore size and the specific surface area and (c) the relationship between the pore volume and the water content for the series of silica gels used in the present study

2.2 Thermal Analysis

The thermal analysis of the series of water-saturated silica gels was performed with a differential scanning calorimeter (DSC) to understand the phase behavior of pore water in the silica gels during a freeze–thaw operation as a function of the pore size in the silica gel. The freezing and thawing temperatures T_f, T_m and freezing and melting enthalpies of water ΔH_f, ΔH_m in each water-saturated silica gel were determined from DSC cooling and heating charts, respectively. The water-saturated silica gels were packed in an aluminum DSC sample cell, and cooling and heating charts were recorded with a DSC200 (Seiko Corporation Instrument). The scanning temperature range and rate in the DSC measurement were 293–243 K and 1 K/min, respectively. Thermal analysis was also performed on the 15 wt% PVA sol sample. The DSC measurements were performed under the same analytical conditions as the case of silica gels. The PVA sample was cooled and heated for five cycles with data acquisition to identify changes in the water behavior (hydration structure) of the PVA.

2.3 Principle of QCM-A

Based on the mathematical equivalency between the mechanical model of the QCM oscillation and LCR electric circuit model under forced oscillation (Figure 2), the viscoelastic property of the viscoelastic soft samples on the QCM were evaluated by admittance analysis of the QCM oscillation. Details of the admittance analysis have been reported elsewhere (Muramatsu, Tamiya & Karube, 1988). The resonance frequency f_s at the maximum conductance G_{max} and the resonance resistance R_1 (i.e., inverse of the G_{max}) of the QCM oscillation were obtained from the admittance analysis through a frequency scan around the resonance frequency. f_s depends on a substantial loading mass and/or elasticity of the load on the QCM (Sauerbrey 1959). R_1 is used as an index for the viscosity of a load. In a QCM, f_s and R_1 shift when the viscoelastic property of the sample changes. f_s also shifts for a viscous sample due to the influence of viscosity on the oscillation of QCM, so f_s is not a simple function of the loading mass in the case of a viscous sample (Kanazawa & Gordon, 1985). Martin, Granstaff, and Frye (1991) reported the independence of f_2 (i.e., the frequency at half of G_{max} and the minimum susceptance B_{min}) from the influence of the load viscosity on the QCM. Thus, the frequency f_2 was also measured in this study to evaluate the mass effect that is free from the influence of the sample viscosity.

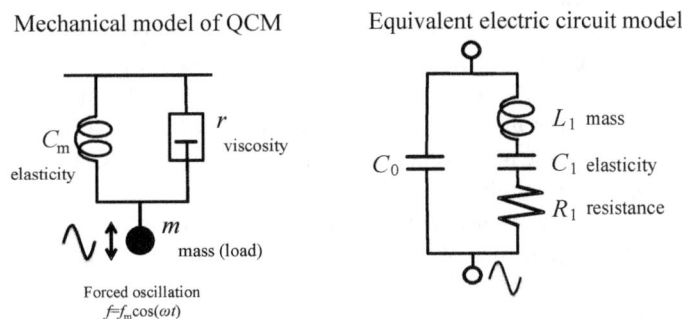

Figure 2. Model comparison between the mechanical oscillation model of QCM and the LCR electric circuit model

2.4 QCM Measurement

The viscoelasticity of the PVA sol during its freeze–thaw process was measured by QCM-A. A 5 MHz QCM with a gold electrode (φ10 mm diameter) purchased from Mitadenpa Co. (Japan) was used in this study. Figure 3 shows a schematic diagram of the QCM system used in this study. The temperature dependence of the frequencies f_s and f_2 and the resonance resistance R_1 of the bare QCM were measured first. Then, 1 μL of the PVA sol sample was dropped onto the surface of the QCM within an area of 1 mm². The QCM was installed in a thin cryo-cell with a Peltier device at the bottom of the cell. To reduce the loss of the aliquot PVA sol sample by drying and to avoid condensation from moisture in the atmosphere on the QCM, the PVA sol sample was covered with a thin polymer film made of poly(propylene), and the entire QCM electrode was covered with a silicon sheet. The film had a reverse dimple with a φ1 mm diameter and 1 mm depth at its center to hold the sample sol. The influence of the thin film on the QCM oscillation was examined in advance and was found to be negligible compared to the changes in f and R_1 induced by the sample on the QCM. The sample temperature was measured by using a φ1 mm thermocouple placed near the edge of the quartz of the QCM. The sample temperature was controlled by using the Peltier device with a proportional–integral–derivative (PID) controller. The sample was cooled from 283 K to 243 K and then heated to 283 K in a stepwise manner. This temperature swing was repeated three times. The cooling and heating rates were 1 K/min. The admittance of the QCM oscillation during the thermal scan cycle was analyzed at each temperature to collect the f_s, f_2, and R_1 data with a network analyzer.

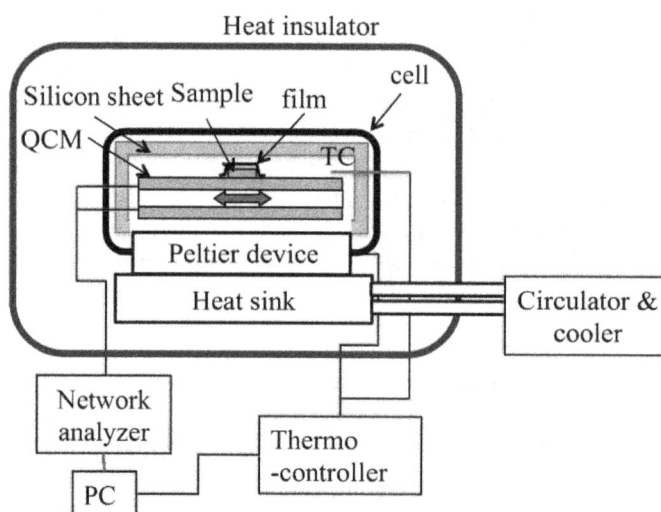

Figure 3. Schematic diagram of the QCM system used in this study

3. Results

3.1 Thermal Behavior of Pore Water in the Silica Gels

Figure 4 indicates the relationship between the pore size of the silica gels and the change in enthalpy with freezing ΔH_f obtained from the cooling charts in the DSC analysis. The water in the silica gels showed supercooled freezing with a single exothermic peak. The freezing temperature T_f decreased monotonically with increasing pore size. ΔH_f in Figure 4 was obtained from DSC time charts because the temperature drift occurred at the freezing point due to the exothermic heat process of the freezing. ΔH_f increased with the pore size and reached near the value for pure water in the silica gel with a pore size of more than 30 nm. This indicates the existence of a large amount of non-freezable bound water in the silica gel with small pores of less than 16 nm. This was because of the increase in the surface area relative to the pore volume with the decreasing pore size. The ratio between the specific surface area S_{BET} and water content in sample A was three times greater than that in sample C, as indicated in Table 1. T_f for pure water obtained by the same method was lower (253 K) than the temperature observed in the silica gels with large pores. Increasing the pore size of the silica gels was found to decrease T_f and increase ΔH_f.

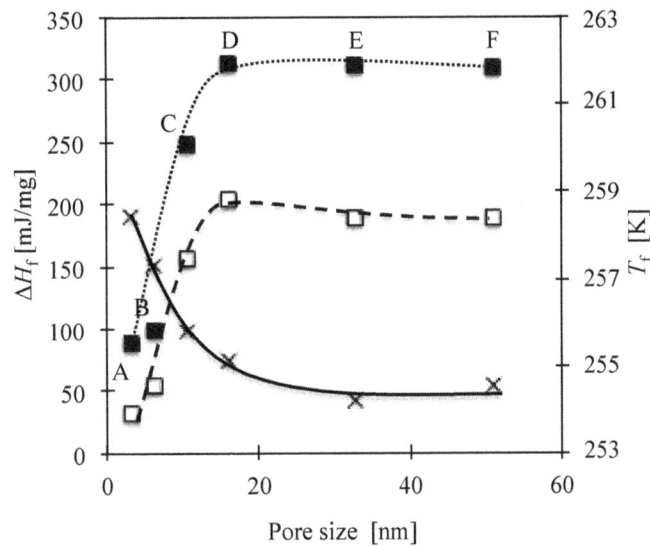

Figure 4. T_f and ΔH_f of water in each silica gel obtained from the cooling charts of DSC. □: weight basis enthalpy, ■: water content basis enthalpy, ×: freezing temperature T_f

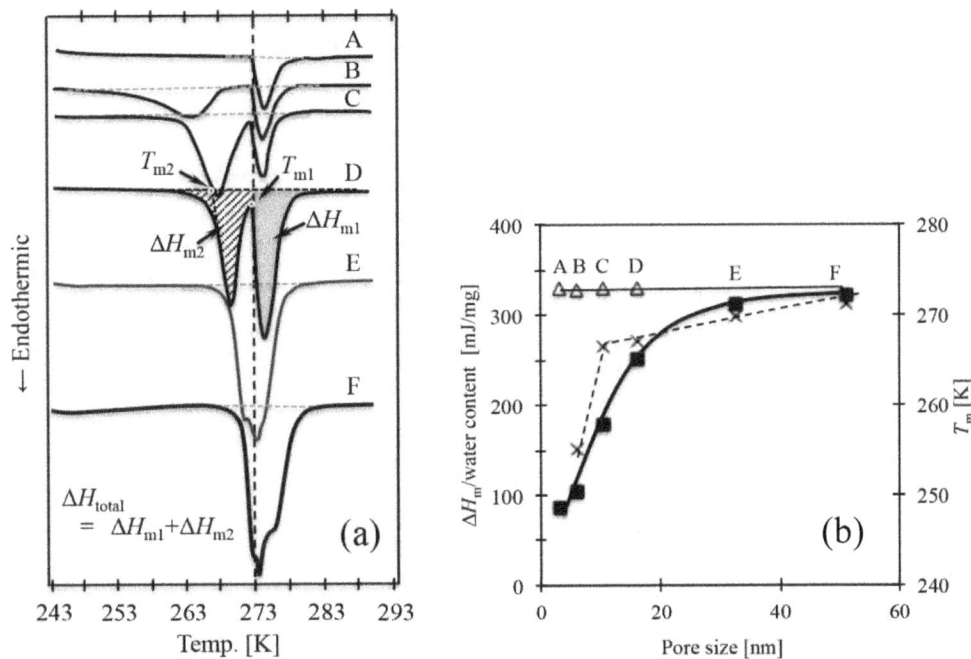

Figure 5. (a) DSC heating charts for the series of silica gels saturated with water. (b) T_m and ΔH_m of water determined from the heating charts of DSC. ■: ΔH_m/water content, △: T_{m1} and ×: T_{m2}

Figure 5(a) indicates the DSC heating charts for the series of water-saturated silica gels. The silica gels frozen at 243 K revealed characteristic endothermic peaks depending on the pore size in the heating process. The endothermic peaks below 273 K indicate the existence of freezable bound water in the silica gels. The ice of strongly bound water starts melting at temperatures below 273 K depending on the strength of the binding. The enthalpy was calculated from the peak area above and below 273 K enclosed by the DSC heating curve and the interpolated baseline as referenced from the peak area of pure water (Watase, Nishinari, & Hatakeyama, 1988; Seida & Nakano, 1996). The melting enthalpy of freezable bound water ΔH_{m2} was calculated from the peak area below 273 K. The melting enthalpy of free water ΔH_{m1} was determined from the area above 273 K. As shown in Figure 5(a), the peak due to melting of the bound water shifted toward a higher temperature with increasing pore size. The endothermic peaks below 273 K indicate the existence of freezable bound water, as mentioned above.

T_{m1} was defined as the melting temperature of free water. T_{m2} was defined as the melting temperature of bound water that could be observed in the DSC analysis. T_{m1} was determined from the inflection point around 273 K in the case of the multi-peak sample. T_{m2} was determined by a conventional method as indicated in Fig.5(a). The endothermic peak attributed to the freezable bound water disappeared in Sample A. Figure 5(b) indicates the melting temperatures T_{m1}, T_{m2} and the melting enthalpy of water ΔH_m ($= \Delta H_{m1} + \Delta H_{m2}$) as a function of the silica gel pore size. T_{m2} and ΔH_m were found to decrease with the pore size of silica gel. The results indicate the existence of some amount of non-freezable bound water in the silica gels with small pores. In the silica gels with pore sizes larger than 30 nm, the total ΔH_m reached close to the melting enthalpy of pure water. This indicates that free water was dominant in the pores larger than 30 nm. The fraction of water that was free from the influence of polar sites on the pore surface was large in the large pores, so T_{m2} and ΔH_{m1} were found to increase with the pore size. The melting temperature T_{m2} of freezable bound water and the melting enthalpy ΔH_m of freezable water can be associated with the pore structure in the case of silica gels. Slight decrease of T_{m1} occurred with the thermal cycle treatment due to a overlapping of the melting peak of bound water in the definition of T_{m1}. The DSC heating charts are a better way to clearly represent the state of water in the present system.

3.2 Thermal Behavior of PVA Sol

Figure 6(a) presents the DSC heating charts of the PVA sol obtained in the freeze–thaw measurements. An endothermic melting peak appeared from around 253 K, which indicates the existence of freezable bound water in the PVA sol. A slight shift in the endothermic peak was observed with the thermal scan cycle. Figure 6(b) summarizes T_f, T_m, and ΔH_m obtained from each DSC chart. The supercooled freezing temperature T_f of the PVA sol shifted toward a lower temperature with the thermal scan cycle. In contrast, the melting temperature T_m determined based on the conventional slope method, as described above, decreased very slightly with the thermal scan cycle. This was due to the tail overlapping of the endothermic melting peak of freezable bound water with the melting peak of free water. Figure 6(b) shows ΔH_m at each freeze–thaw cycle normalized by ΔH_m of the first freeze–thaw cycle $\Delta H_{m,1st}$. The increase in $\Delta H_m / \Delta H_{m,1st}$ indicates the increase in free and freezable bound water with each repetition of the thermal scan cycle treatment. Thus, the melting temperature of the freezable bound water increased because of the decrease in non-freezing bound water with the gelation of PVA. Based on the thermal behavior of water observed in the silica gels, the decrease in T_f of the PVA sol with the thermal scan cycle indicates an increase in free water. The partial densification and/or aggregation of PVA as a result of repeated thermal scan cycle treatment produces a pore structure along with gelation of the sol (Peppas & Stauffer, 1991, Watase, Nishinari, & Nanbu, 1983). The formation of bundled polymers (micellar crystalline aggregates) via hydrogen bonding among hydroxyl groups in PVA results in pore formation with gelation. Based on the knowledge obtained from thermal analysis of the silica gels, the results in Figure 6 indicates the formation and development of a porous structure in the gelling PVA sol by the freeze–thaw thermal scan cycle.

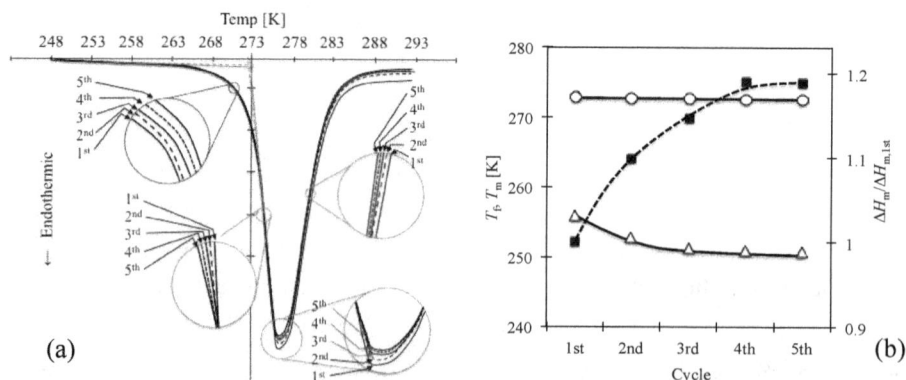

Figure 6. (a) DSC heating charts of the PVA sol, (b) $T_f(\triangle)$, $T_m(\circ)$ and $\Delta H_m / \Delta H_{m, 1st\ cycle}$ ($\blacksquare$) obtained by the DSC analysis for the PVA sol

3.3 QCM Response in the PVA Sol during the Freeze–Thaw Thermal Swing

The f_s and R_1 values showed a slight dependence on the temperature within several hertz in the considered temperature range (243–283 K) for the case of a bare QCM. The thermal shifts in the f_s and R_1 values were negligible compared to the varying ranges of f and R_1 observed in the PVA sol. The use of a thin film to hold the

PVA sol did not significantly affect the oscillation of the QCM. The change in oscillation due to the application of the thin film was within several hertz in the considered temperature range.

Figures 7 (a), (b), and (c) indicate temperature dependences of f_s, f_2, and R_1, respectively, in the QCM analysis of the PVA sol during the freeze–thaw thermal scan cycle. During the cooling process, f_s decreased and R_1 (=1/G) increased with decreasing temperature. In contrast, f_2 slightly decreased, which indicates that the viscosity R_1 change has a significant influence on f_s (Kanazawa & Gordon, 1985). The increase in R_1 during the cooling process was due to the increased density of water with temperature. A drastic shift in f_s, f_2, and R_1 occurred at the freezing temperature of the PVA sol after some supercooling. A large change in the effective load (i.e., penetration depth of the shear wave of the QCM) induced by the phase transition of the water from a liquid to solid (i.e., freezing) caused the behaviors of f_s and R_1. The transition temperature (i.e., point at which the red-shift of f_s occurred) shifted down closer to the freezing temperature of pure water with the thermal scan cycle.

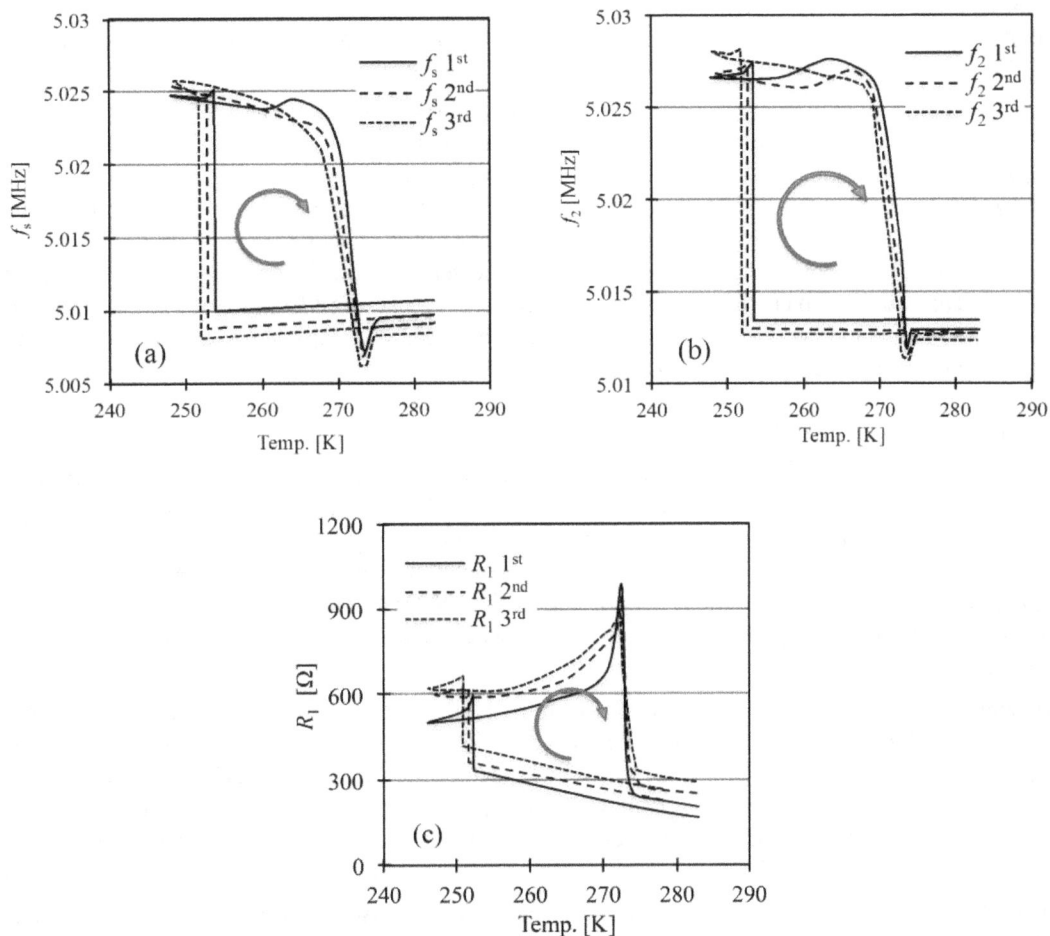

Figure 7. QCM behaviors of the sample during the repeated freeze–thaw temperature scan cycle. (a) f_s, (b) f_2 and (c) R_1

During the heating process, f_s and f_2 decreased with the temperature followed by a large blue-shift after showing a broad peak. The trends were reproducible and depended on the number of thermal scan cycles. The peaks of f_s and f_2 just before the large blue-shift disappeared with the cycle. The formation of crystalline aggregates (bundled polymers) and gelation with the change in hydration structure in the PVA sol were the reasons for such characteristic trends. In contrast, R_1 increased gradually with temperature followed by a large drastic decrease. The f values decreased and the R_1 value increased with the increasing number of thermal scan cycles in the non-freezing phase above 273 K. This indicated the increasing viscoelasticity of the gelling PVA sample. The steep shift of f_s, f_2, and R_1 due to the sample melting began at the temperature where the endothermic peak in the DSC analysis began. The change in viscoelastic property observed in the QCM measurement occurred with the melting of bound and free water of the PVA sol during the thawing process. The changes below 273 K would reflect the melting of freezable bound water in the PVA. Figure 7 represents the first measurements of the soft/wet material during its freeze–thaw process with QCM.

4. Discussion

In the case of the aliquot of pure water on the QCM, the QCM behavior during the freezing and thawing processes of the water was reproducible, and the supercooled freezing temperature and melting temperature did not change with repeated temperature scan cycle measurements. The freezing and melting temperatures were determined from the points of drastic shifts in f_s and R_1 during the QCM measurement. The trend of heating chart for f_s and f_2 of ice from pure water was reproducible as a simple decreasing function of the temperature with a small peak just before the drastic blue-shift due to the ice melting (Seida 2016).

The size and/or amount of samples used in both the DSC analysis and QCM measurements were much larger than the size of the micellar crystalline aggregates of PVA (Hassan & Peppas, 2000) that are produced in the initial stage of the gelation of PVA sol and the pore sizes created in the PVA by the freeze–thaw cycle treatment. From this point of view, knowledge on the gelation mechanism for bulk PVA sol is applicable to this system (Watase & Nishinari, 1985; Peppas & Stauffer, 1991).

The pore size and thermal properties (hydration behavior) of water in the pores correlated well for the silica gels, as indicated above. Increasing the pore size decreased the freezing temperature T_f and increased both the melting temperature of freezable bound water T_{m2} and melting enthalpy of freezable bound water ΔH_{m1} in the silica gels. The freezing water excluded dissolved solutes from its ice crystal, which enhanced the bundle formation and/or micellar crystalline aggregates of PVA and resulted in the porous gel. The elasticity decreased near the melting temperature in the freezing phase of the PVA sample, as expected from the f_2 trend for the heating process shown in Figure 7(b). This would also enable the polymer molecule to be excluded from the ice crystal and enhance the aggregation of PVA. The fractional change in hydration of the PVA also occurred with the progress of the aggregation and/or gelation during the freeze–thaw treatment, as indicated by the trend of ΔH according to the DSC analysis (Figure 6). Thus, the results shown in Figure 6 correspond to the development of a porous structure and gelation with repeated thermal scan cycles.

For the sol of 15 wt% PVA with 80% saponification and a small molecular weight, no obvious temperature shift with the freeze–thaw thermal cycle was observed (Seida, 2015). Bulk gelation does not occur in PVA sol with a small degree of saponification (~80%). Watase and Nishinari (1985) reported that acetate groups can inhibit the formation of gel in PVA. PVA with a small degree of saponification hardly shows bulk gelation even if crystalline PVA aggregates may be produced by the freeze–thaw thermal treatment. Thus, the change in the resonance parameters (i.e., f_s and R_1) during the freeze–thaw thermal scan cycle of the PVA sol, as observed in Figure 7, was attributed to the gelation of the PVA sol.

Rapid freezing produces an internal strain in the freezing phase and/or the interfacial strain between the ice and electrode surface of the QCM. Relaxing the strain would take a long time compared to the scan rate of the temperature in this study. Thus, the strain would affect the f and R_1 trends in the freezing phase of the temperature scan measurement. The scan rate of the temperature is also a potential factor for determining the f and R_1 behaviors in the freezing phase. The scan rate of the temperature in the present study was determined based on the conventional scan rate used in the DSC analysis.

The freezing temperature of the sample observed with the QCM decreased with the progress of gelation; this was associated with the increases in viscoelasticity and/or R_1. The shift in the f_s and R_1 charts for the heating process were also associated with the increased viscoelastic properties with the thermal scan cycles. The increased melting temperature and temperature range of the melting peak of the freezable bound water on the PVA would indicate an increase in free water due to the development of the porous structure. The drastic shift of f_s was due to a large drastic change in the effective load (i.e., the product of the penetration depth δ of the shear wave of the QCM and the density of the load ρ (Kanazawa and Gordon, 1985)). The drastic change in f_2 indicated a drastic change in the penetration depth of the shear wave of QCM in the sample, which means a change in the effective load on the QCM. The steep increase in R_1 during the heating process indicated increased damping of the shear wave energy along with the increased penetration depth of the shear wave. The red-shift of f_s and f_2 with the freezing of the sample may seem to be quite different compared to the generally accepted relation between the loading mass and frequency shift (Sauerbrey, 1956). The large blue-shift of f_s occurred with increases in the water (liquid) phase and penetration depth of the shear wave. The increased effective load with small viscosity observed by the QCM decreased f_s, f_2, and R_1. Because the shear oscillation frequency of a QCM is very high, the freezing phase would behave like viscous body. In the case of a load with a much larger viscosity, f_s and f_2 increase in spite of the large R_1

(Seida, 2016). According to the pore structure-hydration behavior relationship obtained from the results in Figure 5, the decrease of red-shift temperature and the increase of blue-shift temperature of f_s and f_2 occur in the QCM response along with the increase of free water and/or formation of pore structure in the PVA. R_1 becomes large with the increase of free water in the PVA.

The findings obtained from the thermal analysis of the silica gels and the PVA sol with repeated freeze–thaw thermal scan cycles indicate that a QCM can be used to probe the structural formation of a PVA porous network during the gelation of PVA sol.

5. Conclusion

In the present study, the gelation process of PVA gel during repeated freeze–thaw thermal cycles was evaluated based on the thermal behavior of pore water and the viscoelastic behavior of the PVA sol obtained by the cryo-QCM developed in this study. The relationship between pore structure and hydration behavior of pore water was clarified based on the freezing temperature T_f, melting temperature T_m, and freezing and melting enthalpies ΔH_f, ΔH_m of the pore water obtained from the thermal analysis using water saturated porous silica gels with a series of pore size distributions. The non-freezable and freezable bound water decreased with increasing pore size up to a diameter of 16 nm in the silica gels. T_f decreased and T_{m2} increased with the increase of pore size because of the increased amount of free water in the silica gel. The development of a porous structure in the PVA sol during the thermal scan cycle treatment was observed based on the parameters T_f, T_m, and ΔH_m in the thermal analysis. The QCM response of the PVA sol in the repeated freeze–thaw thermal scan cycle was obtained fairly well. The QCM response was interpreted from the viewpoint of hydration behavior referring to the pore structure-hydration behavior relationship observed in the silica gels. The results obtained in this study indicate the viability of the cryo-QCM for clarifying the gelation as well as hydration behaviors of PVA sol during freeze–thaw process in detail from a viscoelastic point of view.

Acknowledgement

This work was supported by JSPS Grant-in-Aid for Scientific Research (C) 23510092, 15K00586. The Silica gels used in this study were supplied from Fuji Silysia Chem. Ltd.

References

Diethelm J. (2015). *The Quartz crystal microbalance in soft matter research, Fundamentals and modeling.* Switzerland. Springer. http://dx.doi.org/ 10.1007/978-3-319-07836-6

Gregg, S. J., & Sing, K. S. W. (1982). *Adsorption, surface area and porosity* (2nd ed.). London. New York. Academic Press.

Hassan, C. H., & Peppas, N. A. (2000). Structure and morphology of freeze/thawed PVA hydrogels. *Macromolecules, 33,* 2472-2479. http://dx.doi.org/10.1021/ma9907587

Iniewski, K. (2012). *Biological and medical sensor technologies (Devices, circuits, and systems).* CRC Press.

Janshoff, A., Galla, H-J., & Steinem, C. (2000). Piezoelectric mass-sensing devices as biosensors -An alternative to optical biosensors?. *Angew. Chem. Int. Ed., 39*(22), 4004-4032. http://dx.doi.org./10.1002/1521 -3773(20001117)39:22%3C4004::AID-ANIE4004%3E3.0.CO;2-2

Kanazawa, K. K., & Gordon II, J. G. (1985). Frequency of a quartz microbalance in contact with liquid. *Anal. Chem., 57,* 1770-1771. http://dx.doi.org/10.1021/ac00285a062

Marx, K. A. (2007). The quartz crystal microbalance and the electrochemical QCM: Applications to studies of thin polymer films, electron transfer systems, biological macromolecules, biosensors, and cells. *Springer Ser. Chem. Sens. Biosens, 5,* 371-424. http://dx.doi.org/ 10.1007/978-3-540-36568-6_11

Martin, S. J., Granstaff V. E., & Frye G. C. (1991). Characterization of a quartz crystal microbalance with simultaneous mass and liquid loading. *Anal. Chem., 63,* 2272-2281. http://dx.doi.org/10.1021/ ac00020a015

Muramatsu M., Tamiya, E., & Karube, I. (1988). Computation of equivalent circuit parameters of quartz crystals in contact with liquids and study of liquid properties. *Anal. Chem., 60*(19), 2142-2146. http://dx.doi.org/10. 1021/ac00170a032

Nakano, Y., Kawabe, K., & Seida, Y. (1998). Detection of multiple phases in ecosensitive polymer hydrogel. *Kobunshi Ronbunshu, 55*(12), 791-795. http://doi.org/10.1295/koron.55.791

Nakano, Y., Itaru, Sato., Seida, Y., & Nakano, Y. (2007). Viscoelastic behavior of thermo-responsive polymer hydrogel with organic adsorbate using quartz crystal microbalance. *Chem. Lett., 36*(10), 1204-1205. http://doi.org/10.1246/cl.2007.1204

Peppas, N. A., & Stauffer, S. R. (1991). Review article: Reinforced uncrosslinked poly(vinyl alcohol) gels produced by cyclic freezing-thawing processes: A short review. *J. Cont. Release, 16,* 305-310. http://dx.doi.org/10.1016/0168-3659(91)90007-Z

Sauerbrey G. (1959). Verwendung von schwingquarzen zur wägung dünner schichten und zur mikrowägung. *Zeitschrift für Physik, 155*(2), 206-222.

Seida, Y. & Nakano, Y. (1996). Effect of salt on the property of adsorption in thermosensitive polymer hydrogels. *J. Chem. Eng. Japan, 29*(5), 767-772. http://doi.org/10.1252/jcej.29.767

Seida, Y., Sato, I., Taki, K., & Nakano, Y. (2007). Sensing of adsorption induced phase behavior of stimuli-sensitive polymer hydrogel using quartz crystal microbalance. *Trans. Mat. Res. Soc. Jpn., 32*(3), 783-786.

Seida, Y. (2013). Viscoelastic behavior of thermo-responsive hydrogel during temperature swing process – Measurement viscosity free mass effect via QCM-A -. *Trans. Mat. Res. Soc. Jpn., 38*(4), 651-654. http://doi.org/10.14723/tmrsj.38.651

Seida, Y. (2015). Demonstration of QCM measurement of viscoelastic phase behavior of PVA sol in its freezing-thawing process. *Macromolecular Symposia, 358*(1), 176-181. http://dx.doi.org/10.1002/masy.201500061

Seida Y., Demonstration of QCM Measurement of Water During Its Freezing-Thawing Process, *J. Chem. Eng. Japan*, submitted 2016.

Tsionsky, V., Daikhin, L., Zagidulin, D., Urbakh, M., & Gileadi, E. (2003). The quartz crystal microbalance as a tool for the study of a "liquidlike layer" at the ice/metal interface. *J. Phys. Chem. B, 107*, 12485-12491. http://dx.doi.org/10.1021/jp030459q

Watase, M., Nishinari, K., & Hatakeyama, T. (1988). DSC study on properties of water in concentrated agarose gels. *Food Hydrocolloids, 2*, 427-438. http://dx,doi.org/10.1016/S0268-005X(88)80043-2

Watase, M., Nishinari, K., & Nanbu, M. (1983). Anomalous increase of the elastic modulus of frozen poly(vinyl alcohol) gels. *Cryo Lett., 4*(3), 197-200.

Watase, M., Nishinari, K. (1985). Rheological and DSC changes in poly(vinyl alcohol) gels induced by immersion in water. *J. Appl. Polym. Sci. Polym. Phys. Ed., 23*, 1803-1811. http://dx.doi.org/10.1002/pol.1985.180230906

Acicular Ferrite Formation and Its Influencing Factors

Denise Loder[1], Susanne K. Michelic[1] & Christian Bernhard[1]

[1] Chair of Ferrous Metallurgy, Montanuniversitaet Leoben, 8700 Leoben, Austria

Correspondence: Susanne K. Michelic, Chair of Ferrous Metallurgy, Montanuniversitaet Leoben, 8700 Leoben, Austria. E-mail: susanne.michelic@unileoben.ac.at

Abstract

Acicular ferrite is a microstructure nucleating intergranularly on non-metallic inclusions and forming an arrangement of fine, interlocking grains. This structure is known to improve steel properties, especially steel toughness, essentially. The formation of acicular ferrite is mainly affected by steel composition, cooling rate, inclusion landscape and austenite grain size. In recent decades, extensive research has been conducted to investigate these factors. The present paper provides an overview of the impact of published results and the state of knowledge regarding acicular ferrite formation. Special attention is paid to the effect of carbon, manganese and titanium addition to steel, as well as the optimum size, number and composition of non-metallic inclusions. In addition, the reactions during the nucleation and growth of acicular ferrite needles are briefly addressed. Further, characteristics of acicular ferrite and bainite are summarized, which should help to distinguish these similar structures.

Keywords: acicular ferrite, cooling rate, non-metallic inclusion, steel

1. Introduction

At the beginning of the 1990s, Takamura and Mizoguchi (1990) introduced the concept of Oxides Metallurgy, which focuses on the use of oxidic inclusions to improve the quality of the final product and stabilize the production process. They described, inter alia, that by controlling the nature and distribution of the inclusions, the formation of intragranular ferrite, the so-called acicular ferrite, can be promoted, which significantly increases the steel's fracture toughness. Although the clear definition of the term 'Oxides Metallurgy' by Takamura and Mizoguchi (1990) was novel, the idea of using acicular ferrite for the optimization of steel properties was not new. Already in 1956, Aaronson and Wells (1956) described a microstructure of intragranularly nucleated, needle-shaped and chaotically arranged ferrite grains; however, they did not name this structure acicular ferrite, but 'Widmannstätten star'. One of the first publications that designated this characteristic structure as acicular ferrite was by Smith, Coldren and Cryderman in 1971. Since then, comprehensive efforts have been devoted to an explanation of acicular ferrite formation. Most of these early studies focused on weld metals (Abson, Dolby, & Hart, 1978; Bonnet & Charpentier, 1983; Cochrane & Kirkwood, 1978; Devillers, Kaplan, Marandet, Ribes, & Riboud, 1983; Düren, 1983; Evans, 1980, 1982, 1983; Ricks, Barritte, & Howell, 1981; Ricks, Howell, & Barritte, 1982; Saggese, Bhatti, Hawkins, & Whiteman, 1983), so using acicular ferrite to improve the toughness of weld fusion zones is quite popular today. In recent years, acicular ferrite has also been of increasing interest to steel producers; the excellent combination of toughness and strength makes acicular ferrite favorable for HSLA (high-strength low-alloy) steels (Bhadeshia, 2001; Sarma, Karasev, & Jönsson, 2009; M. Song, Song B., Hu, Xin, & G.Y. Song, 2015; Sung, Shin, Hwang, C. G. Lee, & S. Lee, 2013). In addition, acicular ferrite is becoming more and more important as a constituent in line pipe steels because of its good sour gas resistance (Garcia-Mateo, Capdevila, Caballero, & Andrés, 2008; Madariaga, Gutierrez, & Bhadeshia, 2001; Ogibayashi, 1994; Xiao et al., 2006; Xiao, Liao, Ren, Shan, & Yang, 2005; Zhao, Shan, Xiao, Yang, & Li, 2002; Zhao, Yang, & Shan, 2003).

Acicular ferrite formation is influenced by four main parameters (Bhadeshia, 2001; Farrar & Harrison, 1987; Grong & Matlock, 1986; Loder & Michelic, 2016; Sarma et al., 2009).

- Austenite grain size (AGS)
- Cooling rate (CR)
- Steel composition
- Non-metallic inclusions (NMI)

Comprehensive research to evaluate the impact of these parameters has been performed in recent decades. However, the effects are not completely clarified in the literature yet, possibly due to their strong interactions. The present work provides an overview of the state of knowledge on acicular ferrite formation and its four main influencing factors.

2. The Formation of Acicular Ferrite

Acicular ferrite nucleates heterogeneously on the surface of NMI during the austenite-ferrite-transition. As transformation continues, the ferrite grains radiate in various directions, creating a chaotic construction of crystallographically disoriented plates of approximately 5 - 15 μm in length and 1 - 3 μm in width. Thereby, the typical fine-grained and interlocking structure of acicular ferrite is formed. Compared with parallel plates, it is much more difficult for cleavage cracks to propagate across this chaotic structure, which leads to a significant increase in mechanical properties, most notably in toughness (Abson et al., 1978; Bhadeshia, 2001; Cai, Zhou, Tong, Yue, & Kong, 2015; Díaz-Fuentes, Iza-Mendia, & Gutiérrez, 2003; Farrar & Harrison, 1987; Fattahi, Nabhani, Hosseini, Arabian, & Rahimi, 2013; Gourgues, Flower, & Lindley, 2000; Jin, Shim, Cho, & Lee, 2003; Lan, Du, & Liu, 2013; Lee, Kim, Kang, & Hwang, 2000; Madariaga et al., 2001; Nako, Hatano, Okazaki, Yamashita, & Otsu, 2014; Ohkita & Horii, 1995; Ricks et al., 1982; Sarma et al., 2009; Wan, Wang, Cheng, & Wu, 2012; Zhang & Farrar, 1996; Zhang, Shintaku, Suzuki, & Komizo, 2012b, 2012a; Zhang, Terasaki, & Komizo, 2010).

The formation of acicular ferrite is diffusionless, so the acicular ferrite plates become supersaturated with carbon and excess carbon is rejected to the remaining austenite shortly after the transformation. Depending on the CR, the carbon-enriched austenite transforms in the ongoing cooling process to pearlite, bainite or martensite. Pearlite and bainite are formed at low and medium CR when the austenite becomes saturated in carbon and cementite precipitates. At a high CR, the remaining austenite transforms to martensite. Occasionally, the austenite does not convert, so islands of retained austenite are found in the final microstructure. In addition to the CR, carbon, silicon and nickel were found to influence the transition of the remaining austenite between the acicular ferrite grains. With increasing carbon content, the fraction of intergranular phases between acicular ferrite needles increases, as observed by Evans (1982, 1983). Increasing the silicon and nickel contents leads to a higher tendency of the intergranular region to transform into martensite, as asserted by Shim et al. (2001) and Evans (1990b). They found that these elements suppress the precipitation of cementite and, consequently, the formation of bainite and pearlite (Bhadeshia, 2001; Capdevila et al., 2006; Capdevila, García-Mateo, Chao, & Caballero, 2009; Evans, 1982, 1983, 1990b; Farrar & Harrison, 1987; Grong & Matlock, 1986; Hossein Nedjad, Zahedi Moghaddam, Mamdouh Vazirabadi, Shirazi, & Nili Ahmadabadi, 2011; Huo, Liu, Zhang, Yan, & Gao, 2013; Kim, Lee, & Kim, 2008; Lan et al., 2013; Madariaga et al., 2001; Madariaga, Gutiérrez, García-de Andrés, & Capdevila, 1999; Rees & Bhadeshia, 1994; Ricks et al., 1982; Shim, Cho et al., 2001; Strangwood & Bhadeshia, 1986; Sugden & Bhadeshia, 1989; Xiao et al., 2005; Yang, Liu, Sun, & Li, 2010; Zhao, Yang, Xiao, & Shan, 2003; Zuo & Zhou, 2015).

Several authors (Aaaronson & Wells, 1956; Barbaro, Krauklis, & Easterling, 1989; Díaz-Fuentes et al., 2003; Jin et al., 2003; Madariaga et al., 2001; Ricks et al., 1981; Ricks et al., 1982; Song et al., 2015; Wu, Inagawa, & Enomoto, 2004; Yang et al., 2010; Zhang & Farrar, 1996; Zhang et al., 2010) separate primary and secondary acicular ferrite. According to them, primary acicular ferrite plates nucleate on NMI, while secondary acicular ferrite forms on primary plates by so-called sympathetic nucleation. Although Wu et al. (2004) support the theory of sympathetic nucleation, they noted that it is hardly possible to distinguish between sympathetic nucleation and impingement of plates in many cases. Additionally, Wan et al. (2012) observed in 3-D reconstructed microstructure models that many acicular ferrite plates, classified in the cross-section as secondary plates, are just grains that impinged or intersected their "primary" plates. Additionally, there is no explanation why secondary acicular ferrite should not form a series of laths on the large surface of primary plates, resulting in packages of parallel laths.

Since the 1970s, much research has been performed to determine the mechanism of acicular ferrite formation. It is a common opinion that NMI are essential for nucleation. The literature describes four main mechanisms of how inclusions promote the nucleation of intragranular ferrite:

- Destruction of crystal structure (Barbaro et al., 1989; Bhadeshia, 2001; Gregg & Bhadeshia, 1997; Lee et al., 2000; Ricks et al., 1981; Ricks et al., 1982; Sarma et al., 2009; Shim, Oh et al., 2001; Zhang et al., 2012a; Zhang et al., 2010)

- Creation of dislocation arrays (Bhadeshia, 2001; Furuhara, Yamaguchi, Sugita, Miyamoto, & Maki, 2003; Grong & Matlock, 1986; Sarma et al., 2009; Zhang et al., 2010)

- Reduction of lattice mismatch (Abson et al., 1978; Cochrane & Kirkwood, 1978; Eijk, Grong, & Hjelen, 1999; Furuhara, Shinyoshi et al., 2003; Furuhara, Yamaguchi et al., 2003; Grong, Kluken, Nylund, Dons, & Hjelen, 1995; Jin et al., 2003; Nako, Okazaki, & Speer, 2015; Sarma et al., 2009; Yamamoto,

Hasegawa, & Takamura, 1996; Zhang et al., 2010)

- Chemical changes in the local matrix (Byun, Shim, Cho, & Lee, 2003; Cai et al., 2015; Chai, Su, Yang, & Xue, 2014; Evans, 1980; Farrar & Harrison, 1987; Gregg & Bhadeshia, 1997; Grong & Matlock, 1986; Hui et al., 2014; Kimura, Nakajima, Mizoguchi, & Hasegawa, 2002; Ogibayashi, 1994; Sarma et al., 2009; Seo, Kim, Evans, Kim, & Lee, 2015; Shim, Cho et al., 2001; Shim, Oh et al., 2001; Shim, Byun et al., 2001; Song et al., 2015; Wan et al., 2012; Yang, Ma, Zhu, Wang, & Zheng, 2012a)

Comprehensive information on the state of knowledge regarding these formation mechanisms were published in Koseki & Thewlis, Loder (2016), Sarma et al. (2009), Thewlis, Whiteman, and Senogles (1997). Hence, detailed information about the mechanisms was not included in the present paper.

3. Acicular Ferrite vs. Bainite

When bainite was first discovered by Davenport, Bain, and Kearny, (1930) (reprinted in Bhadeshia (2010), Davenport, Bain, and Kearny (1970), it was defined as acicular-shaped ferrite. Even still, acicular ferrite is often described as intragranularly nucleated bainite. The similar transformation temperature and comparable transformation mechanism are seen as evidence for the connection of acicular ferrite and bainite. According to Bhadeshia and coworkers (Bhadeshia, (1992, 2001), Rees and Bhadeshia (1994), Sugden and Bhadeshia (1989), a change from bainite to acicular ferrite can be achieved by simple control of the nucleation sites. Nevertheless, a clear distinction between acicular ferrite and bainite is essential because these structures show significant differences, which are summarized in Table 1, and consequently result in different mechanical properties.(Byun et al., 2003; Díaz-Fuentes et al., 2003; Eijk et al., 1999; Gourgues et al., 2000; Sarma et al., 2009; Shim, Cho et al., 2001; Smith et al., 1971; Yang et al., 2010)

Table 1. Characteristics of acicular ferrite and bainite (Bhadeshia, 1992, 2001; Byun et al., 2003; Díaz-Fuentes et al., 2003; Gourgues et al., 2000; Sarma et al., 2009; Shim, Cho et al., 2001; Yang et al., 2010)

	Acicular ferrite	Bainite
Nucleation	intragranular	inter- or intragranular
Nucleation site	point sites (surface of NMI)	mostly austenite grain surface
Transition Temp.	similar	
Grain form	needle-shaped/lenticular plates	fine laths which form sheaves
Growth direction	various directions (star-like)	parallel laths
Parallelism	cannot form parallel laths	laths are parallel within package
Growth mechanism	diffusionless	
Growth stop reason	C-enrichment of austenite or impingement	

Madariaga et al. (1999, 2001) suggested that acicular ferrite plates can grow as so-called morphological packets. These packets are similar to the bainitic morphology, so according to this theory, the only difference between acicular ferrite and bainite is the nucleation site: intragranular particles for acicular ferrite and grain boundaries for bainite. However, this does not conform to the widely used description of acicular ferrite, which includes that acicular ferrite plates radiate in various directions, forming a chaotic structure of crystallographic disoriented plates. Thus, Díaz-Fuentes et al. (2003) recommended that structures of parallel ferrite plates inside the austenite grain should be named intragranularly nucleated bainite and that only chaotically arranged ferrite plates nucleating at point sites should be classified as acicular ferrite.

4. Influence of Austenite Grain Size

During phase transformation, the nucleation of new phases can occur homogenously or heterogeneously. Generally, heterogeneous nucleation is more likely because of the reduced nucleation energy at interphase surfaces. Hence, during the austenite-ferrite-transition, the ferrite grains tend to nucleate heterogeneously on NMI, like acicular ferrite, or on grain boundaries, like Widmannstätten ferrite and grain boundary ferrite. Which nucleation site will be preferred depends on the ratio between the total surface area of NMI and the total surface area of austenite grains. Intragranular nucleation is preferred with increasing AGS or inclusion number, and grain boundary nucleation becomes more likely with decreasing grain size or inclusion number (see Figure 1). However, the AGS cannot be increased unlimitedly without reducing the inclusion number due to pinning effects. (Adrian & Pickering, 1991; Andersen & Grong, 1995; Bernhard, Bragin, & Vanovsek, 2012; Bernhard, Reiter, & Presslinger, 2008; Garbarz, 1995) However, a decrease in inclusions will reduce the nucleation sites for acicular ferrite, so the fraction of acicular ferrite can decrease or stay stable at larger AGS (Barbaro et al., 1989; Bhadeshia, 2001; Bin & Bo, 2012; Byun et al.,

2003; Cochrane & Kirkwood, 1978; Devillers et al., 1983; Díaz-Fuentes et al., 2003; Farrar, Zhang, Bannister, & Barritte, 1993; Garcia-Mateo et al., 2008; Gregg & Bhadeshia, 1997; Grong & Matlock, 1986; Kikuchi et al., 2011; Lee & Pan, 1995; Liu & Olson, 1986; Rees & Bhadeshia, 1994; Ricks et al., 1982; Sarma et al., 2009; Song et al., 2015; Thewlis, 2006; Vanovsek, Bernhard, Fiedler, & Schnitzer, 2013; Wan, Wu, Cheng, & Wei, 2015; Yang, Ma et al., 2015; Yang et al., 2012a, 2012b; Zhang et al., 2012a; Zhang, Terasaki, & Komizo, 2009, 2010).

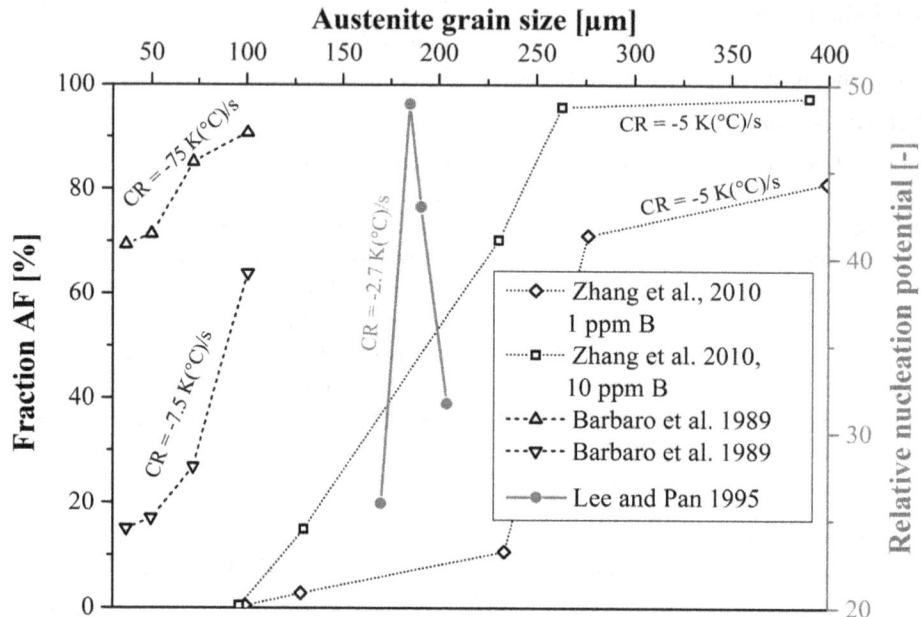

Figure 1. Impact of AGS on the acicular ferrite formation in titanium alloyed low-carbon steels for different CR between 1073,15 K (800 °C) and 773,15K (500 °C) ; adapted from Ref. (Barbaro et al., 1989; Lee & Pan, 1995; Zhang et al., 2010)

Farrar et al. (1993) observed that the AGS also influences the acicular ferrite start temperature. Mostly, ferrite formation starts at austenite grain boundaries. Due to the reduced solubility of carbon in ferrite, the surrounding austenite is enriched in carbon during this transition. In small austenite grains, the intragranular region is also influenced by this effect. The higher carbon content prevents the formation of acicular ferrite and lowers the acicular ferrite start temperature. With increasing AGS, intragranular regions are less enriched in carbon, and the acicular ferrite start temperature remains more stable. The carbon enrichment of austenite grains during grain boundary formation influences the structure of acicular ferrite as well. Due to the larger carbon enrichment of smaller austenite grains, the nucleation rate of acicular ferrite is reduced, and the formed plates can grow larger before they are stopped by impingement. Hence, acicular ferrite grains become coarser with decreasing AGS. (Bhadeshia, 2001; Farrar et al., 1993).

5. Influence of Cooling Rate

It is commonly accepted that the transformation of acicular ferrite is situated between bainite and coarse-grained ferrite; however, the optimum CR to achieve maximum acicular ferrite depends strongly on the production route and the composition of the steel (Farrar & Harrison, 1987; Rees & Bhadeshia, 1994; Sarma et al., 2009; Yang, Wang, Wang, & Song, 2008; Zhang et al., 2012b; Zhao, Yang, Xiao et al., 2003). Due to this strong interaction, a very wide range of CR is described in the literature as suitable for acicular ferrite. Table 2, Table 3 and Table 4 give an overview of recent results in the literature regarding CR for acicular ferrite during heat treatments with or without deformation.

Furthermore, Rees and Bhadeshia (1994) observed that the structure of acicular ferrite becomes finer with increasing CR. A higher CR increases supercooling and therefore the reaction driving force, which leads to a higher nucleation rate. Many acicular ferrite plates nucleate, so the plates have less free space to grow and will become finer.

Table 2. CR between 1073,15 K (800 °C) and 773,15K (500 °C) described in the literature as suitable for the formation of acicular ferrite during heat treatments without deformation

Ref.	CR range [K(°C)/min]	CR opt. [K(°C)/min]	Steel composition [C, Mn, Si, Al, Cu, Ni, Cr, Mo in wt.-%; Ti, Nb, V, B in wt.-ppm]
(Thewlis, 2006)	6 - 498	6 - 30	0.08 C, 2.00 Mn, 0.40 Si, 0.03 Al, 0.25 Mo, 100 Ti
(Thewlis, 2006)	6 - 540	180	0.05 C, 1.50 Mn, 0.30 Si, 0.03 Al, 400 Nb, 800 V
(Thewlis, 2006)	6 - 540	6	0.08 C, 2.00 Mn, 0.40 Si, 0.03 Al, 0.25 Mo, 100 Ti, 10 B
(Capdevila et al., 2006)	6 - 600	18 - 360	0.38 C, 1.44 Mn, 0.62 Si, 0.03 Al, 0.04 Ni, 0.07 Cr, 0.16 Mo, 160 Ti, 1000 V
(Xiao et al., 2005)	12 - 450	-	0.03 C, 1.56 Mn, 0.24 Si, 0.32 Mo, 390 Nb, 190 V
(Zhao et al., 2003)	24 - 3000	1200	0.05 C, 1.94 Mn, 0.30 Si
(Zhao et al., 2003)	24 - 3000	450 - 600	0.08 C, 1.28 Mn, 0.25 Si, 270 Ti, 450 Nb, 530 V
(Capdevila et al., 2006)	30 - 1200	120 - 600	0.36 C, 1.44 Mn, 0.63 Si, 0.02 Al, 0.13 Ni, 0.17 Cr, 0.02 Mo, 220 Ti, 900 V
(Yang et al., 2008)	90 - 150	120	0.39 C, 1.14 Mn, 0.16 Si, 0.01 Al, 250 Ti
(Capdevila et al., 2009)	120 - 360	360	0.10 C, 1.54 Mn, 0.20 Si, 0.22 Cu, 20 Ti, 20 Nb, 600 V
(Lan et al., 2013)	180 - 600	300	0.08 C, 1.59 Mn, 0.25 Si, 0.46 Cr, 0.11 Mo, 1000 Ti+Nb+V, 10 B
(Capdevila et al., 2009)	240 - 600	360	0.08 C, 1.46 Mn, 0.02 Al, 10 Ti, 2500 V
(Jin et al., 2003)	300	-	0.07 C, 1.90 Mn, 160 Ti
(Huo et al., 2013)	400 - 1500	800	0.09 C, 1.29 Mn, 0.29 Si, 0.03 Al, 0.13 Cu, 0.15 Ni, 0.06 Cr, 0.18 Mo, 10 Ti, 300 Nb, 500 V

Table 3. CR between 1073,15 K (800 °C) and 773,15K (500 °C) described in the literature as suitable for the formation of acicular ferrite during heat treatments by HT-LSCM

Ref.	CR range [K(°C)/min]	CR opt. [K(°C)/min]	Steel composition [C, Mn, Si, Al, Cu, Ni, Cr, Mo in wt.-%; Ti, Nb, V, B in wt.-ppm]
(Zhang et al., 2012b)	60 - 240	-	0.15 C, 1.48 Mn, 0.20 Si, 100 Ti, 2 B
(Zhang et al., 2012b)	60 - 240	240	0.15 C, 1.48 Mn, 0.20 Si, 100 Ti, 12 B
(Wang et al., 2012)	300	-	0.07 C, 1.10 Mn, 0.14 Si, 46-240 Ti
(Wan, Wu, Huang, & Wei, 2014)	300	-	0.06 C, 1.60 Mn, 0.21 Si, 0.02 Al, 0.25 Cu+Ni+Cr+Mo, 400 Ti, 390 Nb, 220 V, 5 B
(Wan et al., 2015)	300	-	0.06 C, 1.62 Mn, 0.30 Si, 0.70 Ni+Cr+Mo, 150 Ti+Zr, 550 Nb, 12-20 B, 8-35 Ca

Table 4. CR between 1073,15 K (800 °C) and 773,15K (500 °C) described in the literature as suitable for the formation of acicular ferrite during thermo-mechanical treatments

Ref.	CR range [K(°C)/min]	CR opt. [K(°C)/min]	Steel composition [C, Mn, Si, Al, Cu, Ni, Cr, Mo in wt.-%; Ti, Nb, V, B in wt.-ppm]
(Li, Min, Liu, & Jiang, 2015)	60 - 600	300 - 600	0.05 C, 1.55 Mn, 0.22 Si, 0.03 Al, 0.31 Ni, 140 Ti, 24 Mg, 30 Zr
(Capdevila et al., 2006)	60 - 1200	-	0.36 C, 1.44 Mn, 0.63 Si, 0.02 Al, 0.13 Ni, 0.17 Cr, 0.02 Mo, 220 Ti, 900 V
(Capdevila et al., 2006)	60 - 1200	240 - 600	0.38 C, 1.44 Mn, 0.62 Si, 0.03 Al, 0.04 Ni, 0.07 Cr, 0.16 Mo, 160 Ti, 1000 V
(Wu, Lee, Kim, & Kim, 2012)	60 - 1800	600	0.06 C, 1.80 Mn, 0.25 Si, 0.02 Al, 0.30 Cu, 0.30 Ni, 0.31 Mo, 140 Ti, 400 Nb, 400 V, 10 B
(Xiao et al., 2005)	60 - 2400	300 - 2400	0.03 C, 1.56 Mn, 0.24 Si, 0.32 Mo, 390 Nb, 190 V
(Zhao, Yang, Xiao et al., 2003)	60 - 3000	600	0.03 C, 1.56 Mn, 0.24 Si, 0.32 Mo, 400 Nb, 200 V
(Zhao, Yang, Xiao et al., 2003)	120 - 2400	1200 - 2400	0.08 C, 1.28 Mn, 0.25 Si, 270 Ti, 450 Nb, 530 V
(Shim, Byun et al., 2001)	180	-	0.06 C, 2.46 Mn, 0.20 Si, 90 Ti
(Lan et al., 2013)	180 - 600	300	0.08 C, 1.59 Mn, 0.25 Si, 0.46 Cr, 0.11 Mo, 1000 Ti+Nb+V, 10 B
(Zhao, Yang, Xiao et al., 2003)	210 - 1800	1800	0.05 C, 1.94 Mn, 0.30 Si
(Shim, Byun et al., 2001)	300	-	0.14 C, 1.84 Mn, 0.18 Si, 110 Ti
(Rees & Bhadeshia, 1994)	300	-	0.05 C, 1.08 Mn, 0.46 Si, 2.20 Ni, 0.65 Cr, 0.41 Mo, 220 V
(Rees & Bhadeshia, 1994)	300	-	0.06 C, 2.02 Mn, 0.48 Si, 2.20 Ni, 0.65 Cr, 0.41 Mo, 250 V
(Yang et al., 2010)	300	-	0.03 C, 0.17 Mn, 1.80 Si, 0.02 Al, 0.25 Cu, 0.18 Ni, 0.26 Cr, 110 Ti, 1200 Nb

6. Influence of Steel Composition

Alloying elements can affect the formation of acicular ferrite as solute element in the matrix or because they form active NMI. Table 5 gives an overview of the most important alloying elements' effects. Elements can enhance ($\uparrow$), not influence ($\leftrightarrow$) or decrease ($\downarrow$) the amount of acicular ferrite. Some elements are described to have a positive effect on the formation of acicular ferrite at low contents (L), but a negative one with higher additions (H). About the impact of carbon and aluminium contradictory opinions (?) can be found in literature. An extensive discussion of the listed elements' effects can be found in Loder (2016).

Table 5. Influence of alloying elements on the formation of acicular ferrite

Element	Acts as solute	Acts as NMI	Effect on AF	Literature
C	x	x	? ($\uparrow$ / $\downarrow$)	(Düren, 1983; Evans, 1982, 1983; Farrar & Harrison, 1987; Mu, Mao, Jönsson, & Nakajima, 2016)
O		x	$\uparrow$	(Abson et al., 1978; Düren, 1983; Fox & Evans, 2012; Grong & Matlock, 1986; Liu & Olson, 1986; Saggese et al., 1983)
S		x	L: $\uparrow$ H: $\downarrow$	(Evans, 1986b; Lee & Pan, 1995; Mu, Jönsson, & Nakajima, 2014, 2015)
Ca		x	$\uparrow$	(Lee & Pan, 1995)
Ti		x	$\uparrow$	(Evans, 1986a, 1991b, 1993, 1995; Fox & Evans, 2012; Grong et al., 1995; Ilman, Cochrane, & Evans, 2014, 2015; Saggese et al., 1983; Seo et al., 2015; Vanovsek et al., 2013)
Mn	x	x	L: $\uparrow$ H: $\downarrow$	(Byun et al., 2003; Chai et al., 2014; Chen, Zhang, Liu, Li, & Xu, 2013; Cochrane & Kirkwood, 1978; Evans, 1980; Farrar & Harrison, 1987; Grong & Matlock, 1986; Liu, Kobayashi, Yin, Kuwabara, & Nagai, 2007)
Al		x	? ($\downarrow$ / $\leftrightarrow$)	(Bhadeshia, 2001; Evans, 1990a; Huang, Wang, Jiang, Hu, & Yang, 2016; Inoue et al., 1993; Ogibayashi, 1994; Shim, Cho et al., 2001; Yamada, Terasaki, & Komizo, 2008)
Mg	?	?	$\uparrow$	(Min, Li, Yu, Liu, & Jiang, 2016; Song et al., 2015)
Si	x		$\downarrow$	(Evans, 1986a; Shim, Cho et al., 2001)
B	x		L: $\uparrow$ H: $\downarrow$	(Bhadeshia, 2001; Düren, 1983; Evans, 1996; Grong & Matlock, 1986; Ilman, Cochrane, & Evans, 2012, 2015; Peng, Chen, & Xu, 2001; Takamura & Mizoguchi, 1990; Yamamoto et al., 1996; Zhang et al., 2010)
Nb		x	$\uparrow$	(Chen et al., 2013; Evans, 1991a)
V		x	$\uparrow$	(Capdevila et al., 2009; Evans, 1991c; Garcia-Mateo et al., 2008; Mu et al., 2016)
Cr	x		L: $\uparrow$ H: $\downarrow$	(Bhadeshia, 2001; Capdevila et al., 2006; Evans, 1987, 1989, 1990b; Smith et al., 1971)
Mo	x		L: $\uparrow$ H: $\downarrow$	(Bhadeshia, 2001; Capdevila et al., 2006; Evans, 1987, 1989, 1990b; Smith et al., 1971)
Ni	x		L: $\uparrow$ H: $\downarrow$	(Bhadeshia, 2001; Capdevila et al., 2006; Evans, 1987, 1989, 1990b; Smith et al., 1971)

6.1 Carbon

To date, several publications about the interaction of carbon and acicular ferrite have emerged, but they describe divergent effects: (Düren, 1983; Evans, 1982, 1983; Farrar & Harrison, 1987; Mu et al., 2016)

- Evans (1982, 1983) found a steady increase in the amount of acicular ferrite with increasing carbon content from 0.05 to 0.15 wt.-% when investigating weld metals with 0.65 - 1.72 wt.-% manganese and approximately 0.30 wt.-% silicon.

- Mu et al. (2016) investigated the influence of carbon between 0.01 and 0.40 wt.-% using thermodynamic calculations. The other alloying elements were kept constant at 1.00 wt.-% manganese, 0.40 wt.-% silicon, 100 wt.-ppm titanium and 20 wt.-ppm aluminum. They calculated the driving force for ferrite nucleation and showed a steady decrease with rising carbon content. Furthermore, they demonstrated that the nucleation capability of inclusions is reduced by higher carbon contents; subsequently, the minimum size of the inclusions to act as nuclei is significantly increased (see Figure 2).

- Düren (1983) studied weldments with 0.03 - 0.13 wt.-% carbon, 1.30 wt.-% manganese, 0.35 wt.-% silicon, 0.70 wt.-% nickel, 0.02 wt.-% aluminum, 0.30 wt.-% molybdenum, 250 wt.-ppm niobium and 400 wt.-ppm vanadium. He observed an optimum carbon content for acicular ferrite of 0.07 wt.-%. Below, the acicular ferrite is substituted by coarser ferrite veins, and above, it is substituted by martensite.

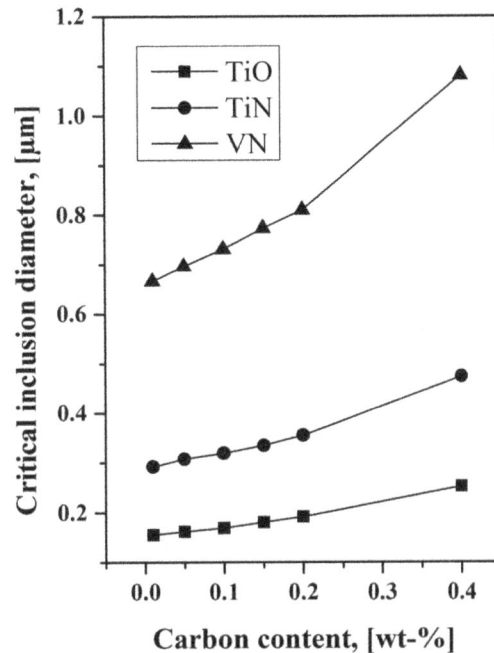

Figure 2. Influence of carbon content on the acicular ferrite formation: increase in the critical inclusion diameter for acicular ferrite nucleation; adapted from (Mu et al., 2016)

Some authors (Chai et al., 2014; Furuhara, Yamaguchi et al., 2003; Gregg & Bhadeshia, 1997) reported that carbon-depleted zones promote acicular ferrite formation. Carbon is an austenite-stabilizing element, so zones with reduced carbon content tend to transform to ferrite. On the one hand, carbon depletion can result from the precipitation of carbide particles, such as VC. On the other hand, Gregg and Bhadeshia (1997) suggested that oxygen, which can be released by oxide particles, e.g., TiO_2 inclusions, reacts with carbon in the surrounding matrix and removes it from solution (Chai et al., 2014; Furuhara, Yamaguchi et al., 2003; Gregg & Bhadeshia, 1997).

6.2 Manganese

The literature describes four main effects of manganese in acicular ferritic steels:

- Suppression of polygonal ferrite and pearlite: Solute manganese suppresses the formation of polygonal ferrite and pearlite, which enhances the acicular ferrite formation. (Farrar & Harrison, 1987)

- Formation of manganese-depleted zones around inclusions: Manganese can be merged in NMI, so manganese-depleted zones around inclusions are formed. These manganese-depleted zones are often described as essential for the nucleation of acicular ferrite. (Chai et al., 2014; Farrar & Harrison, 1987)

- Depression of the austenite-ferrite-transition temperature: Solute manganese decreases the austenite-ferrite-transition temperature. Low phase transition temperatures reduce the driving force for ferrite nucleation. Hence, Chai et al. (2014) asserted that manganese contents that are too high may lead to the suppression of acicular ferrite formation. (Chai et al., 2014)

- Augmentation of yield strength: Solute manganese increases the steel's yield strength. Regarding toughness, this increase in yield strength by manganese can counteract the positive effect of acicular ferrite. Therefore, Grong and Matlock (1986) defined a manganese content of 1.5 wt.-% as optimum for mild and low-alloy steel weld metals.

Thus, it can be concluded that solute manganese negatively influences the formation of acicular ferrite, but if it is bonded in NMI, it becomes supporting. Manganese contents below 1.5 - 2.0 wt.-% are mostly seen as beneficial for acicular ferrite formation, while higher contents are described as having a negative effect on the mechanical properties (Chai et al., 2014; Evans, 1980; Farrar & Harrison, 1987; Grong & Matlock, 1986).

Generally, the formation of manganese-depleted zones is seen as the most important effect of manganese in acicular ferritic steels. Notwithstanding, there are still controversial opinions how this chemical fluctuations in the matrix form. Some authors (Chai et al., 2014; Hui et al., 2014; Kimura et al., 2002; Sarma et al., 2009; Song et al., 2015; Yang et al., 2012a) report that MnS, which precipitates and grows during cooling owing to the change in sulfur solubility, creates such manganese-depleted zones. Others (Byun et al., 2003; Cai et al., 2015; Chai et al., 2014; Gregg & Bhadeshia, 1997; Huang et al., 2016; Ogibayashi, 1994; Seo et al., 2015; Shim, Cho et al., 2001; Shim, Oh et al., 2001; Shim, Byun et al., 2001; Wan et al., 2012) argue that manganese is absorbed by titanium containing inclusions; although, the reason for this manganese absorption is hardly discussed in the literature. One suggestion offered by Gregg and Bhadeshia (1997) asserted that the manganese absorption is associated with the transformation of inclusions from Ti_2O_3 to Ti_3O_5. Another explanation is provided by Shim et al. (2001), who stated that the manganese absorption is caused by the large amount of cation vacancies in titanium oxide and the high solubility of manganese in these cation sites [e.g., > 10 mol.-% at 1473,15 K (1200 °C) in Ti_2O_3]. Seo et al. (2015) described Ti_2O_3 particles as especially favorable for manganese absorption because Ti_2O_3 provides many cation vacancies. Ti^{3+} ions have a similar radius to Mn^{3+} ions, and Ti_2O_3 has nearly the same crystal structure as $MnTiO_3$.

There are two possibilities described in the literature regarding how manganese-depleted zones promote acicular ferrite formation: (Byun et al., 2003; Chai et al., 2014; Chen et al., 2013; Cochrane & Kirkwood, 1978; Liu et al., 2007).

- Reduction of austenite stabilization: Because manganese is an austenite-stabilizing element, the nucleation energy for ferrite is reduced in areas with lower manganese contents.

- Increase in the austenite-ferrite-transition temperature: Composition fluctuations support acicular ferrite nucleation because of a change in the austenite-ferrite-transition temperature. Austenite-stabilizing elements decrease the transition temperature and therefore lead to a reduction in the nucleation driving force. If depleted zones of austenite-stabilizing elements form, the transition temperature is increased, and the driving force for acicular ferrite nucleation is also increased.

6.3 Titanium

Generally, an increase in acicular ferrite with the addition of titanium has been reported in the literature. (Evans, 1986a, 1993, 1995; Fox & Evans, 2012; Grong et al., 1995; Ilman et al., 2015; Saggese et al., 1983; Seo et al., 2015; Vanovsek et al., 2013) The main influence of titanium is to form titanium-containing inclusions, which provide active nucleation sites for acicular ferrite. Bonnet and Charpentier (1983) conducted a comprehensive study of weld metals with 0.04 - 0.09 wt.-% carbon, 0.80 - 2.00 wt.-% manganese, 0.20 - 0.75 wt.-% silicon, 0.05 - 0.15 wt.-% copper, 30 - 140 wt.-ppm titanium and 60 - 120 wt.-ppm vanadium. They found a critical titanium content of 40 - 50 ppm below which it is not possible to form acicular ferrite independent of other conditions (see Figure 3). Similar observations were made by Byun et al. (2003), who determined that without titanium addition, bainite is formed instead of acicular ferrite. With increasing titanium content, the microstructure changed from bainite to acicular ferrite. Additionally, Evans (1991b) and Ilman et al. (2014) reported that without titanium, nearly no acicular ferrite forms in welds with approximately 0.07 wt.-% carbon, 1.50 wt.-% manganese, 0.35 wt.-% silicon and 6 - 255 wt.-ppm titanium. However, a slight increase in the titanium content leads to a boost in acicular ferrite, as shown in Figure 3. The depression of the acicular ferrite at approximately 80 ppm titanium is the result of changes in inclusion morphology.

Nevertheless, deoxidation with titanium is not popular due to its high price. Deoxidation with cheaper elements, such as aluminum, manganese or silicon, and the modification of deoxidation products by titanium are more common in secondary refining. Hui et al. (2014) and Yamada et al. (2008) also proved that inclusions that only show thin layers of TiOx can act as active nuclei. Saggese et al. (1983) demonstrated that only 2 % TiOx in inclusions is necessary to maximize acicular ferrite.

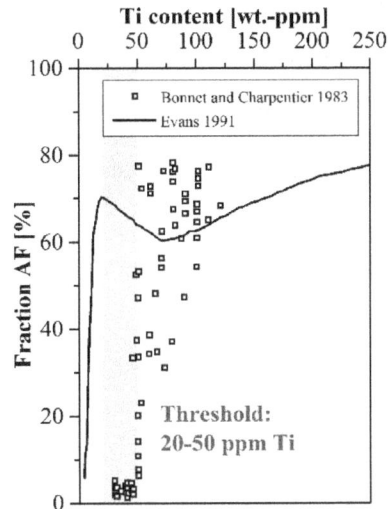

Figure 3. Influence of titanium content on the fraction of acicular ferrite in weld deposits; adapted from Bonnet and Charpentier, (1983) and Evans (1991b)

7. Influence of Non-Metallic Inclusions

7.1 Inclusion Size

The capability of NMI to act as nuclei increases with increasing particle size. This is because the nucleation energy is significantly reduced and larger stress fields, which initiate the austenite-ferrite-transition, are formed.(Abson et al., 1978; Bhadeshia, 2001; Bin & Bo, 2012; Grong & Matlock, 1986; Ricks et al., 1981; Ricks et al., 1982; Sarma et al., 2009). Recent observations (Huang et al., 2016; Lee et al., 2000; Mu et al., 2014, 2015; Mu et al., 2016; Seo et al., 2015; Song et al., 2015) have determined the probability of acicular ferrite nucleation on particles with varying sizes according to:

$$Nucleation\ probability = \frac{Number\ of\ active\ inclusions}{Total\ number\ of\ inclusions}$$

(1)

As illustrated in Figure 4, these authors described a boost in nucleation potential with increasing inclusion size, which is in good accordance with earlier findings of Ricks et al. (1981, 1982). However, it should be noted that in the literature, lower and upper critical sizes for active inclusions are commonly described. The lower critical value is the minimum size beneath which particles are too small to act as active nuclei. The lower critical size was determined to be 0.3 - 0.5 µm in various studies. (Barbaro et al., 1989; Huang et al., 2016; Lee et al., 2000; Mu et al., 2014; Mu et al., 2016) Above the upper critical value, no further significant improvements in the nucleation probability are observable. Ricks et al. (1981, 1982), Lee et al. (2000) and Mu et al. (2014, 2016), who investigated inclusions smaller than 1.5 µm, defined 1 µm as the upper critical value. Above 1 µm, the probability curve flattens out. Huang et al. (2016), who considered inclusions up to 7 µm, observed 1.5 µm as the upper critical value but described a sharp decrease in the nucleation probability above 4 µm, resulting in inert behavior of inclusions larger than 6.5 µm. Similar results were gained by Song et al. (2015) and Wang et al.(2012), who found inclusions with sizes of 1 - 3 µm and 1 - 2 µm, respectively, as the most appropriate nuclei. Inclusions below or above these ranges were less active. Seo et al. (2015) determined the upper critical value to be approximately 0.6 - 0.7 µm and showed that the value decreased with increased titanium contents, which may result from the high nucleation potential of titanium-containing inclusions (Barbaro et al., 1989; Huang et al., 2016; Lee et al., 2000; Liu & Olson, 1986; Mu et al., 2014, 2015; Mu et al., 2016; Ricks et al., 1981; Ricks et al., 1982; Seo et al., 2015; Song et al., 2015; Wang et al., 2012).

The demand for larger inclusions is in contrast to the requirements of high-toughness steels in which inclusions should be as small as possible. On the one hand, it is therefore important to consider the findings of Ref. (Huang et al., 2016; Lee et al., 2000; Mu et al., 2014; Ricks et al., 1981), which showed that an increase in inclusion size above 1.0 - 1.5 µm does not have a further beneficial effect on the nucleation probability of acicular ferrite. Hence, an inclusion size of 1 µm can be suggested as optimum for acicular ferrite because the nucleation energy is minimized, but the material properties are not degraded by large inclusions. On the other hand, the nature of inclusions is much more important than their size or number, so acicular ferrite microstructures can also be produced by appropriate modifications of the inclusions' composition without increasing their content (Bhadeshia, 2001; Huang et al., 2016; Lee et al., 2000; Mu et al., 2014, 2015; Sarma et al., 2009).

Several authors reported that the inclusion size also affects the structure of nucleating ferrite grains: Song et al. (2015) observed, for example, that decreasing the inclusion size also decreased the size of the acicular ferrite laths. Fattahi et al. (2013) and Zhang and Farrar (1996) described that the inclusion size influences how ferrite nucleates on NMI: Smaller inclusions are engulfed by the ferrite plates ('boxing in'). With increasing inclusion size, more needle-shaped ferrite grains will emanate from the particle. Large inclusions act as nuclei for several acicular ferrite laths, and the typical star-like structure of acicular ferrite is developed (Fattahi et al., 2013; Zhang & Farrar, 1996).

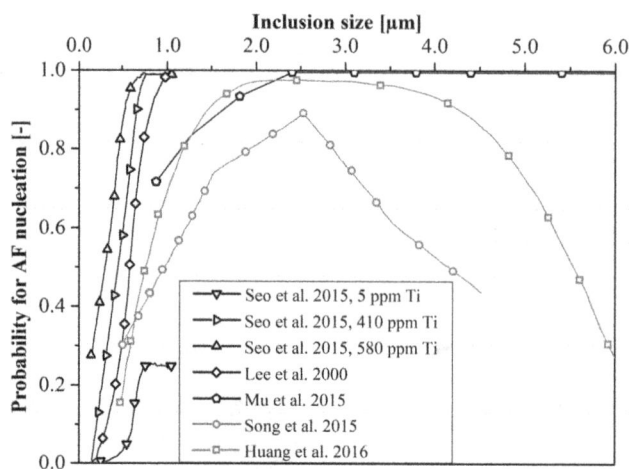

Figure 4. Impact of inclusion size on the probability for acicular ferrite nucleation; adapted from Ref. (Huang et al., 2016; Lee et al., 2000; Mu et al., 2015, 2015; Seo et al., 2015; Song et al., 2015)

7.2 Inclusion Number

NMI are indispensable for acicular ferrite. Several authors (Abson et al., 1978; Bin & Bo, 2012; Garcia-Mateo et al., 2008; Thewlis, 2006) reported that few active inclusions inhibit the nucleation of acicular ferrite. Andersson et al. (2011) calculated that at least ten active inclusions have to be inside an austenite grain of 10 μm in size to provide efficient conditions for acicular ferrite. Nevertheless, higher numbers of inclusions limit the AGS by pinning effects. This worsens the conditions for acicular ferrite nucleation and downgrades the steel's mechanical properties. (Abson et al., 1978; Andersson et al., 2011; Bin & Bo, 2012; Cochrane & Kirkwood, 1978; Garcia-Mateo et al., 2008; Grong & Matlock, 1986; Kikuchi et al., 2011; Lee & Pan, 1995; Liu & Olson, 1986; Sarma et al., 2009; Thewlis, 2006).

Liu and Olson (1986) reported that the number of active inclusions influences the structure of acicular ferrite. If the density of active particles is low, fewer ferrite plates are formed, so they can grow continuously before their growth is stopped by impingement with other plates. With an increasing amount of active particles more acicular ferrite needles are nucleated, so the free space for growth is reduced, and the growing plates are stopped earlier, which results in a finer structure.

7.3 Inclusion Type

Barbaro et al. (1989) asserted that all types of inclusions became active at a critical particle size, but almost all other research has come to the conclusion that the potential of NMI strongly depends on chemical composition. Some inclusion types are seen as active, and some as inactive, strongly dependent on the observed system of the inclusion landscape, steel composition, CR and AGS.

Table 6 - Table 10 give an overview of inclusion types described as active or inactive in the literature. The listed types are divided into three classes: inclusions that are generally described as active, inclusions that are generally described as inactive and inclusions whose effect is seen contradictory in the literature. There are several reports in the literature (e.g., Bin & Bo, 2012; Nako et al., 2015; Thewlis, 2006) about the use of rare earth elements to create active nuclei for acicular ferrite, but rare earth inclusions are not considered in this overview.

Table 6. Literature review on the potential of oxidic inclusions

		ACTIVE	INACTIVE
OXIDES	**(Al,Mg,Zr)O$_x$**	(Li et al., 2015; Min et al., 2016)	
	(Al,Si)O$_x$	(Sarma et al., 2009)	
	(Mn,Al,Si)O$_x$	(Devillers et al., 1983; Inoue et al., 1993; Sarma et al., 2009; Zhang & Farrar, 1996)	
	(Ti,Mn)O$_x$	(Byun et al., 2003; Fox & Evans, 2012; Kikuchi et al., 2011; Lee & Pan, 1995; Nako et al., 2014; Shim, Oh et al., 2001; Shim, Byun et al., 2001)	
	(Ti,Mn,Al)O$_x$	(Huang et al., 2016; Ilman et al., 2015; Inoue et al., 1993; Lee & Pan, 1995; Yamada et al., 2008)	
	(Ti,Mn,Al,Si)O$_x$	(Wan et al., 2012; Wang et al., 2012; Zhang & Farrar, 1996)	
	(Ti,Mn,Si)O$_x$	(Nako et al., 2014; Zhang & Farrar, 1996)	
	(Ti,Si)O$_x$	(Zhang & Farrar, 1996)	
	TiO$_x$	(Bhadeshia, 2001; Fox & Evans, 2012; Gregg & Bhadeshia, 1997; Hossein Nedjad & Farzaneh, 2007; Ilman et al., 2012, 2015; Kikuchi et al., 2011; Lee & Pan, 1995; Mu et al., 2016; Shim, Cho et al., 2001; Zhang & Farrar, 1996; Zhang et al., 2010)	(Düren, 1983; Shim, Oh et al., 2001)
	(Ti,Al)O$_x$	(Ilman et al., 2015; Wang et al., 2012)	(Huang et al., 2016; Wang et al., 2012)
	SiO$_2$	(Zhang & Farrar, 1996)	(Liu et al., 2007; Shim, Oh et al., 2001)
	Al$_2$O$_3$		(Bhadeshia, 2001; Devillers et al., 1983; Evans, 1990a; Fox & Evans, 2012; Gregg & Bhadeshia, 1997; Huang et al., 2016; Kikuchi et al., 2011; Min et al., 2016; Shim, Cho et al., 2001; Shim, Oh et al., 2001; Wang et al., 2012; Yamada et al., 2008; Zheng, Wang, Li, Shang, & He, 2012)
	FeO$_x$		(Sarma et al., 2009)
	(Mn,Al)O$_x$		(Gregg & Bhadeshia, 1997; Yamada et al., 2008)
	(Mn,Si)O$_x$		(Byun et al., 2003; Shim, Oh et al., 2001)
	(Mn,Si,Fe)O$_x$		(Liu et al., 2007)

Table 7. Literature review on the potential of sulfidic inclusions.

		ACTIVE	INACTIVE
SULPHIDES	**CuS$_y$**	(Zhang & Farrar, 1996)	
	(Mn,Cu)S$_y$	(Capdevila et al., 2006; Díaz-Fuentes et al., 2003; Madariaga et al., 2001; Zhang & Farrar, 1996)	
	FeS	(Liu et al., 2007)	
	(Fe,Cu)S$_y$	(Liu et al., 2007)	
	(Mn,Fe,Cu)S$_y$	(Liu et al., 2007)	
	MnS	(Bhadeshia, 2001; Kikuchi et al., 2011; Lee & Pan, 1995; Sarma et al., 2009; Yang, Xu et al., 2015; Zhang & Farrar, 1996; Zhang et al., 2012a; Zhang et al., 2010)	(Furuhara, Yamaguchi et al., 2003; Gregg & Bhadeshia, 1997; Huang et al., 2016; Hui et al., 2014; Mu et al., 2014; Sarma et al., 2009; Shim, Oh et al., 2001; Zheng et al., 2012)

Table 8. Literature review on the potential of oxysulfidic inclusions

		ACTIVE	INACTIVE
	$(Al,Mn,Si)O_xS_y$	(Inoue et al., 1993)	
	$(Mn,Al,Mg)O_xS_y$	(Min et al., 2016)	
	$(Mn,Al,Mg,Zr)O_xS_y$	(Li et al., 2015; Min et al., 2016)	
	$(Mn,Al,Si,Cu)O_xS_y$	(Zhang & Farrar, 1996)	
	$(Ti,Mn)O_xS_y$	(Andersson et al., 2011; Bhadeshia, 2001; Fattahi et al., 2013; Huang et al., 2016; Lee & Pan, 1995; Mu et al., 2014, 2015; Wang et al., 2012; Zheng et al., 2012)	
OXYSULPHIDES	$(Ti,Mn,Al)O_xS_y$	(Cai et al., 2015; Lee & Pan, 1995; Yang et al., 2008)	
	$(Ti,Mn,Al,Ca)O_xS_y$	(Lee & Pan, 1995)	
	$(Ti,Mn,Al,Mg)O_xS_y$	(Song et al., 2015)	
	$(Ti,Mn,Al,Si)O_xS_y$	(Barbaro et al., 1989)	
	$(Ti,Mn,Al,Si,Zr)O_xS_y$	(Wan et al., 2015)	
	$(Ti,Mn,Ca)O_xS_y$	(Lee & Pan, 1995)	
	$(Ti,Mn,Si)O_xS_y$	(Hui et al., 2014)	
	$(Ti,Mn,Si,Cu)O_xS_y$	(Evans, 1991b; Ilman et al., 2012, 2014)	
	$(Mn,Al)O_xS_y$	(Yang et al., 2008)	(Li et al., 2015; Min et al., 2016; Wang et al., 2012; Yamada et al., 2008)
	$(Ti,Mn,Cu)O_xS_y$	(Díaz-Fuentes et al., 2003)	(Díaz-Fuentes et al., 2003)
	$(Al,Ca)O_xS_y$		(Eijk et al., 1999)
	$(Mn,Al,Si)O_xS_y$		(Devillers et al., 1983)
	$(Mn,Si)O_xS_y$		(Byun et al., 2003; Seo et al., 2015)
	$(Mn,Si,Cu)O_xS_y$		(Evans, 1991b; Ilman et al., 2014)
	$(Si,Cu)O_xS_y$		(Liu et al., 2007)

Table 9. Literature review on the potential of nitridic and oxynitridic inclusions

		ACTIVE	INACTIVE
(OXY)NITRIDES	TiO_xN	(Jin et al., 2003)	
	$(Ti,Al)O_xN$	(Ilman et al., 2015)	
	$(Ti,B)O_xN$	(Ilman et al., 2012; Zhang et al., 2012a)	
	VN	(Capdevila et al., 2009; Garcia-Mateo et al., 2008; Mu et al., 2016)	
	TiN	(Bhadeshia, 2001; Grong et al., 1995; Ilman et al., 2014, 2015; Jin et al., 2003; Kikuchi et al., 2011; Mu et al., 2016; Zhang et al., 2010)	(Düren, 1983; Shim, Oh et al., 2001; Zheng et al., 2012)

Table 10. Literature review on the potential of other inclusion types in addition to those mentioned before

		ACTIVE	INACTIVE
OTHERS	$(Mn,Al,V)O_xS_yN$	(Ogibayashi, 1994)	
	$(Ti,Mn,Al)O_xS_yN$	(Eijk et al., 1999; Wan et al., 2014)	
	$(Ti,Mn,Cu,B)O_xS_yN$	(Ilman et al., 2012)	
	$(Ti,Mn,Al,Si)O_xS_yN$	(Barbaro et al., 1989; Mu et al., 2016)	
	$(Mn,V)S_yCN$	(Capdevila et al., 2009; Furuhara, Shinyoshi et al., 2003)	
	$(Ti,Mn)O_xS_yN$	(Mu et al., 2014; Ogibayashi, 1994)	
	$(Ti,Mn)S_yN$	(Jin et al., 2003)	
	V(C,N)	(Capdevila et al., 2009; Garcia-Mateo et al., 2008)	(Capdevila et al., 2009)
	VC		(Capdevila et al., 2009)

8. Conclusions

The formation of acicular ferrite is mainly influenced by austenite grain size, cooling rate, steel composition and non-metallic inclusions. In recent decades, extensive work was conducted to evaluate the impact of these parameters, and important knowledge was gained. The present work aimed at displaying the current state of knowledge regarding acicular ferrite formation and its influencing factors. The key aspects can be summarized as follows:

- Austenite grain size: A steel's capability for acicular ferrite increases with increasing austenite grain size. However, the positive effect of large austenite grains is diminished at austenite grain sizes above 150 μm.

- Cooling rate: The optimum cooling rate to produce a maximum fraction of acicular ferrite essentially depends on the steel's composition and its production route. Deformation and thermomechanical treatment significantly change the optimum cooling rates compared to the cooling without deformation.

- Steel composition:
  - The addition of titanium and manganese is essential to the production of acicular ferrite. Titanium forms titanium oxides, which were found to be highly active nucleation sites for acicular ferrite. Manganese merges in inclusions and forms manganese-depleted zones, which stimulate the nucleation of acicular ferrite.
  - The adjustment of appropriate carbon, oxygen and sulfur contents is important to provide sufficient conditions for acicular ferrite.
  - Low to moderate concentrations of manganese, boron, chromium, nickel and molybdenum are described in the literature to promote the formation of acicular ferrite, while higher amounts of these elements negatively influence the formation of acicular ferrite.
  - Magnesium, niobium, vanadium and calcium are asserted to enhance the formation of acicular ferrite.
  - Additions of aluminum and silicon are generally seen as negative for acicular ferrite.

- Non-metallic inclusions:
  - The nucleation potential of acicular ferrite significantly increases with inclusion size. Nevertheless, an inclusion size of 1 μm is defined as optimum because of mechanical properties.
  - The number of inclusions should be high enough to provide a sufficient number of nucleation sites, but with respect to the requirements of high-quality steels, it should be as low as possible.
  - Various inclusion types have already been investigated in the literature, but inconsistent information about some inclusion types still prevails. This may result from the interaction between inclusion type and steel composition. Generally, titanium- and manganese-containing inclusions are described in particular as favorable for acicular ferrite nucleation.

Acknowledgments

Financial support by the Federal Ministry for Transport, Innovation and Technology (bmvit) and the Austrian Science Fund (FWF): [TRP 266-N19] is gratefully acknowledged.

References

Aaaronson, H. I., & Wells, C. (1956). Sympathetic nucleation of ferrite. *Journal of Metals*, 1216–1223.

Abson, D. J., Dolby, R. E., & Hart, P. H. (1978). The role of nonmetallic inclusions in ferrite nucleation in carbon steel weld metals. *Proceedings of Conference on 'Trends in Steels and Consumables for Welding'*, 75–101.

Adrian, H., & Pickering, F. B. (1991). Effect of titanium additions on austenite grain growth kinetics of medium carbon V–Nb steels containing 0·008–0·018%N. *Materials Science and Technology*, 7(2), 176–182. http://dx.doi.org/10.1179/026708391790194860

Andersen, I., & Grong, Ø. (1995). Analytical modelling of grain growth in metals and alloys in the presence of growing and dissolving precipitates—I. Normal grain growth. *Acta Metallurgica et Materialia*, 43(7), 2673–2688. http://dx.doi.org/10.1016/0956-7151(94)00488-4

Andersson, M., Janis, J., Holappa, L., Kiviö, M., Naveau, P., Brandt, M., & Van der Eijk, C. (2011). *Grain size control in steel by means of dispersed non-metallic inclusions - GRAINCONT: Final report. Research Fund for Coal and*

Steel series: Vol. 24993. Luxembourg: Publications Office of the European Union. Retrieved from http://bookshop.europa.eu/en/grain-size-control-in-steel-by-means-of-dispersed-non-metallic-inclusions-grainco nt -pbKINA24993/

Barbaro, F. J., Krauklis, P., & Easterling, K. E. (1989). Formation of acicular ferrite at oxide particles in steels. *Materials Science and Technology, 5*(11), 1057–1068. http://dx.doi.org/10.1179/026708389790340888

Bernhard, C., Bragin, S., & Vanovsek, W. (2012). Near surface austenite grain growth in different casting processes for steel: Experiments and modeling. *Proceedings of 5th Congress on 'Science and Technology of Steelmaking'*, 1–8.

Bernhard, C., Reiter, J., & Presslinger, H. (2008). A model for predicting the austenite grain size at the surface of continuously-cast slabs. *Metallurgical and Materials Transactions B, 39*(6), 885–895. http://dx.doi.org/10. 1007/s11663-008-9197-8

Bhadeshia, H. K. D. H. (1992). Modelling of steel welds. *Materials Science and Technology, 8*(2), 123–133. http://dx.doi.org/10.1179/mst.1992.8.2.123

Bhadeshia, H. K. D. H. (2001). *Bainite in steels: Transformations, microstructure and properties* (2nd ed.). London: IOM Communications Ltd.

Bhadeshia, H. K. D. H. (2010). A personal commentary on "Transformation of austenite at constant subcritical temperatures". *Metallurgical and Materials Transactions A, 41*(6), 1351–1390. http://dx.doi.org/10.1007/ s11661-010-0250-2

Bin, W., & Bo, S. (2012). In situ observation of the evolution of intragranular acicular ferrite at Ce-containing inclusions in 16Mn steel. *steel research international, 83*(5), 487–495. http://dx.doi.org/10.1002/srin.2 01100266

Bonnet, C., & Charpentier, J.-P. (1983). Effect of deoxidation residues in wire and of some particular oxides in CS fused fluxes on the microstructure of submerged-arc weld metals. *Proceedings of Conference on 'The Effects of residual, impurity, and microalloying elements on weldability and weld properties'*. (Paper 8), 1–31.

Byun, J.-S., Shim, J.-H., Cho, Y. W., & Lee, D. N. (2003). Non-metallic inclusion and intragranular nucleation of ferrite in Ti-killed C–Mn steel. *Acta Materialia, 51*(6), 1593–1606. http://dx.doi.org/10.1016/S1359-6454 (02)00560-8

Cai, Z., Zhou, Y., Tong, L., Yue, Q., & Kong, H. (2015). Effect of Ti-Al-O inclusions on the formation of intragranular acicular ferrite. *Materials Testing, 57*(No. 7-8), 649–654. http://dx.doi.org/10.3139/120.110760

Capdevila, C., García-Mateo, C., Chao, J., & Caballero, F. G. (2009). Effect of V and N precipitation on acicular ferrite formation in sulfur-lean vanadium steels. *Metallurgical and Materials Transactions A, 40*(3), 522–538. http://dx.doi.org/10.1007/s11661-008-9730-z

Capdevila, C., Ferrer, J. P., García-Mateo, C., Caballero, F. G., López, V., & García de Andrés, C. (2006). Influence of deformation and molybdenum content on acicular ferrite formation in medium carbon steels. *ISIJ International, 46*(7), 1093–1100. http://dx.doi.org/10.2355/isijinternational.46.1093

Chai, F., Su, H., Yang, C.-F., & Xue, D.-M. (2014). Nucleation behavior analysis of intragranular acicular ferrite in a Ti-killed C-Mn steel. *Journal of Iron and Steel Research International, 21*(3), 369–374. http://dx.doi.org/10.1016/S1006-706X(14)60057-1

Chen, Y., Zhang, D., Liu, Y., Li, H., & Xu, D. (2013). Effect of dissolution and precipitation of Nb on the formation of acicular ferrite/bainite ferrite in low-carbon HSLA steels. *Materials Characterization, 84*, 232–239. http://dx.doi.org/10.1016/j.matchar.2013.08.005

Cochrane, R. C., & Kirkwood, P. R. (1978). The effect of oxygen on weld metal microstructure. *Proceedings of Conference on 'Trends in Steels and Consumables for Welding'*. (Paper 35), 103–121.

Davenport, E. S., Bain, E. C., & Kearny, N. J. (1930). Transformation of austenite at constant subcritical temperatures. *Transactions of the Metallurgical Society of AIME, 90*, 117–154.

Davenport, E. S., Bain, E. C., & Kearny, N. J. (1970). Transformation of austenite at constant subcritical temperatures: Metallurgical Classics Communications. *Metallurgical Transactions, 1*(12), 3503–3530.

Devillers, L., Kaplan, D., Marandet, B., Ribes, A., & Riboud, P. V. (1983). The effect of low level concentrations of some elements on the toughness of submerged-arc welded CMn steel welds. *Proceedings of Conference on 'The Effects of residual, impurity, and microalloying elements on weldability and weld properties'*. (Paper 1), 1–11.

Díaz-Fuentes, M., Iza-Mendia, A., & Gutiérrez, I. (2003). Analysis of different acicular ferrite microstructures in low-carbon steels by electron backscattered diffraction. Study of their toughness behavior. *Metallurgical and Materials Transactions A*, *34*(11), 2505–2516. http://dx.doi.org/10.1007/s11661-003-0010-7

Düren, C. F. (1983). Toughness properties in two-pass welds. *Proceedings of Conference on 'The Effects of residual, impurity, and microalloying elements on weldability and weld properties'*. (P34), 1–10.

Eijk, C. van der, Grong, Ø., & Hjelen, J. (1999). Quantification of inclusion-stimulated ferrite nucleation in wrought steel using the SEM-EBSD technique. *Proceedings of Conference on 'Solid-Solid Phase Transformations'*, 1573–1576.

Evans, G. M. (1980). The effect of manganese on the microstructure and properties of C-Mn all-weld metal deposits. *Welding Research Supplement*, 67.

Evans, G. M. (1982). The effect of carbon on the microstructure and properties of C-Mn all-weld deposits: Der Einfluss des Kohlenstoffs auf Mikrogefüge und Eigenschaften von C- und Mn-haltigen reinen Schweißgütern. *OERLIKON-Schweißmitteilungen*, *40*(99), 17–31.

Evans, G. M. (1983). The effect of carbon on the microstructure and properties of C-Mn all-weld metal deposits. *Welding Research Supplement*, 313.

Evans, G. M. (1986a). The effect of silicon on the microstructure and properties of C-Mn all-weld deposits: Der Einfluss von Silizium auf Mikrogefüge und Eigenschaften von C- und Mn-haltigen reinen Schweißgütern. *OERLIKON-Schweißmitteilungen*, *44*(110), 19–33.

Evans, G. M. (1986b). The effects of sulphur and phosphorus on the microstructure and properties of C-Mn all-weld deposits: Der Einfluss von Schwefel und Phosphor auf Mikrogefüge und Eigenschaften von C- und Mn-haltigen reinen Schweißgütern. *OERLIKON-Schweißmitteilungen*, *44*(111), 22–35.

Evans, G. M. (1987). The effect of molybdenum on the microstructure and properties of C-Mn all-weld metal deposits: Der Einfluss von Molybdän auf Mikrogefüge und Eigenschaften von C- und Mn-haltigen reinen Schweißgütern. *OERLIKON-Schweißmitteilungen*, *45*(115), 10–27.

Evans, G. M. (1989). The effect of chromium on the microstructure and properties of C-Mn all-weld metal deposits: Der Einfluss von Chrom auf Mikrogefüge und Eigenschaften C- und Mn-haltiger reiner Schweißgüter. *OERLIKON-Schweißmitteilungen*, *47*(120), 17–34.

Evans, G. M. (1990a). The effect of aluminium on the microstructure and properties of C-Mn all-weld deposits: Der Einfluss von Aluminium auf Mikrogefüge und Eigenschaften C- und Mn-haltiger reiner Schweißgüter. *OERLIKON-Schweißmitteilungen*, *48*(124), 15–31.

Evans, G. M. (1990b). The effect of nickel on the microstructure and properties of C-Mn all-weld metal deposits: Der Einfluss von Nickel auf Mikrogefüge und Eigenschaften C- und Mn-haltiger reiner Schweißgüter. *OERLIKON-Schweißmitteilungen*, *48*(122), 18–35.

Evans, G. M. (1991a). The effect of niobium in manganese containing MMA weld deposits: Der Einfluss von Niob in Mn-haltigem Schweißgut. *OERLIKON-Schweißmitteilungen*, *49*(127), 24–39.

Evans, G. M. (1991b). The effect of titanium on the microstructure and properties of C-Mn all-weld deposits: Der Einfluss von Titan auf Mikrogefüge und Eigenschaften C- und Mn-haltiger reiner Schweißgüter. *OERLIKON-Schweißmitteilungen*, *49*(125), 22–33.

Evans, G. M. (1991c). The effect of vanadium in manganese containing MMA weld deposits: Der Einfluss von Vanadium in Mn-haltigem Schweißgut. *OERLIKON-Schweißmitteilungen*, *49*(126), 18–33.

Evans, G. M. (1993). Microstructure and properties of ferritic steel welds containing Al and Ti. *IIW Doc. II-A-901-93*.

Evans, G. M. (1995). Microstructure and properties of ferritic steel welds containing Al and Ti. *Welding Research Supplement*, 249.

Evans, G. M. (1996). Microstructure and properties of ferritic steel welds containing Ti and B. *Welding Research Supplement*, 251.

Farrar, R. A., & Harrison, P. L. (1987). Acicular ferrite in carbon-manganese weld metals: An overview. *Journal of Materials Science*, *22*(11), 3812–3820. http://dx.doi.org/10.1007/BF01133327

Farrar, R. A., Zhang, Z., Bannister, S. R., & Barritte, G. S. (1993). The effect of prior austenite grain size on the transformation behaviour of C-Mn-Ni weld metal. *Journal of Materials Science, 28*(5), 1385–1390. http://dx.doi.org/10.1007/BF01191982

Fattahi, M., Nabhani, N., Hosseini, M., Arabian, N., & Rahimi, E. (2013). Effect of Ti-containing inclusions on the nucleation of acicular ferrite and mechanical properties of multipass weld metals. *Micron, 45*, 107–114. http://dx.doi.org/10.1016/j.micron.2012.11.004

Fox, A. G., & Evans, G. M. (2012). A comparative study of the non-metallic inclusions in C-Mn steel weld metals containing titanium or aluminium. *Presentation at 9th Conference on 'Trends in Welding Research'*.

Furuhara, T., Shinyoshi, T., Miyamoto, G., Yamaguchi, J., Sugita, N., Kimura, N., ... Maki, T. (2003). Multiphase crystallography in the nucleation of intragranular ferrite on MnS+V(C,N) complex precipitate in austenite. *ISIJ International, 43*(12), 2028–2037. http://dx.doi.org/10.2355/isijinternational.43.2028

Furuhara, T., Yamaguchi, J., Sugita, N., Miyamoto, G., & Maki, T. (2003). Nucleation of proeutectoid ferrite on complex precipitates in austenite. *ISIJ International, 43*(10), 1630–1639. http://dx.doi.org/10.2355/isijinternational. 43.1630

Garbarz, B. (1995). The effect of some continuous casting parameters and microalloying elements on the effectiveness of controlling of austenite grain size. *Journal of Materials Processing Technology, 53*(No. 1-2), 147–158. http://dx.doi.org/10.1016/0924-0136(95)01971-G

Garcia-Mateo, C., Capdevila, C., Caballero, F. G., & Andrés, C. G. de. (2008). Influence of V precipitates on acicular ferrite transformation: Part 1: The role of nitrogen. *ISIJ International, 48*(9), 1270–1275. http://dx.doi.org/10.2355/isijinternational.48.1270

Gourgues, A.-F., Flower, H. M., & Lindley, T. C. (2000). Electron backscattering diffraction study of acicular ferrite, bainite, and martensite steel microstructures. *Materials Science and Technology, 16*(1), 26–40. http://dx.doi.org/10.1179/026708300773002636

Gregg, J. M., & Bhadeshia, H. K. D. H. (1997). Solid-state nucleation of acicular ferrite on minerals added to molten steel. *Acta Materialia, 45*(2), 739–748. http://dx.doi.org/10.1016/S1359-6454(96)00187-5

Grong, Ø., Kluken, A. O., Nylund, H. K., Dons, A. L., & Hjelen, J. (1995). Catalyst effects in heterogeneous nucleation of acicular ferrite. *Metallurgical and Materials Transactions A, 26*(3), 525–534. http://dx.doi.org/10.1007/BF02663903

Grong, Ø., & Matlock, D. K. (1986). Microstructural development in mild and low-alloy steel weld metals. *International Materials Reviews, 31*(1), 27–48. http://dx.doi.org/10.1179/imr.1986.31.1.27

Hossein Nedjad, S., & Farzaneh, A. (2007). Formation of fine intragranular ferrite in cast plain carbon steel inoculated by titanium oxide nanopowder. *Scripta Materialia, 57*(10), 937–940. http://dx.doi.org/10.1016/j.scriptamat.2007.07.016

Hossein Nedjad, S., Zahedi Moghaddam, Y., Mamdouh Vazirabadi, A., Shirazi, H., & Nili Ahmadabadi, M. (2011). Grain refinement by cold deformation and recrystallization of bainite and acicular ferrite structures of C–Mn steels. *Materials Science and Engineering: A, 528*(3), 1521–1526. http://dx.doi.org/10.1016/j.msea.2010.10.064

Huang, Q., Wang, X., Jiang, M., Hu, Z., & Yang, C. (2016). Effects of Ti-Al complex deoxidization inclusions on nucleation of intragranular acicular ferrite in C-Mn steel. *Steel Research International, 87*(4), 445–455. http://dx.doi.org/10.1002/srin.201500088

Hui, K., YunFu, S., YaHui, Z., Qiang, Y., Lianhai, T., & ZhengYu, C. (2014). Effects of titanium oxide precipitates on the acicular ferrite nucleation in carbon structural steel. *Materials Testing, 56*(2), 131–135. http://dx.doi.org/10.3139/120.110534

Huo, J., Liu, Y., Zhang, D., Yan, Z., & Gao, Z. (2013). Isochronal phase transformations of low-carbon high strength low alloy steel upon continuous cooling. *Steel Research International, 84*(2), 184–191. http://dx.doi.org/10.1002/srin.201100277

Ilman, M. N., Cochrane, R. C., & Evans, G. M. (2012). Effect of nitrogen and boron on the development of acicular ferrite in reheated C-Mn-Ti steel weld metals. *Welding in the World, 56*(No. 11-12), 41–50. http://dx.doi.org/10.1007/BF03321394

Ilman, M. N., Cochrane, R. C., & Evans, G. M. (2014). Effect of titanium and nitrogen on the transformation characteristics of acicular ferrite in reheated C–Mn steel weld metals. *Welding in the World, 58*(1), 1–10. http://dx.doi.org/10.1007/s40194-013-0091-x

Ilman, M. N., Cochrane, R. C., & Evans, G. M. (2015). The development of acicular ferrite in reheated Ti–B–Al–N type steel weld metals containing various levels of aluminium and nitrogen. *Welding in the World, 59*(4), 565–575. http://dx.doi.org/10.1007/s40194-015-0231-6

Inoue, T., Ohara, M., Tomita, Y., Tsuda, Y., Tanabe, K., Koyama, K.,. . . Isoda, S. (1993). Development of heavy steel plates with excellent electron beam weldability. *Nippon Steel Technical Report, 58*, 17–25.

Jin, H.-H., Shim, J.-H., Cho, Y. W., & Lee, H.-C. (2003). Formation of intragranular acicular ferrite grains in a Ti-containing low carbon steel. *ISIJ International, 43*(7), 1111–1113. http://dx.doi.org/10.2355/isijinternational.43.1111

Kikuchi, N., Nabeshima, S., Yamashita, T., Kishimoto, Y., Sridhar, S., & Nagasaka, T. (2011). Micro-structure refinement in low carbon high manganese steels through Ti-deoxidation, characterization and effect of secondary deoxidation particles. *ISIJ International, 51*(12), 2019–2028. http://dx.doi.org/10.2355/isijinternational.51.2019

Kim, Y. M., Lee, H., & Kim, N. J. (2008). Transformation behavior and microstructural characteristics of acicular ferrite in linepipe steels. *Materials Science and Engineering: A, 478*(No. 1-2), 361–370. http://dx.doi.org/10.1016/j.msea.2007.06.035

Kimura, S., Nakajima, K., Mizoguchi, S., & Hasegawa, H. (2002). In-situ observation of the precipitation of manganese sulfide in low-carbon magnesium-killed steel. *Metallurgical and Materials Transactions A, 33*(2), 427–436. http://dx.doi.org/10.1007/s11661-002-0103-8

Koseki, T., & Thewlis, G. Inclusion assisted microstructure control in C-Mn and low alloy steel welds. *Materials Science and Technology, 21*(8), 867–879. Retrieved from ftp://info.metallurgy.ac.at/M2CC%20NEU/2005_Koseki.pdf

Lan, H. F., Du, L. X., & Liu, X. H. (2013). Microstructure and mechanical properties of a low carbon bainitic steel. *steel research international, 84*(4), 352–361. http://dx.doi.org/10.1002/srin.201200186

Lee, J.-L., & Pan, Y.-T. (1995). The formation of intragranular acicular ferrite in simulated heat-affected zone. *ISIJ International, 35*(8), 1027–1033. http://dx.doi.org/10.2355/isijinternational.35.1027

Lee, T.-K., Kim, H. J., Kang, B. Y., & Hwang, S. K. (2000). Effect of inclusion size on the nucleation of acicular ferrite in welds. *ISIJ International, 40*(12), 1260–1268. http://dx.doi.org/10.2355/isijinternational.40.1260

Li, X., Min, Y., Liu, C., & Jiang, M. (2015). Study on the formation of intragranular acicular ferrite in a Zr-Mg-Al deoxidized low carbon steel. *Steel Research International.* http://dx.doi.org/10.1002/srin.201500167

Liu, S., & Olson, D. (1986). The role of inclusions in controlling HSLA steel weld microstructures. *Welding Journal, 65*(6), 139.

Liu, Z., Kobayashi, Y., Yin, F., Kuwabara, M., & Nagai, K. (2007). Nucleation of acicular ferrite on sulfide inclusion during rapid solidification of low carbon steel. *ISIJ International, 47*(12), 1781–1788. http://dx.doi.org/10.2355/isijinternational.47.1781

Loder, D. (2016). *On the systematic investigation of acicular ferrite on laboratory scale* (Doctoral thesis). Montanuniversitaet Leoben, Leoben.

Loder, D., & Michelic, S. K. (2016). Systematic investigation of acicular ferrite formation on laboratory scale. *Materials Science and Technology.* (HTTP://DX.DOI.ORG/ 10.1080/02670836.2016.1165902). http://dx.doi.org/10.1080/02670836.2016.1165902

Madariaga, I., Gutierrez, I., & Bhadeshia, H. K. D. H. (2001). Acicular ferrite morphologies in a medium-carbon microalloyed steel. *Metallurgical and Materials Transactions A, 32*(9), 2187–2197. http://dx.doi.org/10.1007/s11661-001-0194-7

Madariaga, I., Gutiérrez, I., García-de Andrés, C., & Capdevila, C. (1999). Acicular ferrite formation in a medium carbon steel with a two stage continuous cooling. *Scripta Materialia, 41*(3), 229–235. http://dx.doi.org/10.1016/S1359-6462(99)00149-9

Min, Y., Li, X., Yu, Z., Liu, C., & Jiang, M. (2016). Characterization of the Acicular Ferrite in Al-Deoxidized Low-Carbon Steel Combined with Zr and Mg Additions. *steel research international*, n/a-n/a. http://dx.doi.org/10.1002/srin.201500440

Mu, W., Jönsson, P. G., & Nakajima, K. (2014). Effect of sulfur content on inclusion and microstructure characteristics in steels with Ti2O3 and TiO2 additions. *ISIJ International*, *54*(12), 2907–2916. http://dx.doi.org/10.2355/isijinternational.54.2907

Mu, W., Jönsson, P. G., & Nakajima, K. (2015). Effect of the inclusion size and sulfur content on the intragraunlar ferrite transformation in steels with Ti2O3 additions. *Proceedings of 6th Congress on 'Science & Technology of Steelmaking'*, 767–771.

Mu, W., Mao, H., Jönsson, P. G., & Nakajima, K. (2016). Effect of carbon content on the potency of the intragranular ferrite formation. *steel research international*, *87*(3), 311–319. http://dx.doi.org/10.1002/srin.201500043

Nako, H., Hatano, H., Okazaki, Y., Yamashita, K., & Otsu, M. (2014). Crystal orientation relationships between acicular ferrite, oxide, and the austenite matrix. *ISIJ International*, *54*(7), 1690–1696. http://dx.doi.org/10.2355/isijinternational.54.1690

Nako, H., Okazaki, Y., & Speer, J. G. (2015). Acicular ferrite formation on Ti-rare earth metal-Zr complex oxides. *ISIJ International*, *55*(1), 250–256. http://dx.doi.org/10.2355/isijinternational.55.250

Ogibayashi, S. (1994). Advances in technology of oxide metallurgy. *Nippon Steel Technical Report*, *61*, 70–76.

Ohkita, S., & Horii, Y. (1995). Recent development in controlling the microstructure and properties of low alloy steel weld metals. *ISIJ International*, *35*(10), 1170–1182. http://dx.doi.org/10.2355/isijinternational.35.1170

Peng, Y., Chen, W., & Xu, Z. (2001). Study of high toughness ferrite wire for submerged arc welding of pipeline steel. *Materials Characterization*, *47*(1), 67–73. http://dx.doi.org/10.1016/S1044-5803(01)00155-3

Rees, G. I., & Bhadeshia, H. K. D. H. (1994). Thermodynamics of acicular ferrite nucleation. *Materials Science and Technology*, *10*(5), 353–358. http://dx.doi.org/10.1179/mst.1994.10.5.353

Ricks, R. A., Howell, P. R., & Barritte, G. S. (1982). The nature of acicular ferrite in HSLA steel weld metals. *Journal of Materials Science*, *17*(3), 732–740. http://dx.doi.org/10.1007/BF00540369

Ricks, R., Barritte, G., & Howell, P. (1981). The influence of second phase particles on diffusional phase transformations in steels. *Proceedings of Conference on 'Solid-Solid Phase Transformations'*, 463–468.

Saggese, M. E., Bhatti, A. R., Hawkins, D. N., & Whiteman, J. A. (1983). Factors influencing inclusion chemistry and microstructure in submerged arc-welds. *Proceedings of Conference on 'The Effects of residual, impurity, and microalloying elements on weldability and weld properties'*. (Paper 15), 1–11.

Sarma, D. S., Karasev, A. V., & Jönsson, P. G. (2009). On the role of non-metallic inclusions in the nucleation of acicular ferrite in steels. *ISIJ International*, *49*(7), 1063–1074. http://dx.doi.org/10.2355/isijinternational.49.1063

Seo, K., Kim, Y.-M., Evans, G. M., Kim, H. J., & Lee, C. (2015). Formation of Mn-depleted zone in Ti-containing weld metals. *Welding in the World*, *59*(3), 373–380. http://dx.doi.org/10.1007/s40194-014-0207-y

Shim, J.-H., Oh, Y.-J., Suh, J.-Y., Cho, Y., Shim, J.-D., Byun, J.-S., & Lee, D. (2001). Ferrite nucleation potency of non-metallic inclusions in medium carbon steels. *Acta Materialia*, *49*(12), 2115–2122. http://dx.doi.org/10.1016/S1359-6454(01)00134-3

Shim, J.-H., Byun, J.-S., Cho, Y. W., Oh, Y.-J., Shim, J.-D., & Lee, D. N. (2001). Mn absorption characteristics of Ti2O3 inclusions in low carbon steels. *Scripta Materialia*, *44*(1), 49–54. http://dx.doi.org/10.1016/S1359-6462(00)00560-1

Shim, J.-H., Cho, Y. W., Shim, J.-D., Oh, Y.-J., Byun, J.-S., & Lee, D. N. (2001). Effects of Si and Al on acicular ferrite formation in C-Mn steel. *Metallurgical and Materials Transactions A*, *32*(1), 75–83. http://dx.doi.org/10.1007/s11661-001-0103-0

Smith, Y. E., Coldren, A. P., & Cryderman, R. L. (1971). Manganese-molybdenum-niobium acicular ferrite steels with high strength and toughness. *Proceedings of Conference on 'Toward Improved Ductility and Toughness'*, 119–142.

Song, M.-M., Song, B., Hu, C.-L., Xin, W.-B., & Song, G.-Y. (2015). Formation of acicular ferrite in Mg treated Ti-bearing C–Mn steel. *ISIJ International, 55*(7), 1468–1473. http://dx.doi.org/10.2355/isijinternational. 55.1468

Strangwood, M., & Bhadeshia, H. K. D. H. (1986). The mechanism of acicular ferrite formation in steel weld deposits. *Proceedings of Conference on 'Advances in Welding Science and Technology'*, 209–213.

Sugden, A. A. B., & Bhadeshia, H. K. D. H. (1989). Lower acicular ferrite. *Metallurgical Transactions A, 20*(9), 1811–1818. http://dx.doi.org/10.1007/BF02663212

Sung, H. K., Shin, S. Y., Hwang, B., Lee, C. G., & Lee, S. (2013). Effects of cooling conditions on microstructure, tensile properties, and charpy impact toughness of low-carbon high-strength bainitic steels. *Metallurgical and Materials Transactions A, 44*(1), 294–302. http://dx.doi.org/10.1007/s11661-012-1372-5

Takamura, J.-I., & Mizoguchi, S. (1990). Roles of oxides in steels performance: Metallurgy of oxides in steels - 1. *Proceedings of 6th Congress on 'Iron and Steel'*, 591–597.

Thewlis, G. (2006). Effect of cerium sulphide particle dispersions on acicular ferrite microstructure development in steels. *Materials Science and Technology, 22*(2), 153–166. http://dx.doi.org/10.1179/026708306X81432

Thewlis, G., Whiteman, J. A., & Senogles, D. J. (1997). Dynamics of austenite to ferrite phase transformation in ferrous weld metals. *Materials Science and Technology, 13*(3), 257–274. http://dx.doi.org/10.1179/mst. 1997.13.3.257

Vanovsek, W., Bernhard, C., Fiedler, M., & Schnitzer, R. (2013). Effect of titanium on the solidification and postsolidification microstructure of high-strength steel welds. *Welding in the World, 57*(5), 665–674. http://dx.doi.org/10.1007/s40194-013-0063-1

Wan, X. L., Wang, H. H., Cheng, L., & Wu, K. M. (2012). The formation mechanisms of interlocked microstructures in low-carbon high-strength steel weld metals. *Materials Characterization, 67*, 41–51. http://dx.doi.org/10.1016/j.matchar.2012.02.007

Wan, X., Wu, K., Cheng, L., & Wei, R. (2015). In-situ observations of acicular ferrite growth behavior in the simulated coarse-grained heat-affected zone of high-strength low-alloy steels. *ISIJ International, 55*(3), 679–685. http://dx.doi.org/10.2355/isijinternational.55.679

Wan, X., Wu, K., Huang, G., & Wei, R. (2014). In situ observations of the formation of fine-grained mixed microstructures of acicular ferrite and bainite in the simulated coarse-grained heated-affected zone. *steel research international, 85*(2), 243–250. http://dx.doi.org/10.1002/srin.201200313

Wang, X., Hu, Z., Jiang, M., Wang, W., Yang, C., & Li, S. (2012). Investigation on non-metallic inclusions acting as nucleation sites of intra-granular acicular ferrites in Al and Ti deoxidized HSLA steels. *Proceedings of 5th Congress on 'Science and Technology of Steelmaking'*, 1–10.

Wu, K., Inagawa, Y., & Enomoto, M. (2004). Three-dimensional morphology of ferrite formed in association with inclusions in low-carbon steel. *Materials Characterization, 52*(2), 121–127. http://dx.doi.org/10.1016/j.matchar.2004.04.004

Wu, X., Lee, H., Kim, Y. M., & Kim, N. J. (2012). Effects of processing parameters on microstructure and properties of ultra high strength linepipe steel. *Journal of Materials Science & Technology, 28*(10), 889–894. http://dx.doi.org/10.1016/S1005-0302(12)60147-9

Xiao, F., Liao, B., Ren, D., Shan, Y., & Yang, K. (2005). Acicular ferritic microstructure of a low-carbon Mn–Mo–Nb microalloyed pipeline steel. *Materials Characterization, 54*(No. 4-5), 305–314. http://dx.doi.org/10.1016/j.matchar.2004.12.011

Xiao, F.-R., Liao, B., Shan, Y.-Y., Qiao, G.-Y., Zhong, Y., Zhang, C., & Yang, K. (2006). Challenge of mechanical properties of an acicular ferrite pipeline steel. *Materials Science and Engineering: A, 431*(No. 1-2), 41–52. http://dx.doi.org/10.1016/j.msea.2006.05.029

Xiong, Z., Liu, S., Wang, X., Shang, C., Li, X., & Misra, R. (2015). The contribution of intragranular acicular ferrite microstructural constituent on impact toughness and impeding crack initiation and propagation in the heat-affected zone (HAZ) of low-carbon steels. *Materials Science and Engineering: A, 636*, 117–123. http://dx.doi.org/10.1016/j.msea.2015.03.090

Yamada, T., Terasaki, H., & Komizo, Y. (2008). Microscopic observation of inclusions contributing to formation of acicular ferrite in steel weld metal. *Science and Technology of Welding and Joining, 13*(2), 118–125. http://dx.doi.org/10.1179/174329308X271797

Yamamoto, K., Hasegawa, T., & Takamura, J.-I. (1996). Effect of boron on intra-granular ferrite formation in Ti-oxide bearing steels. *ISIJ International, 36*(1), 80–86. http://dx.doi.org/10.2355/isijinternational.36.80

Yang, J., Ma, Z., Zhu, K., Wang, R., Xu, G., & Wang, J. (2015). Improvement of heat affected zone toughness of steel plate for large heat input welding with fine inclusions at Baosteel. *Proceedings of 6ᵗʰ Congress on 'Science & Technology of Steelmaking'*, 723–726.

Yang, J., Ma, Z., Zhu, K., Wang, R., & Zheng, Q. (2012a). Development of oxide metallurgy technology in Baosteel. *Proceedings of Conference on 'Asia Steel'*. (Z001), 1–4.

Yang, J., Ma, Z., Zhu, K., Wang, R., & Zheng, Q. (2012b). Development of oxide metallurgy technology in Baosteel. *Baosteel Technical Research, 9*(3), 41–46.

Yang, J., Xu, L., Zhu, K., Wang, R., Zhou, L., & Wang, W. (2015). Improvement of HAZ toughness of steel plate for high heat input welding by inclusion control with Mg deoxidation. *steel research international, 86*(6), 619–625. http://dx.doi.org/10.1002/srin.201400313

Yang, J.-H., Liu, Q.-Y., Sun, D.-B., & Li, X.-Y. (2010). Microstructure and transformation characteristics of acicular ferrite in high niobium-bearing microalloyed steel. *Journal of Iron and Steel Research, International, 17*(6), 53–59. http://dx.doi.org/10.1016/S1006-706X(10)60114-8

Yang, Z., Wang, F., Wang, S., & Song, B. (2008). Intragranular ferrite formation mechanism and mechanical properties of non-quenched-and-tempered medium carbon steels. *Steel Research International, 79*(5), 390–395.

Zhang, D., Shintaku, Y., Suzuki, S., & Komizo, Y. (2012a). In situ observation of phase transformation in low-carbon, boron-treated steels. *Metallurgical and Materials Transactions A, 43*(2), 447–458. http://dx.doi.org/10.1007/s11661-011-0892-8

Zhang, D., Shintaku, Y., Suzuki, S., & Komizo, Y. (2012b). Effect of cooling rate on phase transformation in the low-carbon boron-treated steel. *Journal of Materials Science, 47*(14), 5524–5528. http://dx.doi.org/10.1007/s10853-012-6444-9

Zhang, D., Terasaki, H., & Komizo, Y. (2009). In situ observation of phase transformation in Fe–0.15C binary alloy. *Journal of Alloys and Compounds, 484*(1-2), 929–933. http://dx.doi.org/10.1016/j.jallcom.2009.05.074

Zhang, D., Terasaki, H., & Komizo, Y. (2010). In situ observation of the formation of intragranular acicular ferrite at non-metallic inclusions in C–Mn steel. *Acta Materialia, 58*(4), 1369–1378. http://dx.doi.org/10.1016/j.actamat.2009.10.043

Zhang, Z., & Farrar, R. A. (1996). Role of non-metallic inclusions in formation of acicular ferrite in low alloy weld metals. *Materials Science and Technology, 12*(3), 237–260. http://dx.doi.org/10.1179/026708396790165704

Zhao, M.-C., Shan, Y.-Y., Xiao, F. R., Yang, K., & Li, Y. H. (2002). Investigation on the H2S-resistant behaviors of acicular ferrite and ultrafine ferrite. *Materials Letters, 57*(1), 141–145. http://dx.doi.org/10.1016/S0167-577X(02)00720-6

Zhao, M.-C., Yang, K., & Shan, Y.-Y. (2003). Comparison on strength and toughness behaviors of microalloyed pipeline steels with acicular ferrite and ultrafine ferrite. *Materials Letters, 57*(No. 9-10), 1496–1500. http://dx.doi.org/10.1016/S0167-577X(02)01013-3

Zhao, M.-C., Yang, K., Xiao, F.-R., & Shan, Y.-Y. (2003). Continuous cooling transformation of undeformed and deformed low carbon pipeline steels. *Materials Science and Engineering: A, 355*(No. 1-2), 126–136. http://dx.doi.org/10.1016/S0921-5093(03)00074-1

Zheng, C., Wang, X., Li, S., Shang, C., & He, X. (2012). Effects of inclusions on microstructure and properties of heat-affected-zone for low-carbon steels. *Science China Technological Sciences, 55*(6), 1556–1565. http://dx.doi.org/10.1007/s11431-012-4812-y

Zuo, X., & Zhou, Z. (2015). Study of pipeline steels with acicular ferrite microstructure and ferrite-bainite dual-phase microstructure. *Materials Research, 18*(1), 36–41. http://dx.doi.org/10.1590/1516-1439.256813

13

A Review on a Straight Bevel Gear Made from Composite

Haidar F. AL-Qrimli[1], Karam S. Khalid[2], Ahmed M. Abdelrhman[1], Roaad K. Mohammed A[3] & Husam M. Hadi[4]

[1] Department of Mechanical Engineering, Curtin University of technology, Miri, Sarawak, Malaysia

[2] Department of Mechanical Engineering, Ibra College of Technology, Ibra, Oman

[3] Oil Products Distribution Company, Ministry of Oil, Iraq

[4] SIG Combibloc Obeikan, Iraq

Correspondence: Ahmed M. Abdelrhman, Department of Mechanical Engineering, Curtin University of technology, Miri, Sarawak, Malaysia. E-mail: ahmed.mohammed@curtin.edu.my

Abstract

The purpose of this work is to present a clear fundamental thought for designing and investigating straight bevel gear made of composite material. Composite materials have the advantage of being light, producing low noises, and extra loading capacities. Due to these properties, it is highly preferable over conventional materials. A comparison between different types of material used in a gear structure will be shown. The outcome shows that a new form of cheap material may be useful for designing a new type of lighter and stiffer gear, designed for robotic arm applications or any power transmission application.

Keywords: orthotropic materials, matrix composite, gear, torque

1. Introduction

Metals and alloys cannot forever fulfill the demand that is continuously created and increasing in today's market and technological advances. In order to emulate and supplement metals and alloys, combining existing materials will allow us to meet the ever demanding performance that is required for numerous applications. Recent advances in material processing technology have propelled composite materials to the forefront of material technological development. Due to its most prominent advantage such as high specific strength, high specific modulus, and special electrical properties, composite materials have attracted considerable attention of the engineering community, and are touted as potential substitutes for metal, such as in athletic equipment, automotive parts, aircraft and aerospace structural parts, transportation, oil installations and nuclear industries. Specifically, a number of composite transmission gears can be found in many different locations, for example, gear pumps, watches, devices, and washing machines. This is due to the fact that composites have good properties such as excellent corrosion resistance, high strength to weight ratio, high impact resistance and design flexibility. These properties make parts manufactured from composite materials high-quality, durable and cost-effective products (Mallick, 2007, Deborah, 2010, Autar, 2006, Valery et al., 2007, Denial et al., 2003). Although at this point it is also worth noting that although composite materials are able to withstand high impact resistance extremely well, there is often poor tolerance to accidental low velocity impacts which could limit their use in certain industrial applications. The types of damage which can occur are sometimes very difficult to detect visually and yet can also seriously affect the structural integrity of the materials. Composite are flexible in terms of modifications, fatigue resistance and its ability to absorb impact energy, making them very attractive to current industrial needs. The literature on isotropic metallic gears of both experimental and theoretical studies is massive, however, the author has managed to locate a few references regarding gears made from composite materials. In a general sense, the word composite means constituted of two or more different parts. In practice, the term composite material or composite is used in a more restrictive sense, as a material constituted by the assemblage of two or more materials of different natures, with complementary properties leading to a material which has better properties than the properties of the composite components considered separately (Ezhil & Paul, 2015).

During the last century, there have been major developments in the technology of power transmission by gears. In general, the materials used are mild steel and other conventional materials to construct the gear structure. However, these materials require extensive lubrication, which is quite costly, the lack of which will produce noise and vibration, which will in turn induce gear failure. These failures, and how to solve them using composites will be discussed in next section. Composite materials are an area of research that plays a key role by reviewing the mesh stiffness of orthotropic material gears. It is capable of rebuilding the macro structures of materials in order to create new mechanical properties with the combination of low noise, lightweight, high strength and low usage of lubricants. As a result, it can be useful in power transmission applications. The metallic gear is heavy, and cannot be used without using a proper lubrication system. Moreover, plastic gears are lighter, has low loading capability, and weaker impact loading. Hence, composite gears are desired or even necessary in order to meet industrial requirements. This is a part of orthotropic material that can solve their entire gear problem without sacrificing strength, because they are lighter materials, making them essential for many industrial applications. The authors will attempt to understand the direction of potential stress and deformation variation for a straight bevel gear made from composite epoxies material (Peter, 2005; Sufyan et al., 2007, Haidar et al., 2015).

2. Effectiveness of Orthotropic Materials

The failure of metallic gears and other conventional materials can be caused by breakage and surface failure; however, composite materials are capable of circumventing these failures modes. Failures due to breakage are classified into three modes; the first mode involves a sudden breakage occurring at the root due to impact loading of the gearing. Composite materials address this problem by reinforcing the gear tooth with strong fibers, the combination of which will create materials with more load potential. The second failure mode due to breakage is fatigue at the tooth root due to constant repeated loading, and by utilizing composite materials; the stiffness of the gear tooth is increased and performs on par with another tooth body. And finally, the tooth corner breakage failure mode is mostly due to the uneven distribution of loads from axis misalignment, which is mostly caused by tooth alignment errors. This problem can be solved by die-casting the gear body using composite material in order to avoid manufacturing (generation) the gear tooth profile by rack cutters.

Wear and pitting on gear surfaces is another common type of failure. Pitting is caused by heavy contact pressure and the effect of lubrication. This failure can be avoided by making the tooth profiles smoother by casting them from composite material. Using composites make the surface of the gear tooth harder and stiffer, and the rigidity (higher yielding point) of composite material makes the teeth contact of the meshing gear Hartezian contact work with less lubricant. Improper material combination, poor surface finish and insufficient lubrication may cause wear failure (adhesive and abrasive). However, the hardness of composite material reduces the wear in gear teeth. Therefore, composite material is able to solve the entire gear problem, and create new fields of research of new types of material needed or required, without sacrificing strength. This will lead to extensive study in this field, and the formation of new macrostructures with excellent load ability, strength, lighter weight, less noise, less usage of oil and more wear resistance, by carefully selecting the type of matrix and reinforcement and studying the types of the arrangement that will address the gear tooth shape problem (Mohammed et al., 2007, Rongxian et al., 2010, Min et al., 2011).

3. Composite Gear Categories

Generally, any gear made from two or more materials with different properties and boundaries is considered a composite gear. This is further clarified below:

3.1 Metal Matrix Composite (MMC) Gear

Materials consisting of metal alloys reinforced with continuous fibres, particulates or whiskers are classified as MMC. The reinforced alloys by these constituents produce superior and unique physical and mechanical characteristics, which makes MMC suitable for a variety for applications. Its use is especially widespread in engineering due to its high strength, high power transmission, and medium weight, however, MMC are relatively more costly than polymers or ceramics. Generally, the reinforcement is done to tailor the specific applications. An example is the addition of SiC fibres, where it improves elastic stiffness and strength of metals, while reducing electrical conductivity. Ganesan and Vijayarangan (1993) conducted an investigation and compared the performance of a spur gear made of MMC materials to that made of conventional steel materials, and analyzed its static behavior in three dimensions using finite element analysis. They concluded that the behavior and performance of both MMC material gears and the mild steel gear are quite similar. The safety factor for the metal matrix composite showed the lowest values compared to other materials. Cedergren (2003) and his team demonstrate a method using a gear wheel to determine the porosity distribution within a complex powder compacted 3D structures using a dynamic 3D dilatant finite strain finite element program. The analysis showed that the porosity distribution in the powder compacted gears is dependent on the number of the gear teeth, thickness of the inner ring, the pressure angle, and is an excessive influence in the values of the porosity, which will induce small changes in its geometrical parameters. They stated that the minimum porosity distribution in the metal matrix composite gear is 3%, and this is achieved by supplying a maximum compaction during molding. However this ratio is not enough to prevent the gear tooth from initiation and propagation of cracks. The two modes of failure seen for the gears are tooth root bending fatigue and contact fatigue. Ramesh and Ganesan (1993) used metal matrix composites (MMC) in a railway wheel and compared them with steel and fiber reinforcement plastic (FRP) wheels from a static point of view. They concluded that the deflection in MMC is much lower than that of FRP, but the fundamental frequency of the MMC wheels are higher than those of steel or FRP wheels. Junichi Nozawa et al. (2009) studied the tribology of a metal spur gear and a hybrid gear, in order to reduce noise with greaseless metal gear and plastic gear with smaller rate of tooth failure. The test was at rotation speeds of 1000 rpm. They discovered that the noise is suddenly amplified when a single polymer sheet was spontaneously removed from the gear's surface; and their research indicates that this is due to its low adhesive strength against a shear.

3.2 Polymeric Matrix Composites Gear

Many researchers have studied, fabricated, and discussed the behavior of composite gears as a polymer matrix and glasses or graphite fibre, which named a (PMC) used in engineering applications as gear due to their medium level strength and low density. Kozo (1986) investigated the effect of different types of composite material on the strength of plastic gears. They used glass and carbon fibres in the gear tooth surfaces as reinforcement in order to improve bending strength, epoxy resin as the matrix material. Experimental static and dynamic tests were carried out to evaluate the effect of fibre reinforcements, and it shows the reinforcement is effective in improving the strength of plastic gears. Masaya et al. (1999, 2000, 2003) improved the performance of a plastic gear by using five kinds of carbon fibres and blending it with poly-ether-ether-ketone (PEEK). They evaluated the load capabilities and the wear properties of the reinforcement material used, and the results showed that the carbon fibre reinforced composite gear have superb affinity with PEEK. In 2007 Melick examined the effect of steel and plastic gear transmission with numerical and analytical methods by studying the influence of the stiffness of the gear material on the bending of the gear teeth, the consequences on contact path, load sharing, stresses and kinematics. The tooth bending of the plastic gear teeth results in an increase in the contact path length and considerable changes to load sharing. The root stresses are in dependent of its modulus, and were different from plastic gears from changes in load sharing. Also, the modulus dependent contact stresses, has significant influence on the preliminary and prolonged contact path, inducing very high stress peaks. Furthermore, the FEA results showed that the kinematics of the plastic gears changed dramatically. Hoskins and his group in 2011 presented a study of the dynamic performance of noise emission of polymeric gears. They tested five types of materials; Polyoxymethylene, Unreinforced polyetheretherketone, Carbon–fiber reinforced PEEK, Polyamide, and Glass fiber reinforced, and suggested an optimization of material combinations and the improvement of the techniques that minimizes transmission errors in polymeric gear trains, to give the opportunity of design development.

Senthilvelan and Gnanamoorthy (2007) experimentally studied the effect of various rotational speeds and stress on the performance of unreinforced Nylon 6 spur gears and glass fiber reinforced Nylon 6 spur gears. The result showed that speed has no influence on the gear life on both under low stress levels. At low stress levels, gear tooth root cracking and gear wear were observed, while at high stress levels, the rotational speed influenced the performance of both Nylon 6 and the glass fiber Nylon 6, due to the increase in the gear's surface temperature, which weakens the materials, performance reduction and gear life reduction. It was concluded that the Glass fiber reinforced Nylon 6 gears showed higher performance over unreinforced Nylon 6 gears due its better mechanical strength and resistance to thermal deformation. They also reported and described the computer-aided simulation of unreinforced glass fibre reinforced Nylon 6/6 gear to understand the gating and fibre orientation effect on part shrinkage. Detailed metrological inspection of the molded gears was conducted, and the results were correlated with fibre orientation in the gear and the simulations results (Senthilvelan & Gnanamoorthy, 2008). Mao (2007) carried out experimental investigations and modeled polymer composite (glass fibre reinforced nylon with PTFE). The design method is based on the relationship between polymer composite gears wear rate and surface temperature. A similar test was conducted on non lubricated metal gears, and it was discovered that the polymer gear wear rate dramatically increased when the load reaches a critical rate for a specific geometry. Furthermore, he reported that there are two failure modes that might occur in the polymer composite gears, which is fatigue, or wear. Polymer composite gears have been used in many applications such as the automotive industry, office machines and textile machinery, as well as other gearing industry. It is preferred due to the economical and technical advantages such as its minimal need of grease or lubrication, low production cost, low density, high resilience and internal damping capacity (KMao, 2007).

Huseyin Imrek experimented with plastic spur gears made of Nylon 6 by modifying the width of the gear teeth, shown in Figure 1. He investigated the performances of both the modified and unmodified gears with different loadings with a fixed rotational speed of 1000 rpm.

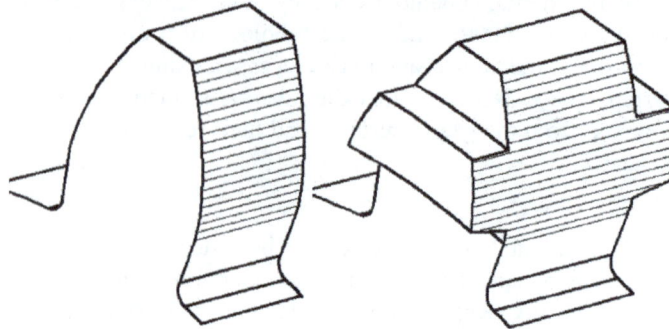

Figure 1. Unmodified and modified gear profile models (HuseyinImrek, 2009)

The purpose of the investigation is to improve the poor heat conduction of gears under high loads and speeds, due to the fact that it decreases the strength and the performance of the gears. From the study, the results showed that the modified gears exhibited lesser tooth temperatures, and its wear rates were significantly reduced compared to unmodified gears. He concluded that the teeth width modification reduced the wear rates on the tooth profile and increased the gear's performance, but the reduction in the tooth's surface temperature is 10-15 °C, which is considered quite slight (Huseyin, 2009).

Hayrettin (2009) created a modified Polyamide gear tooth by drilling cooling holes at different locations on the gear tooth in order to improve heat distribution in the gear. This experiment showed that this method of heat distribution is effective as the service life of the gear significantly improved compared to standard spur polyamide gears. Figure 2 showed the new design of the gear, and where the failure occurs. However, mass transfer occurred between the holes, so the positions and dimension of the drilled cooling holes must be carefully selected. In general, the thermal conductivity of polyamide gears is abysmal. Furthermore, Zhu and Li (2011) established a finite element model on a straight bevel gear by drilling holes in gears to reduce its weight. Nevertheless, they concluded that straight bevel gears with holes would weaken the mesh impact beneath the vibration compared to the normal straight bevel gear under the same loading and environment conditions. Vilmos Simon (2009) presents an optimal tooth modification for a spiral bevel gear that improves load distribution and decreases the maximum tooth contact pressure.

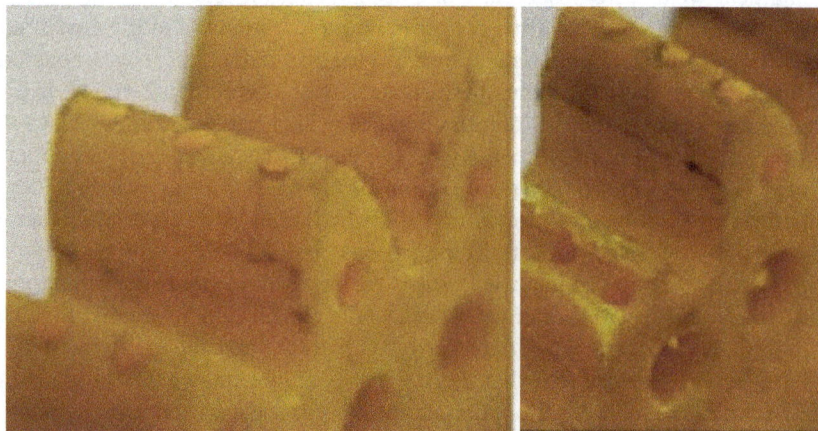

Figure 2. The damage in newly designed gears (Hayrettin, 2009)

3.3 Fibre Reinforcement Composite Gear

Fibre reinforcement results in high specific strength and stiffness compared to metal matrix composite, depending on different types of fibre (glass, carbon and Kevlar) and their arrangement in the matrix when acting as laminates, which are mostly used in weight-sensitive industry gear applications, offering the highest specific strength and stiffness. However, the growing use of fibre reinforcement composites in many gear transmission applications consist of embedded laminated layers of unidirectional fibers in a matrix, which are very directionally dependent in terms of strength. Composite materials are suitable when complex shape components with lower manufacturing expenses are needed. Hasim Pihtili (2009) investigated the wear resistance and behavior of woven glass fiber and composite materials under different loads, speeds and sliding distances. Antonio and Marcos (1993) used a polyethylene terephtalate to replace a spur gear. The module of the spur gear equals to 2mm/tooth, with a pitch diameter 34mm, and a pressure angle of 20° and the width of 18 mm. Nabi and Ganesan (1993) studied the effect of teeth proximity stiffness while analytically evaluating the root stresses of a gear using the cyclic symmetry approach. In their analysis, an attempt had been made to compare the root stresses with and without considering the effects of the adjacent gear teeth stiffness's using the concept of cyclic symmetry. Furthermore, they analyzed the performance of a composite spur gear made of Glasses/Epoxy using the above approach, and it was discovered that root stresses of cyclic symmetry is less for orthotropic spur gears.

4. Gear Stress Analysis Studies

The manufacturing and designing of bevel gears is currently considered an important field of research, and it is essential applications in robotic transmissions, automobile gears, and in other manufacturing processes. Investigations and experiments in this area are numerous, and current researches gravitate toward enhancing the endurance of the gear drives while simultaneously decreasing the noise. One of the main causes of gear tooth failures is the large tensile stress on the root fillets of a loaded gear tooth. These stresses have the tendency to trim down the gear's service life, and result in catastrophic failure under peak loading conditions. Enormous efforts have been made by previous investigators to relate tensile fillet stresses observed in statically and dynamically loaded gear teeth to the geometric appearance of the tooth. There are a number of approaches used in the past to verify the stresses and deflection in the gear's teeth, Wilfred Lewis made where the first effort to find the tooth root stresses. He based his analysis on a cantilever beam, and assumed that failure will occur at the weakest point of the beam, with him assuming it to be at the cross-section at the base of the gear. At the same time, Heinrich Hertz researched the contact pressures of the teeth, and his research was based on the elastic contact of two cylindrical bodies that determines the contact pressure between a gear and a pinion. With this tendency and experimental studies on bending stress analysis for the gears, the American Gear Manufacturers Association (AGMA) published their own standards based on Lewis's equation. It is more accurate, and calculates all the geometrical factors, which are important in calculating the bending stresses for the gears. These geometrical factors take into account the loading position and the fillet radius tooth of the tip and base. Alexander (2003) detail an engineering method that balances bending stress in the pinion and the gear. In general, the pinion and the gear have diverse tooth shapes and widths, constituting different materials. They presented an equation for equal bending safety factors for maximum bending stresses in both the pinion and the gear, in order to provide equally strong teeth to the pinion and gear. The equation is:

$$Smax1 - kb \cdot Smax2 < \delta s \tag{1}$$

Where Smax1 and Smax2 are the maximum bending stresses in the fillet area of the pinion and the gear, kb is the bending stress balance coefficient reflecting the difference of material properties and the number of tooth load cycles for the pinion and the gear, and δs is the permissible balance tolerance (typically less than 1%). In their research, different pressure angles were used (25° and 28°), and the results showed that the maximum bending stress is much lower than the standard gear pressure angle (20°). Furthermore, they reported that the bending stress reduction in turn leads to size and weight reduction, longer life, and higher load application for gear, however, the friction effects were ignored in their research. Hasan et at. (2006) analytically studied the elastic–plastic stress analysis on an orthotropic rotating annular disc. The disc is made from metal matrix curvilinear reinforced steel fibers, and they used different angular velocities to enable them to see the separation of the plastic region. The results showed that the radial displacements and the plastic flow at the inner surface have higher values than those at the outer surface. In 2004, Glodez developed a new model to determine the service life of mechanical elements in contact that is showing signs of pitting. They use the finite element method to determine the stress field in the contact area between the gears, and the simulation of fatigue and cracks on the contacting surfaces. The contacting mechanical elements contain crack initiation and propagation required for pit formation. The required number of stress cycles N for the pitting rate of the gear is determined by:

$$N = No + Nf \tag{2}$$

Where No is the number of stress cycles required for the fatigue crack initiation, and Nf is the number of stress cycles required for the initial crack to propagate to the critical crack length. Ahmad and Ahmet (2007) proposed two dynamic models in order to study the interaction between a gear's surfaces wear and its dynamic response. The model is made up of the influence of the worn surface profiles on dynamic tooth forces, its transmission error, as well as the influence of dynamic tooth forces on the wear profiles. Kumar et al. (2008) investigated the use of asymmetric toothed gear to develop fillet capacity in bending. They also analyzed the maximum fillet stress in order to help improve fillet capacity in bending. In their study, they used a non-standard asymmetric rack cutter for the pinion and the gear. They concluded that the asymmetric gear drive enhances the fillet load capacity of the pinion and the gear at higher-pressure angles.

Michele et al. (2005) and har group described a procedure for analyzing a gear under torque. it's a three step procedure, beginning with manufacturing simulation for the tooth profile of the face gear, followed by the unloaded kinematics simulation to determine the possible contact regions, and finally the computation of the load sharing between all the teeth that is in contact with each other. This stage involves the computation of instantaneous pressure distribution, meshing stiffness and the loaded transmission error. They present a model of a gear, based on analytical simulation, and conducted experimental tests using strain gages and compare with numerical results. A modified bevel gear was used to carry out the strain gage measurements, with the strain gages fitted on the gear with no respect to the nodes of the model. As a result, the stress at the similar position of the strain gauge fitted was not estimated in the experiment.

5. Finite Element Analysis

Finite Element (FE) is a viable engineering method that is capable of solving structural related problems without actually having to construct the structure itself. The method involves discretization of a structure into smaller individual units (i.e. finite elements), which are then reassembled in order to accurately portray the distortion of each finite element under various loading conditions. Each element has an assumed displacement field, and it is imperative that the selection of appropriate elements of the correct size and distributions is made to ensure accuracy of prediction (the FE 'mesh') (Matthews et al., 2003).

A finite-element analysis allows for the exploration of variables and loadings when no analytical model is available. Modelling of physical objects and the discretisation into finite elements allows for a numerical solution that closely predicts the physical behaviour of the actual material, provided that the material model is appropriate and accurate. By using the numerical model, numerous variables can be examined in order to limit the range of practical variables for actual physical testing. The most reliable technique for determining stress and deflection is the finite-element method (FEM). Arafa and Megahed (1999) constructed an FEA model of a spur gear to gather more information on the gear's mesh stiffness. The analysis involves quasi-static meshing conditions, where its compliance is evaluated at discrete meshing positions; with it assumed to be homogeneously isotropic. Their FEA results prove that the deflection of a loaded tooth is proportional to the load location along the tooth profile, and not on the mating tooth geometry, which is determined by the number of teeth, with it being based on a single tooth contact only. Faydor conducted an experimental and numerical investigation on a steel spiral bevel gear. Their main goal was to reduce the levels of noise and vibration of the gear, and they manage to achieve this by the application of tooth contact analysis and stress analysis using the finite element method (Faydor et al., 2006). The gear tooth deflection under an applied load is an important factor that affects the gear design, and it is essential to calculate the tooth profile and the pattern of the load. A student and some other researchers investigated the gear tooth deflection in Laval University Quebec, and they presented a finite-element model for a spur and a straight bevel gear (Gagnon et al., 1997).

Yi and Chia (2011) investigated contact stress for a non parallel axes concave conical involute gear pairs, with the first model being a bevel gear, and the second model a helical gear, using a finite-element analysis. They used the commercial software (ABAQUS), which is able to determine the contact stress of the two 3D deformable gears in order to evaluate the stress distribution on the gear tooth surfaces. The boundary conditions that they used to generate their FE model were that they considered the material isotropic, the stress is in the elastic region of the material, and they neglect the thermal stress, heat generation and the friction loss. In their FEA simulation, they applied 50Nm torque, and they fixed the gear, and gave the pinion a static degree of freedom. However, the FEA model did show the failure mode(s) that can occur. They calculate the Hertzian contact stress from Hertzian formulae and compared the results with FEA simulation.

In 2008, Chao Lin and his team used the finite element software (ABAQUS) to establish a 3DFinite element model of a noncircular gear transmission. They choose ABAQUS because it is one of the most advanced large-scale general finite element software in the world, and has powerful functions for nonlinear analysis, dynamic stress, and contact problem fields. The material used for the oval gear is 45 steel, with a Young's modulus of (2.0×10 N /mm2), Poisson's ratio of (0.3), and a Density of (7.85×10 kg /m2).

6. Linear and Nonlinear FEA Limitations

The linear and nonlinear analyses are two types of FEA. Their main divergence is that in the nonlinear analysis, the displacement is assumed to be large, while in the linear analysis, the displacement is assumed to be very small. The other main difference is that the nonlinear analysis considers the elastic and the plastic behavior of the material and its output to be not necessarily linear, while in the linear analysis, only the elastic linear zone will be considered. Comparing the two analyses, the linear analysis problems are easier to solve, with a low computational rate. Besides that, special load cases and boundary conditions can be utilized in linear analysis, which is inapplicable in nonlinear analysis. Nonlinear analysis is considered as the modeling of real world systems, while linear FE is the idealization of such systems. This idealization can be reasonably satisfactory in some cases, but for special cases, nonlinear modeling is the only option, and this includes straight bevel gear simulations. The sources of nonlinearity depend on geometric nonlinearity, in which the change in geometry is taken into consideration in setting the strain-displacement relations. Material nonlinearity response and possible deformation can occur. The boundary condition nonlinearity, in which the applied force and displacement depends on the deformation of the structure, is also another parameter to that effect. In nonlinear and also in linear FEA, many assumptions are made, due to the lack of information on some physical parameters, or due to the uncertainties about their actual values, such as the exact value of the material's yield strength, and the exact value of the part's thickness. This may lead to a deviation between the results from test experiments and FE models. The sources of deviation can be classified as follows:

1) *Deviation due to material:* Consider that in FEA, isotropic and orthotropic materials are assumed to have homogeneous properties, whereas in reality, there are many factors that may lead to non-homogeneous material properties, for example, imperfections in the microstructure, voids ...etc.

2) *Deviation due to geometry:* FEA assumes that the model's dimensions are uniform along the gear part, whereas this may not be the case in the real world due to the manufacturing processes deviations, which can lead to non-uniform sizing.

3) *Deviation due to load:* Consider that in FEA, a longitudinal axial loading is assumed to be completely perpendicular to the gear's tooth surface, and it is also assumed that the load and torque is perfectly aligned with the gear axis, whereas in reality, these precision cannot be guaranteed.

4) *Deviation due to measurement:* In reality, physically measured quantities are subject to the measuring tool's tolerances, and are also subject to the variation between different samples. This may cause a difference between the measured and the actual values.

The aforementioned sources can lead to a large difference between idealized finite element simulations and reality, especially in nonlinear finite element analysis. The intentional introduction of slight limitation within the nonlinear FE model helps in accounting for the difference between reality and idealization.

7. Summary

As mentioned in the literature, there is a demand for composite materials, as it is important for robotics applications due to the heavy mechanical parts made from conventional materials. Gears are one of the heavy parts in the mechanical structure of industrial robotic arms, it is obvious that combining more than one material to fabricate a composite structure is more efficient. The advantage of using composite gears is the inability of metals and their alloys in meeting today's gears' requirements. For example, carbon fiber matrices has better mechanical properties, higher specific strength (strength to density ratio), higher specific modulus, and better fatigue and wear resistance compared to metals. They also have higher specific energy absorption, which is imperative in industrial parts design, due to the fact that it will reduce weight, cost, noise and vibrations. Moreover, polymeric matrix composites (PMC) have limited transmission gearing and low loading abilities. In addition, it has been specified that the PMC have low surface temperature, and beyond a specific critical temperature value, sharp failures will occur. As a result, PMC gears have low load carrying capabilities, short service life, and poor heat resistance, rendering it inefficient for use in industrial applications. As a result, the present work will provide evidence that a new form of cheap material may be useful for designing a new type of lighter and stiffer gear, designed for robotic arm applications or any power transmission application.

Acknowledgement

The researchers whould like to thank Curtin University for their support of this work.

References

Abdelrhman, A. M., AL-Qrimli, H. F., Hadi, H. M., Mohammed, R. K., & Sultan, H. S. (2016). Times Three Dimensional Spur Gear Static Contact Investigations Using Finite Element Method. *Modern Applied Science, 10*(5), 145-150. http://dx.doi.org/10.5539/mas.v10n5p145

Al-Qrimli, H. F., Oshkour, A. A., Ismail, F. B., & Mahdi, F. A. (2015). Material design consideration for gear component using functional graded materials. *International Journal of Materials Engineering Innovation, 6*(4), 243-265. http://dx.doi.org/10.1504/IJMATEI.2015.072849

Al-Shhyab, A., & Kahraman, A. (2007). A nonlinear torsional dynamic model of multi-mesh gear trains having flexible shafts. *JJMIE, 1*(1), 31-41.

Arafa, M. H., & Megahed, M. M. (1999). Evaluation of spur gear mesh compliance using the finite element method. *Proceedings of the Institution of Mechanical Engineers, Part C: Journal of Mechanical Engineering Science, 213*(6), 569-579.

Autar, K. K. (2006). *Mechanics of Composite Materials* (2nd ed.). Taylor and Francis Group, LLC.

Avila, A. F., & Duarte, M. V. (2003). A mechanical analysis on recycled PET/HDPE composites. *Polymer Degradation and Stability, 80*(2), 373-382. http://dx.doi.org/10.1016/S0141-3910(03)00025-9

Çallıoğlu, H., Topcu, M., & Tarakcılar, A. R. (2006). Elastic–plastic stress analysis of an orthotropic rotating disc. *International journal of mechanical sciences, 48*(9), 985-990. http://dx.doi.org/10.1016/j.ijmecsci.2006.03.008

Cedergren, J., Sørensen, N. J., & Melin, S. (2003). Numerical investigation of powder compaction of gear wheels. *International journal of solids and structures, 40*(19), 4989-5000. http://dx.doi.org/10.1016/S0020-7683(03)00250-6

Chen, Y. C., & Liu, C. C. (2011). Contact stress analysis of concave conical involute gear pairs with non-parallel axes. *Finite Elements in Analysis and Design, 47*(4), 443-452. http://dx.doi.org/10.1016/j.finel.2010.12.005

Deborah D. L. C. (2010). *Composite Materials Science and Applications* (2nd ed.). Springer-Verlag London Limited.

Düzcükoğlu, H. (2009). Study on development of polyamide gears for improvement of load-carrying capacity. *Tribology International, 42*(8), 1146-1153. http://dx.doi.org/10.1016/j.triboint.2009.03.009

Gagnon, P., Gosselin, C., & Cloutier, L. (1997). Analysis of spur and straight bevel gear teeth deflection by the finite strip method. *Journal of Mechanical Design, 119*(4), 421-426.

Ganesan, N., & Vijayarangan, S. (1993). A static analysis of metal matrix composite spur gear by three-dimensional finite element method. *Computers & structures, 46*(6), 1021-1027. http://dx.doi.org/10.1016/0045-7949(93)90088-U

Garoushi, S., Yokoyama, D., Shinya, A., & Vallittu, P. K. (2007). Fiber-reinforced composite resin prosthesis to restore missing posterior teeth: A case report. *Libyan Journal of Medicine, 2.*

Gay, D., Hoa, S. V., & Tsai, S. W. (2003). *Composite materials design and applications* (pp. 6-84). CRC Press printed in United States of America. Retrieved from http://dlia.ir/Scientific/e_book/Technology/Engineering_Civil_Engineering_(General)/TA_401_492_Materials_of_Engineering_/020470.pdf

Glodež, S., Aberšek, B., Flašker, J., & Ren, Z. (2004). Evaluation of the service life of gears in regard to surface pitting. *Engineering fracture mechanics, 71*(4), 429-438. http://dx.doi.org/ 10.1016/S0013-7944(03)00049-3

Guingand, M., De Vaujany, J. P., & Jacquin, C. Y. (2005). Quasi-static analysis of a face gear under torque. *Computer methods in applied mechanics and engineering, 194*(39), 4301-4318. http://dx.doi.org/10.1016/j.cma.2004.10.010

Hayajneh, M. T., Tahat, M. S., & Bluhm, J. (2007). A study of the effects of machining parameters on the surface roughness in the end-milling process. *Jordan Journal of Mechanical and Industrial Engineering, 1*(1), 1-5.

Hoskins, T. J., Dearn, K. D., Kukureka, S. N., & Walton, D. (2011). Acoustic noise from polymer gears–A tribological investigation. *Materials & Design, 32*(6), 3509-3515. http://dx.doi.org/10.1016/j.matdes.2011.02.041

Ikegami, K., Kikushima, K., & Shiratori, E. (1986). Effects of material constitutions on the strength of fiber reinforced plastic gears. *Composites science and technology*, *27*(1), 43-61. http://dx.doi.org/10.1016/0266-3538(86)90062-X

İmrek, H. (2009). Performance improvement method for Nylon 6 spur gears. *Tribology International*, *42*(3), 503-510. http://dx.doi.org/ 10.1016/j.triboint.2008.08.011

Kapelevich, A. L., & Shekhtman, Y. V. (2003). Direct gear design: Bending stress minimization. *Gear Technology*, *20*(5), 44-47.

Kumar, V. S., Muni, D. V., & Muthuveerappan, G. (2008). Optimization of asymmetric spur gear drives to improve the bending load capacity. *Mechanism and Machine Theory*, *43*(7), 829-858. http://dx.doi.org/10.1016/j.mechmachtheory.2007.06.006

Kurokawa, M., Uchiyama, Y., & Nagai, S. (1999). Performance of plastic gear made of carbon fiber reinforced poly-ether-ether-ketone. *Tribology International*, *32*(9), 491-497. http://dx.doi.org/10.1016/S0301-679X(99)00078-X

Kurokawa, M., Uchiyama, Y., & Nagai, S. (2000). Performance of plastic gear made of carbon fiber reinforced poly-ether-ether-ketone: Part 2. *Tribology International*, *33*(10), 715-721. http://dx.doi.org/10.1016/S0301-679X(00)00111-0

Kurokawa, M., Uchiyama, Y., Iwai, T., & Nagai, S. (2003). Performance of plastic gear made of carbon fiber reinforced polyamide 12. *Wear*, *254*(5), 468-473. http://dx.doi.org/ 10.1016/S0043-1648(03)00020-6

Lin, C., Cheng, K., Qin, D., Zhu, C., Qiu, H., & Ran, X. (2008). Design and finite element mode analysis of noncircular gear (pp. 703-710). *6th International Conference on Manufacturing Research (ICMR08)*. Brunel University, UK.

Litvin, F. L., Fuentes, A., & Hayasaka, K. (2006). Design, manufacture, stress analysis, and experimental tests of low-noise high endurance spiral bevel gears. *Mechanism and Machine Theory*, *41*(1), 83-118. http://dx.doi.org/ 10.1016/j.mechmachtheory.2005.03.001

Mallick, P. K. (2007). Fiber reinforced Composites Materials, Manufacturing, and Design (3rd ed.). Taylor and Francis Group.

Mao, K. (2007). A new approach for polymer composite gear design. *Wear*, *262*(3), 432-441. http://dx.doi.org/10.1016/j.wear.2006.06.005

Matthews, F. L., Davies, G. A. O., Hitchings, D., & Soutis, C. (2003). *Finite element modelling of composite materials and structures*. Woodhead Publishing.

Nabi, S. M., & Ganesan, N. (1993). Static stress analysis of composite spur gears using 3D-finite element and cyclic symmetric approach. *Composite Structures*, *25*(1-4), 541-546.

Nozawa, J. I., Komoto, T., Kawai, T., & Kumehara, H. (2009). Tribological properties of polymer-sheet-adhered metal hybrid gear. *Wear*, *266*(9), 893-897. http://dx.doi.org/ 10.1016/j.wear.2008.12.008

Ou, R., Zhao, H., Sui, S., Song, Y., & Wang, Q. (2010). Reinforcing effects of Kevlar fiber on the mechanical properties of wood-flour/high-density-polyethylene composites. *Composites Part A: Applied Science and Manufacturing*, *41*(9), 1272-1278. http://dx.doi.org/ 10.1016/j.compositesa.2010.05.011

Peter Morgan, Carbon Fibers and Their Composites. (2005). CRC press is an imprint of Taylor and Francis Group LCC, No. 10: 0-8247-0983-7, printed in United States America on acid-free paper 10987654321.

Pihtili, H. (2009). An experimental investigation of wear of glass fibre–epoxy resin and glass fibre–polyester resin composite materials. *European polymer journal*, *45*(1), 149-154. http://dx.doi.org/10.1016/j.eurpolymj.2008.10.006

Ramesh, T. C., & Ganesan, N. (1993). Studies on metal matrix composite railroad wheels. *Computers & structures*, *47*(2), 259-263. http://dx.doi.org/ 10.1016/0045-7949(93)90375-N

Senthilvelan, S., & Gnanamoorthy, R. (2007). Effect of rotational speed on the performance of unreinforced and glass fiber reinforced Nylon 6 spur gears. *Materials & design*, *28*(3), 765-772. http://dx.doi.org/ 10.1016/j.matdes.2005.12.002

Senthilvelan, S., & Gnanamoorthy, R. (2008). Influence of reinforcement on composite gear metrology. *Mechanism and machine theory*, *43*(9), 1198-1209. http://dx.doi.org/10.1016/j.mechmachtheory.2007.09.002

Simon, V. (2009). Head-cutter for optimal tooth modifications in spiral bevel gears. *Mechanism and Machine Theory, 44*(7), 1420-1435. http://dx.doi.org/ 10.1016/j.mechmachtheory.2008.11.007

Su, M., Gu, A., Liang, G., & Yuan, L. (2011). The effect of oxygen-plasma treatment on Kevlar fibers and the properties of Kevlar fibers/bismaleimide composites. *Applied Surface Science, 257*(8), 3158-3167. http://dx.doi.org/ 10.1016/j.apsusc.2010.10.133

Valery, V., Vasiliev, E., & Morozov, V. (2007). Advanced Mechanics of Composite Materials. Elsevier Ltd.

Van Melick, I. H. (2007). Tooth-Bending Effects in Plastic Spur Gears Influence on Load Sharing, Stresses and Wear, Studied by FEA. *Journal Gear Technology, 58*, 58-66.

Vannan, E., & Vizhian, P. (2015). Corrosion Characteristics of Basalt Short Fiber Reinforced with Al-7075 Metal Matrix Composites. *Jordan Journal of Mechanical and Industrial Engineering, 9*(2), 121-128. Retrieved from http://jjmie.hu.edu.jo/vol9-2/JJMIE%2086-14-01%20Proof%20Reading-corrected.pdf

Xiaoyuan, Z., Zongde, F., & Wei, L. (2011, April). Research on vibration reduction performance of holey straight bevel gear. In *Consumer Electronics, Communications and Networks (CECNet), 2011 International Conference on* (pp. 4371-4374). IEEE.

14

Development of New Structural Materials with Improved Mechanical Properties and High Quality of Structures through New Methods

Salokhiddin Nurmurodov[1], Alisher Rasulov[1], Nodir Turakhodjaev[2], Kudratkhon Bakhadirov[1], Lazizkhan Yakubov[2], Khusniddin Abdurakhmanov[2] & Tokhir Tursunov[2]

[1] Materials Science and Materials Technology department, Tashkent State Technical University, Tashkent, Uzbekistan

[2] Machine Building Department, Tashkent State Technical University, Tashkent, Uzbekistan

Correspondence: Kudratkhon Bakhadirov, Tashkent State Technical University, University str. 2, 100057, Tashkent, Uzbekistan. E-mail: bahadirov@gmail.com

Abstract

Up-to-date science and technology requires further development and wide introduction of new high- performance processes to produce refractory metals. These may include plasma chemical technology of high dispersed powders production. Practical implementation of plasma chemical method in producing and processing of high dispersed powders is in its initial stage. Along with this at the present time the demand for processing of structural materials with improved physical and mechanical properties is now steadily increasing. Such materials have low machinability due to high hardness and durability at high temperatures which results in heavy wear of a cutting tool. To improve the efficiency when processing hard-to-cut materials it is necessary to enhance the tool's durability; this can be provided by application of new grades of hard alloys received from tungsten nanopowders. New alloy, obtained by the new developed technology, has higher degree of hardness and wear resistance compared with existing alloys and will be intended for hard materials processing.

Keywords: tungsten oxide, mechanical properties of materials, powder materails, plasma apparatus, powders granulation

1. Introduction

Plasma chemical technology in comparison with the traditional one has a number of essential advantages, namely, large productivity, energy savings, environmental cleanliness and the possibility of complete mechanization and automation.

Tungsten ultrafine powders (UFP) have become common use in traditional technology production of goods and semi-finished products. Basing on them the materials have been obtained which due to their physical and mechanical properties surpasses by far serial production.

It is now known that in the world of science based on nanotechnology can obtain defect-free large-sized construction materials. Investment is sufficiently developed. For example, the annual rate of 9 in the world for the nanotechnology industries - 10 billion dollars.. The United States, including $ 4 - $ 5 billion, in Japan 2 - 3 billion dollars.

In this direction we have will drive scientific research scientist of the Institute of Metallurgy and Engineering named after AA Baikov (Russian) prof. Tsvetkov Yu.V. in the development of nano-dispersed, aimed at physical and chemical bases of processes of plasma recovery and synthesis, scientists Institute of problems of chemical physics RAS (Russian) prof. Andrievsky RA in the field of nanostructures in materials science, physics and chemistry of solids, scientists of the Institute of Thermal Physics, RAS (Russia) Perepechko LV Ulanova and IM in the field of technology of various refractory metals and compounds and producing nanostructures of hard alloys with nizkoinduktsionnyh transformer installations in the laboratory of nationally Roskilde (USA) competitor Gleiter G. concept in nanocrystalline materials, nanostructures and nanocomposites nanophases in of nationally Argon Laboratory (USA) R.Zigelom applicant to obtain ultrafine powders through condensation evaporation, the Japan Institute of metallurgy (Japan) and scientists R.Bringer U.Herrom changes in surface structure modification of nanomaterials. In addition, the American University of Science and Technology (USA), the University of

Nagoya (Japan), University of Technology in English (England), the Belarusian National Technical University (Belarus), the Institute issues materialavedeniya name Frantsevich IN (Ukraine) and other research centers, higher education institutions are working to obtain a powder of refractory metals and the development of technology is carried out on the basis of these powders in order to manufacture products of multifunctional purposes.

As a result of receiving the UDP has been mastered and made of refractory metals are tools for quality and superior durability of existing analogues.

The development of scientific school in the field of physical chemistry and technology of plasma processes in metallurgy and processing of materials owned by the Russian scientist, the founder of which is Rykalin NN, and the further development and practical implementation of the CIS countries is absolute merit Tsvetkova Yu.

A significant role in the study nanoworld irresistible played, at least two things: the creation of a scanning tunneling microscope and the discovery of a new form of existence of carbon in nature. New methods and tools for the study materials it possible to observe the structure of the atomic-molecular structure of single crystal surfaces in the nanometer size range, thanks to the use of quantum tunneling effect theory.

2. Method

In the Figure 1 plasma reduction apparatus (PRA-300) is a prototype specimen of serial plasma devices and operates at the workshop Nr.2 in JSC "Uzbekistan refractory and heat-resistant metals" in one - shift operating regime.

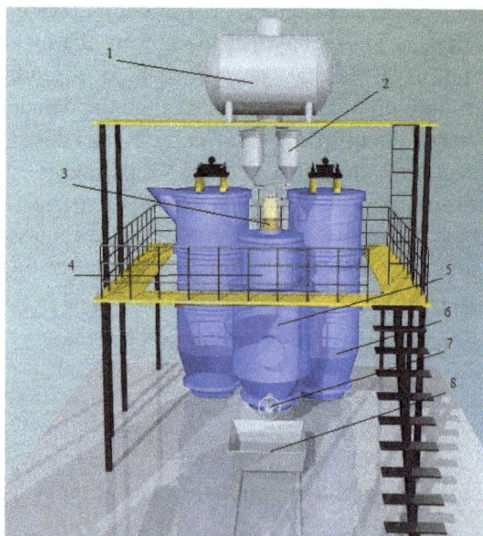

Figure 1. Plasma reduction apparatus (PRA - 300)

1 - plasma generator; 2 - reactor - 1 p; 3 - settling chamber, 4 - filters block, 5 - powder feeders; 6 - hopper for raw material; 7 - receiving hopper, consisting of a container and bogey; 8 – auger

Raw material is fed by air transport into the hopper (6) from which under its own weight it continuously proceeds into four feeders (5). From feeders the raw material is moved by transporting gas through the input unit into a plasma jet into reactor 2 where the raw material's mixing with the plasma jet, heating, melting - evaporation, chemical reaction of reduction and powder condensation occur. Moving along the reactor, the powder particles collide with each other forming conglomerates that fall out of steam and gas flow and accumulate at the bottom of the settling chamber (3). The smallest particles of the powder are lifted out with the steam and gas flow to the filters (4), where the mixture clearing from the powder takes place. Purified steam gas flow is directed to a bleeder or recycle hydrogen shop collector for regeneration and further use. The apparatus provides separation of powders obtained into PRA and submicron. There are two points of load: 1) the settling chamber, and 2) the filter. Powder fraction collected in the settling chamber contains a number of sub-oxides, and therefore oxygen and water vapor content is 5 - 10% (weight). The average particle's size falls in the range of 0,8 - 1,0 mm. Tungsten anhydride in the apparatus PRA - 300 restores under the following parameters (Table 1).

Table 1. Tungsten anhydride recovery parameters

№	Parameter Name	Rate
1	Arc current, kA	from 0,45 to 0,55
2	Arc voltage, V	from 380 to 410
3	Consumption of hydrogen through the plasma jet, m^3/ h	from 60 to 70
4	Consumption of hydrogen for tungsten anhydride transportation, m^3/h	from 2 to 4
5	Water rate for the plasma jet, m^3/ h	from 2.15 to 3.6
6	Water pressure in the high pressure header, Pa kgc/cm^2	$8,73 \times 10^5 - 11,77 \times 10^5$ 9-12
7	The temperature difference of softened water at the inlet and outlet of the plasma jet, ^{0}C, not more	20
8	Gas pressure in the unit, Pa, not more than kgc/c^2, not more	$0,98 \times 10^4$ 0,1
9	Softened water pressure at the inlet to the water distribution comb, Pa, kgc/cm^2	 $78,48 \times 10^4$ 8
10	Industrial water pressure in the water pipeline of low pressure, Pa kgc/cm	$16.62 \times 10^4 - 39,24 \times 10^4$ 2-4
11	Hydrogen pressure in the gas pipeline at the inlet to the gas distributing comb, Pa., no less kgc/cm^2, not less	 $3,9 \times 10^4$ 0,4
12	Softened water pressure at the outlet from plasma jet, Pa, not less, kgc/cm^2, no less	 $(-1,76) \times 10^4$ (-1,18)

The powders obtained have found application in the following types of products:

• powders brand 1.9 - 2.2 for ceramic metallization;

• tungsten powder brands PVV, PVO, PV1, PVPV and others for the production of technical tungsten powders.

The new type of plasma chemical reactor for hydrogen reduction of tungsten and molybdenum oxides has a distinctive feature of energy supply to the reaction zone. The energy is introduced not only in the form of plasma jet, but also as an additional stream, heated to a high temperature gas entering the reaction zone through the porous, permeable wall, heated by an electrical air heater.

The new type of plasma chemical reactor has been created, based on the following considerations. The process of plasma chemical reduction in the standard reactor, where d/D is equal to 1/10, is divided during no more than 0.03 seconds, and the plasma flow at free discharge in a large volume quickly loses thermal energy store, therefore some part of the powder that gets into the peripheral area of the jet remains underreduced. Thus, it is necessary to extend stay time of tungsten oxides particles in the hot area.

The constructive scheme of plasma chemical reactor is shown in Figure 2. Plasma jet with diameter of 30-40 mm is crimped by porous molybdenum tube with inner diameter of 40-50 mm, which is located coaxially in the porous tube made of stainless steel. Between the porous tubes there is a molybdenum heater for heating the flowing gas.

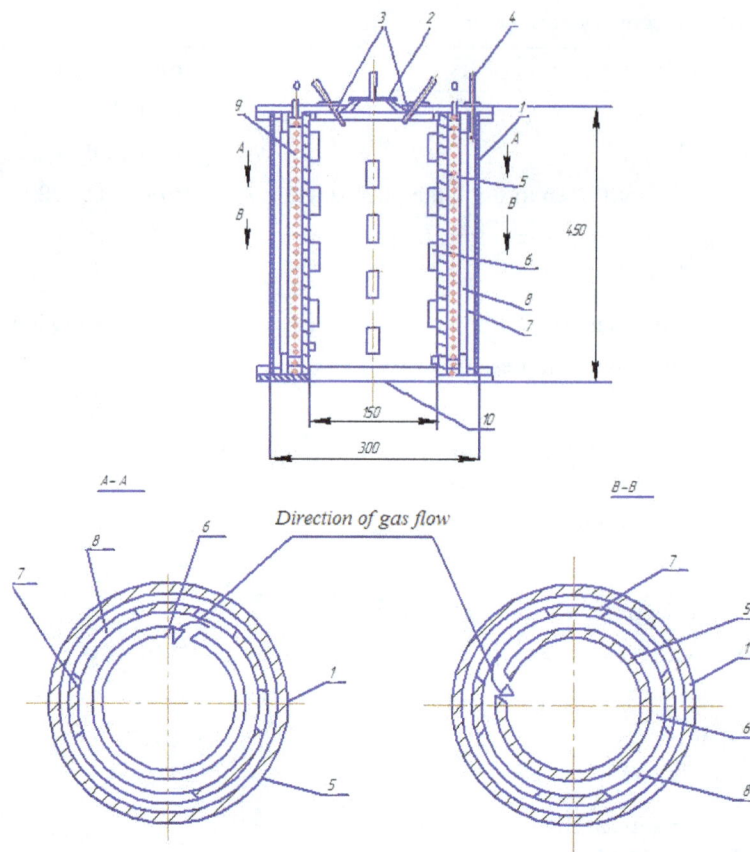

Figure 2. Plasma chemical reactor with a high degree of raw material processing

1 - porous molybdenum cylinder, 2 – porous stainless steel cylinder; 3 - plasmatron A-26 (PG-2, I); 4 - case

The above described construction is placed into a sealed case, where additional hydrogen is blasted, when heated between the porous tubes, crimps the plasma jet.

Technological researches of the new reactor were conducted on the strengthened plasma laboratory apparatus. Plasma capacity was kept within the range of 45-55 W, calorifer capacity was 16-18 kW, hydrogen consumption through the plasma generator was 20 m^3/h, through the heater was from 40 to 60 m^3/h. Consumption of tungsten oxide was 6-10 kg/h. Hydrogen temperature heated by the calorifer was 1500 -1600 C^0. Thus, the total capacity is not more than 75 kW, and the total hydrogen consumption is up to 80 m^3/h, while plasma jet capacity under the traditional scheme is 100 kW at a flow rate of hydrogen 75 m^3/h.

3. Results and Discussions

The study of the obtained powders showed that the average grain size due to Fisher is 0.07 - 0.09 microns with the content of oxygen and moisture vapor 1.5% wt (Table 2).

Table 2. The grain size due to Fisher and the mass fraction of oxygen in powders obtained in plasma-chemical reactor

№/№	Extraction point	The grain size due to Fischer, micron	Mass fraction of oxygen and moisture vapor,%
1	W plasma from the filter	0,08	1,5
2	W plasma from the filter	0,09	1,5
3	W plasma from the settling chamber	0,09	1,5
4	W plasma from the filter	0,07	1,4

X-ray phase analysis showed the presence of β- tungsten up to 50%, I - Tungsten up to 35%, and the rest is tungsten oxide without amorphous phases.

In the process of powders production in the new reactor in the settling chamber under the reactor practically no powder was found; it testifies that the entire obtained tungsten oxide was reduced and UFP tungsten entered the filters.

Thus, proceeding from the analysis of technological tests of the new plasma chemical reactor one can come to the following conclusions:

• raw material processing degree increases up to 95%;

• significantly smaller and more active UFP has been received;

• the actual number of amorphous tungsten in the powders produced has been revealed;

• process control is increasing, including the dispersion of the powder;

• uniformity of grain size is increasing;

• coefficient of reduction process activity is increasing.

The process of tungsten anhydride reduction in the plasma chemical reactor occurs in the plasma jet of hydrogen due to the reactions:

$$WO_3 + H_2 = WO_{2.90} + H_2O,$$

$$WO_{2.90} + H_2 = WO_{2.72} + H_2O,$$

$$WO_{2.72} + H_2 = WO_2 + H_2O,$$

$$WO_2 + H_2 = W + H_2O,$$

The apparatus has two points of powder discharge (Figure 1): settling chamber; filter with the powder accumulation in the settling chamber and filter it is discharged into the intake reservoirs. Depending on the variety of raw material and places of discharge one can obtain:

plasma reduction powder - unannealed (from the filter);

plasma reduction powder - unannealed powder (from the settling chamber);

Unannealed plasma reduction powder from the filter of black colour.

Unannealed plasma reduction powder from the settling chamber of gray-black colour.

The powder should not have cakes and balls.

Dispersity of the powder is characterized by an average particle size due to Fisher and must be not more than 0, 4 microns. The powder mass loss at hydrogen calcinations is not more than 3%.

Before-reduced plasma powder has the following characteristics:

- oxygen content is 0.5%;

- average grain size due to Fisher is 0,2-0,4 micron.

Tungsten plasma powders are designed for the production of tungsten powders of various grades to improve the quality of products, to intensify the sintering process, fine-grained carbides and hard alloys.

The application of tungsten UFP obtained as an alloying component of many high-temperature and heat-resistant materials and the products received from them allowed to improve the quality of these products, to reduce the sintering temperature and lower energy consumption during the operation "Welding."

The study used plasma powders of tungsten, selected from the filter of the industrial apparatus PRA-300. Plasma tungsten powders specification is given in Table. 3.

Table 3. Plasma tungsten powders specification

Powder		Oxygen content in %	Bulk density, g/cm^3	Specific surface area, m^2/g,
Tungsten	1	2,0	0,70	4,0
	2	2,4	0,82	5,8
	3	3,0	0,90	8,0

The oxygen content was calculated using the weight method, annealing powder in hydrogen stream at 900 °C.

Weight by volume was determined using volumeter and the specific surface was determined by the argon thermal desorption method.

Powders granulation was done in the system of "drunken barrel". Pressing of industrial samples was done on the presses of P474A, P807 brands. Testing regimes of extraction, sintering, welding was carried out on the industrial equipment of shop 2 (PRA-300, CEP-214, CTN-1, 6).

Powders were used in the experiment that included the gases under mass- spectrometry data: O_2, H^2, W_2, H_2O, CO, CO_2.

Compact half-finished products preparing. The experiment was carried out using initial tungsten finely dispersed powders with introduction of a plasticizer (alcohol, glycerol, 1:1) into their composition. Mixing of powders with a plasticizer and at the same time their granulation was done out on vibromixer.

Formation of finely dispersed powders is associated with sharp increase in resistance to the punch, due to friction along the walls. Therefore, to obtain rods without compacting crack at pressing is practically impossible. Using methods of impact compaction does not give positive results too. The value of bulk density is shown in Table 4.

Table 4. Density of bulk depending on powders' grain size

Powder	Ultrafine			Standard
	initial	sifted	granulated	sifted
Bulk density W, g/cm^3	0,7-0,9	1,4-1,8	2,0-2,2	3,4-3,65

Quality rods are obtained from granular tungsten powders, the compaction process of which can be represented as follows: in the initial period of pressing the compaction occurs due to the movement of granules associated with the destruction of bridges and arches formed at the free fill of powder. This results in an increase in stress at the particles contact places, which in turn causes irreversible elastic deformation and brittle fracture of grains. Further compaction occurs under the usual scheme by filling the voids by the particles formed by the destruction of the granules.

Compacts with uniform density are obtained by hydrostatic pressing of powders in elastic shells due to the development of plastic deformation by brittle solids under all-round compression.

Specific features of formation and sintering of ultrafine tungsten powders. Recrystallized mechanism of sintering, based on the formation of the nonequilibrium vacancies excess concentration at recrystallization, can be resulted in considerable activation of self-diffusive processes. Shrinkage in the initial period of sintering occurs under two mechanisms of mass transfer: particles grain boundary sliding and diffuse - viscous flow (liquid-like coalescence).

Shrinkage in this case will depend on the initial porosity of compacts; at that the largest one will not necessarily be observed in the compacts with minimal porosity. Obtained data show that the most favorable values of pressing efforts for ultrafine tungsten powders are in the range 100-200 MPa.

Production of rods from ultrafine tungsten powders. 5% of plasticizer was added into superfine tungsten powder; powder granulation was carried out for an hour. Granulated powder was pressed into rods by section 12x12x500 at effort of 200 MPa. The density of compacts was 10.2 g/cm^3. The rods have been sintered in a furnace TSEP214 at 1373 ° K for two hours. Rods' welding was carried out under the existing technology at various welding currents. Standard density can already be obtained at a welding current equal to 2.6 kA. The results of the welding rods from ultrafine tungsten powders are shown in Table 5.

Table 5. Rods density change depending on welding current

Wn/n	Welding current, kA	Rods density, g/cm^3
1	2,2	17,0
2	2,4	17,8
3	2,6	17,9
4	2,8	18,0
5	3,0	18,2

Currently, the basic requirements for durable cutting materials include high hardness and uniform finely dispersed structure.

To equip the cutting tool the industry produces alloys of the type "M" (VKZ-M VK6-M, VK10-M) with a grain size of WC-phase 1.8 mm without alloy additions.

The developed new alloy has higher degree of hardness and wear resistance compared with existing alloys and will be intended for hard materials processing.

The most efficient means to create such an alloy are the use of tungsten nano-powders and WC-intensive grinding phase. The paper shows that the researches done revealed that the developed alloy is different from the standard fine alloys by its ultra-finely grained structure. Thus, the volume of WC-phase grain up to 1 micron is 80-85%, and in alloys VK10-M - 65-75%. The average grain size of WC-phase is 1.1 microns, and in the alloy VK10-M - 1, 3-1, 5 microns.

Acknowledgments

First of all we would like to gratitude staff of the JSC "Uzbekistan refractory and heat-resistant metals" and JSC "Metallurgy combinat of Uzbekistan" for prowiding us an opportunity to carry out our experimental works, their support with raw materials for the experiments and most important, their sugessions in planning of the experiments.

References

Bakhadirov, K. G. (2013). Study of orientation of crystallographic lattices of aluminum sheet after asymmetric rolling. *Composition Materials Journal,* 18-21. Uzbekistan, 2013, #1.

Kalamazov, R. U. (2004). *Nanocrystalline structures in materials* (p. 98, in Russian). - Tashkent: Tashkent State Technical University.

Nurmurodov, S. D. (2012). *Theoretical and technological aspects of construction materials on the basis of fine powders of refractory metals. Monograph* (p. 136, in Russian). - Tashkent, Tashkent State Technical University.

Nurmurodov, S. D. (2013). *Plasma-chemical reactor №IAP No. 04732* (in Russian).

Nurmurodov, S., & Norkulov, A. (2010). *Thermophysical basics of structure in the cast bimetallic composites (in Russian): Monography* (p. 160, in Russian). - Tashkent: Fan va technology.

Saidaxmedov, R. Kh., & Bakhadirov, K. G. (2014). Mechanical properties of material after rolling and heat treatment. *Journal of Technical University of Gabrovo, 47.*

Samokhin, A. V., Alexeev, N. V., Kornev, S. A., & Tsvetkov, Yu. V. (2009). WC nanosized composition synthesis and characterization. *19th International Symposium on Plasma Chemistry (ISPC-19), Bochum, Germany.*

15

Thermal and Nanoindentation Behaviours of Layered Silicate Reinforced Recycled GF-12 Nanocomposites

A. Shalwan[1], A. I. Alateyah[2], B. Aldousiri[3] & M. Alajmi[1]

[1] Manufacturing Engineering Technology Department, College of Technological Studies, Public Authority for Applied Education and Training, Kuwait City 13092, Kuwait

[2] Mechanical Engineering Department, Qassim University, P.O.B.6677, Buraydah, Saudi Arabia

[3] Advanced Polymer and composites (APC) Research Group, Department of Mechanical & Design Engineering, University of Portsmouth, PO1 3DJ, UK

Correspondence: A. Shalwan, Manufacturing Engineering Technology Department, College of Technological Studies, Public Authority for Applied Education and Training, Kuwait City 13092, Kuwait.
E-mail: ama.alajmi1@paaet.edu.kw

Abstract

This work is an attempt to improve the thermal and nanoindentation behaviours of recycled Glass-Filled Polyamide-12 (GF-12) by adding layered Silicate (Nanoclay) as a reinforced filler. Differential Scanning Calorimetry (DSC) and Nanoindentation tests were conducted to study the effect of various loading levels (0-7 wt. %) of Nano-layered silicate on the thermal and nanohardness behaviours of GF-12 and its nanocomposites. Wide Angle X-ray Diffraction (WAXD) was employed to characterise the nanostructure of material and determine the intercalation/exfoliation for layered silicate in a GF-12 matrix. This study reveals that the layer silicate has a positive effect on the hardness results. Nanoindentation results showed remarkable improvement when layered silicate was added to recycled GF-12. The Glass transition (T_g) and crystallization (T_c) temperatures showed a slight improvement over the base polymer by the incorporation of layered silicate. Moreover, the enhancement of crystallization was obvious with the addition of clay loading, which functioned as a nucleating agent that could increase the rate of crystallization.

Keywords: Nanocomposites, polymers, Differential Scanning Calorimetry (DSC), nanoindentation

1. Introduction

Polymer nanocomposites (PNCs) have recently emerged as substitutes for conventional materials and have aroused considerable interest for their use in numerous industrial applications, such as automobiles, food packaging and rapid manufacturing (RM) (Aldousiri, Shalwan, & Chin, 2013; Alexandre & Dubois, 2000; Fischer, 2003; Sinha Ray & Okamoto, 2003; Youssef, 2013). This is mainly due to the superior properties of PNCs in mechanical, thermal, physical and chemical terms to conventional composite materials. They have greater tensile strength, impact resistance, elastic modulus, flexural strength, heat resistance and chemical resistance (Shalwan & Yousif, 2013; Thomas & Pothan, 2009; Youssef, 2013).

As the literature shows, the enhancement of the PNCS' performance results from the ability of nanometer particles to disperse in a polymer matrix (Alexandre & Dubois, 2000; Fischer, 2003; Kiliaris & Papaspyrides, 2010; Shalwan & Yousif, 2014). In other words, the performance of PNCs greatly depends on the characteristics of nanometer particles and the structural characterisation of nanocomposites, for instance their microcomposite, intercalated or exfoliated structure. In this context, layered silicate has shown remarkable performance as an additive to polymers, which has led to the enhancement of its mechanical and thermal properties, due to the capacity of the silicate to separate or break down into individual layers (Djebara, El Moumen, Kanit, Madani, & Imad, 2016; Kotal & Bhowmick, 2015; Liu, Ping Lim, Chauhari Tjiu, Pramoda, & Chen, 2003). Many investigators have investigated the production of new nanocomposite materials with improved mechanical and thermal properties (Djebara et al., 2016; Hu et al., 2006; Liu et al., 2003; Yu, Zhao, Chen, Juay, & Yong, 2007; Zeng, Yu, Lu, & Paul, 2005). For example, Liu et al. (Liu et al., 2003) studied the mechanics of Nylon 11/organoclay nanocomposites containing (0, 1, 2, 4 and 8 wt.%) organoclay. Dynamic Mechanical Analyser

(DMA) tests showed that the storage modulus noticeably increased by 100% when the clay loading went up to 8 wt. %, compared with the neat Nylon 11.

In addition, Hu et al. (Hu et al., 2006) used the nanoindentation technique to investigate the mechanical properties (i.e. hardness and modulus) of Nylon 11 and its nanocomposites. It was found that the hardness and modulus of the nanocomposites steadily increased with increasing clay content. With the inclusion of 5 wt. % clay into the PA11 matrix, the hardness and elastic modulus were improved by about 30%. Moreover, Shen et al. (Shen, Phang, Liu, & Zeng, 2004) investigated the mechanical properties of PA66/clay nanocomposites using the nanoindentation technique. It was obvious that the hardness and the elastic modulus were gradually enhanced with increasing clay concentration. Moreover the creep behaviour of the nanocomposites was considerably increased by increasing the clay content. In addition, Shen et al.(Shen, Tjiu, & Liu, 2005) studied the nanoindentation behaviour of PA6/clay. It was clear that the modulus and hardness of the nanocomposites with 2.5 wt. % of clay were enhanced by 74% and 80%, respectively compared with the neat PA6.

Likewise, Aldousiri et al.(Aldousiri, Dhakal, Onuh, Zhang, & Bennett, 2011) carried out a study of the nanoindentation of spent polyamide-12 and the PA-12/nanocomposite. It was found that the addition of layered silicate led to increases of up to 116% and 73%, respectively of nano-hardness and modulus, compared to the unreinforced polymer. The authors attributed the enhancement in properties to the changes in the properties matrix caused by the high aspect ratio of layered silicate.

However, characterizing and studying the structures of nanocomposites is considered the first step in understanding the behaviour of nanocomposites. X-ray diffraction (XRD) and wide angle x-ray diffraction (WAXD) are the most widely used techniques for evaluating nanocomposite systems (Hussain, Hojjati, Okamoto, & Gorga, 2006). In addition, the XRD method provides an understanding of the distances between the layers in the silicates structure, using Bragg's Law (Alexandre & Dubois, 2000). The degree of dispersion of the nanoclay in the polymer can be discovered by using XRD technique to determine the d-spacing in nanocomposite material (Giannelis, 1996).

Any improvement in thermal properties will also give nanocomposites many more opportunities to be applied industrially, in the construction and automotive sectors, for example (Pinnavaia, 2000; Yu et al., 2007; Zeng et al., 2005). Suzhu et al. (Yu et al., 2007) found that layered-silicate loading was affected in the crystalline and thermal behaviors of polyamide layered-silicate nanocomposites. Moreover, layered-silicates have led to slight reductions in the crystalline and melting temperature of polyamide layered-silicate nanocomposites.

Selective Laser Sintering (SLS) is one of the rapid prototyping techniques that are commonly used in Rapid Manufacturing to build parts from powder. A wide range of polymers can be used in this process, such as thermoplastic polymers (polyamide, polycarbonate), which can be used to manufacture aerospace products and the automotive and packaging industries. However, those products which are wasted in the processing must in the end be replaced by new products made from virgin raw materials and the disposal of the old products can represent a serious environmental issue.

This paper reports on a study of the effect of various loading levels of Nano-layered silicates used to reinforce the thermal and nanoindentation behaviours of recycled Glass-Filled Polyamide-12.

2. Experimental

2.1 Materials

In this experiment, Recycled Glass-filled Polyamide-12 (GF-12) was used as the matrix material (mix ratio 45 wt.% virgin, 55 wt.% recycled) provided by the Centre for Rapid Design and Manufacturing (CRDM), Buckinghamshire, U.K. Glass-filled polyamide-12 is widely used in Selective Laser Sintering (SLS). The filler material used was Nano-layered Silicate, commercially known a LK-PA-CR1. Both polymer and filler were supplied as fine white powders with particles of similar size (RGF ≈ 48μm, LK-PA-CR1 ≈ 50μm).

2.2 Sample Preparation

The Recycled Glass-filled Polyamide-12 (GF-12) and Nano-layered Silicate were dried separately in a fan assisted oven at 100 °C for 1 h in order to remove the residual moisture, because both these materials absorb water. A dry grinding process was employed to mix various weight percentages (3, 5 and 7 wt. %) of Nano-layered Silicate with GF-12, using a high speed rotary Kenwood blender at a rotary speed of 800 rpm for 10 min. Test specimens were prepared using an MCP Mini Moulder 12/90 HSP with a temperature profile of 240°–260 °C with a screw speed of 200 rpm.

2.3 Wide Angle X-ray Diffraction (WAXD)

Injection moulded samples were examined and analysed using the wide angle X-ray diffraction technique to verify clay dispersion and characterise the nanostructure of the material. WAXD analysis was performed using a Philips X-ray diffractometer with Cu Ka radiation generated (λ = 1.542) at 20 mA and 40 kV. The d-spacing values (i.e. the values of the basal distance between the clay layers) were calculated using Bragg's Law:

$$n\lambda = 2d\sin\theta \tag{1}$$

where n is an integer, d is the inter layer d-spacing and λ is the wave length.

2.4 Differential Scanning Calorimetry (DSC)

DSC measurements were employed to investigate the effect of the layered silicates on the melting and crystallization behavior. Samples of recycled GF-12 and GF-12/nanocomposites were studied using TA Instruments. The encapsulated sample pan and an empty reference pan were positioned inside a DSC machine and heated and cooled at the rate of 20 ^{0}C/min up to 250 ^{0}C. The specimen for the DSC test was in the form of powder obtained by scraping the bulk sample with a knife. The sample mass was approximately 10 mg. The DSC test was made in an inert nitrogen atmosphere. A minimum of three samples were used in this test and the standard deviation was determined for each group.

2.5 Nanoindentation Testing

The Nanoindentation tests were carried out in a NanoTest apparatus (Micro Materials, UK) as shown in Figure 1. A Berkovich diamond indenter tip manufactured by Micro Materials was used to measure the nanohardness properties. Nine symmetrical indentations (in the form of a matrix, 30 µm apart) were made on each specimen. The Nanoindentation specimens were prepared by injection moulding and then cut into pieces sized approximately 15 × 15 × 3 mm.

Figure 1. Schematic view of the NanoTest system

The parameters used for all measurements were as follows:

Initial load: 0.1 mN

Maximum load for all indents: 3.2 mN

Loading and unloading rate (strain rate): 2.00 mN S^{-1}

Dwell time or holding time at maximum load: 5 s

Hardness and elastic modulus measurement

In indentation testing, the hardness is defined as the indentation load divided by the projected contact area. The hardness (H) is determined from the peak load (P_{max}) and the projected area of contact (A,) using the empirical relation:

$$H = \frac{P_{max}}{A} \tag{2}$$

To obtain the elastic modulus, the unloading portion of the depth-load curve is analyzed according to a relation, which depends on the contact area, as explained elsewhere [11,12]. The indentation modulus E indent can be calculated from the slope of the tangent for the calculation of indentation hardness and is comparable to Young's modulus of materials. The theoretical modulus was calculated by Equation (3) [10]:

$$E_{indent} = \frac{1-(v_s)^2}{\frac{1}{E_r} - \frac{1-(v_i)^2}{E_i}} \qquad (3)$$

v_s Poisson's ratio of the test piece; (for polymer approximately 0.2)

v_i Poisson's ratio of the indenter (for diamond 0.07)

E_i Young's modulus for the indenter (for a diamond 1141 GPa).

E_r Modulus of the indentation contact (reduced modulus)

3. Result and Discussion

3.1 Wide Angle X-ray Diffraction (WAXD)

Figure 2 shows the results for the nanoscale dispersion of layered silicate in the GF-12 and GF-12/nanocomposites obtained by wide angle x-ray diffraction. The peak position of WAXD was used to calculate the interlayer distances of samples, the distance between the basal layers of nanoclay. Specifically, the d-spacing values were calculated using Bragg's Law for all theGF-12 and GF-12/nanocomposite samples; they summarised in Table 1.

It can be concluded that whenever the angle of the X- ray beam decreases, the spacing between the silicate layers grows, which could indicate enhancement of the ability of polymers to move in between the layers of silicate in the nanocomposite sample. In other words, the d-spacing value also indicates the degree of intercalation/exfoliation for the layered silicate in the polymer matrix. Many investigators have found that the d-spacing has a significant effect on and plays an important role in the ability of a polymer to enter between layers of clay. Put differently, d-spacing could be used to indicate the dispersion of clay into a polymer and to determine the structural level of the nanocomposite sample (whether an immiscible, intercalated or exfoliated nanocomposite structure) (Ginzburg, Singh, & Balazs, 2000; Krishnamoorti, Vaia, & Giannelis, 1996).

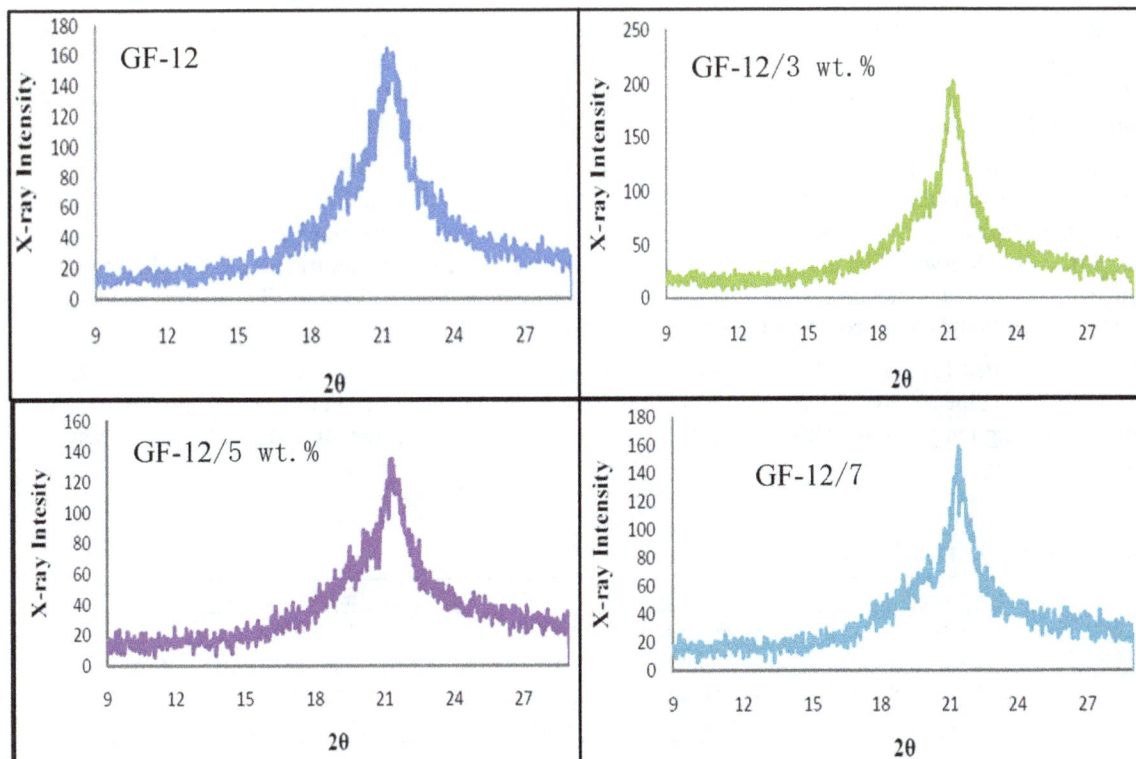

Figure 2. X-ray diffraction patterns for the GF-12 and GF-12 nanocomposites

Table 1. WAXD results obtained from different clay loadings of nanocomposites

Specimen	RGP	RGPS /3wt.%	RGPS /5wt.%	RGPS /7 wt.%
2θ value	21.48	21.24	24.24	21.36
d-spacing (nm)	4.13	4.18	4.18	4.15

The peak positions, as shown in Figure 2, were obtained from X-ray diffraction to find the interlayer distance of each sample. For the un-reinforced GF-12 sample, the peak 2θ value position of the 2θ value was 21.48° and the interlayer d-spacing value was 4.13 nm. The peak 2θ values for 3% and 5 wt. % of the nanoclay loadings were 21.24° and 21.24° respectively and the inter layer d-spacing values were 4.18 and 4.18 nm, respectively. In the same way, for a 7 wt. % nanoclay loading a 2θ value of 21.36° and an interlayer d-spacing value of 4.15 nm were recorded. It can be concluded that as the angle in nanocomposite samples becomes smaller, the inter layer d-spacing increases and that this is related to the intercalated structures (Alexandre & Dubois, 2000; Sinha Ray & Okamoto, 2003).

3.2 Differential Scanning Calorimetry (DSC)

The DSC characterization method was used in to investigate the change in the glass transition temperature (T_g) of the neat and corresponding nanocomposites as well as to study the effect on the crystallization (T_c) and melting temperature (T_m) of adding layered silicate to the polymer matrix. All the DSC results are shown in Figures 3-5 and summarized in Table 2. From the DSC results, it was found that the addition of layered silicate into the polymer matrix resulted in a slight improvement in the T_g, compared to the base polymer. As shown in Figure 3, the neat matrix exhibited 47°C and for nanocomposites of 3%, 5%, and 7 wt. % showed 47.20°C, 47.33° C, and 47.50°C respectively. The improvement of T_g can be attributed to the well dispersed layered silicate within the polymer matrix, which played an important part in increasing the T_g results, as demonstrated by the results of XRD. When the amount of clay loading goes up, the T_g goes down, compared to other wt. % nanocomposites. This effect can be traced to the lower adhesion of the layered silicate and polymer matrix, in which the layered silicate does not insulate and protect the polymer matrix from heat, as shown in the values of XRD. Thus, the heat can easily reach the polymer matrix and increase the polymer's mobility. In addition, the formation of an interphase between the clay sheets can be important in reducing the T_g value when the matrix material is close to the layered silicate surface. In turn, various properties from the bulk material are introduced. The phenomenon of interphase formation is attributed to the plasticization of the polymer surfactants (Alateyah, Dhakal, & Zhang, 2013). Another explanation of the lower reduction of the T_g value at a higher clay loading is the increased free volume of resin. In conclusion, it is worth mentioning that many factors can be important in changing the T_g value, for example, the curing temperature and time, the modification of the layered silicate, and the intercalation level between sheets. The T_g results were found to be in close agreement with the work reported in other studies (Alateyah, Dhakal, & Zhang, 2014; Bakar, Kostrzewa, Hausnerova, & Sar, 2010; Yasmin, Luo, Abot, & Daniel, 2006).

The DSC test can also show important information about the crystallization and melting temperature by scan cycles of cooling and heating. As shown in Figures 4 & 5, up to 5 wt.%, the improvement of crystallinity in the nanocomposites was almost proportional to the clay loading, but at higher clay loading, i.e. 7 wt.%, a reduction of crystallinity was observed. The enhancement of the crystallinity value may be traced to the addition of layered silicate, which acts as a nucleating agent which can increase the rate of crystallization. The melting temperatures of the neat polymer and the nanocomposites were almost the same.

The GF-12 exhibited 177.63° C. The nanocomposites samples of 3%, 5%, and 7 wt.% showed 177.37°C, 177.21° C, and 177.47°C respectively. The crystallite size can be reduced by the addition of layered silicate, which in turn affects the melting temperature. This finding was in close agreement with the study by Kim and Creasy (Kim & Creasy, 2004).

Table 2. DSC results for EX and its nanocomposites

Sample	Tg (ºC) (SD)	Tc (ºC) (SD)	Tm (ºC) (SD)
GF-12	47.00 (±0.12)	152.80 (±0.22)	177.63 (±0.09)
GF-12/ 3 wt.%	47.20 (±0.10)	154.20 (±0.13)	177.37 (±0.1)
GF-12/ 5 wt.%	47.33 (±0.11)	155.19 (±0.14)	177.21 (±0.12)
GF-12/ 7 wt.%	47.50 (±0.13)	152.75 (±0.11)	177.47 (±0.2)

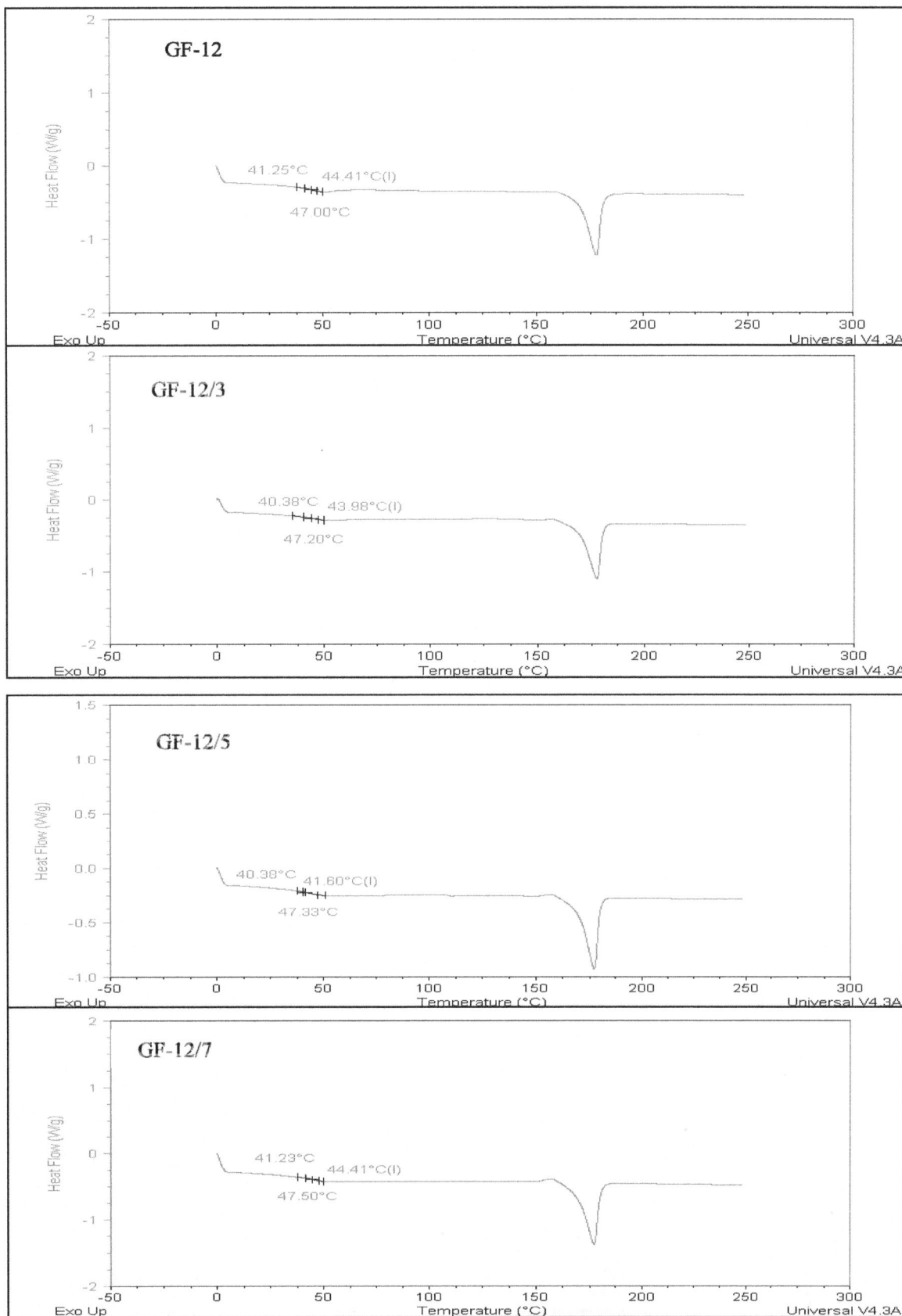

Figure 3. Glass Transition Temperature (T_g) curve of the GF-12 samples

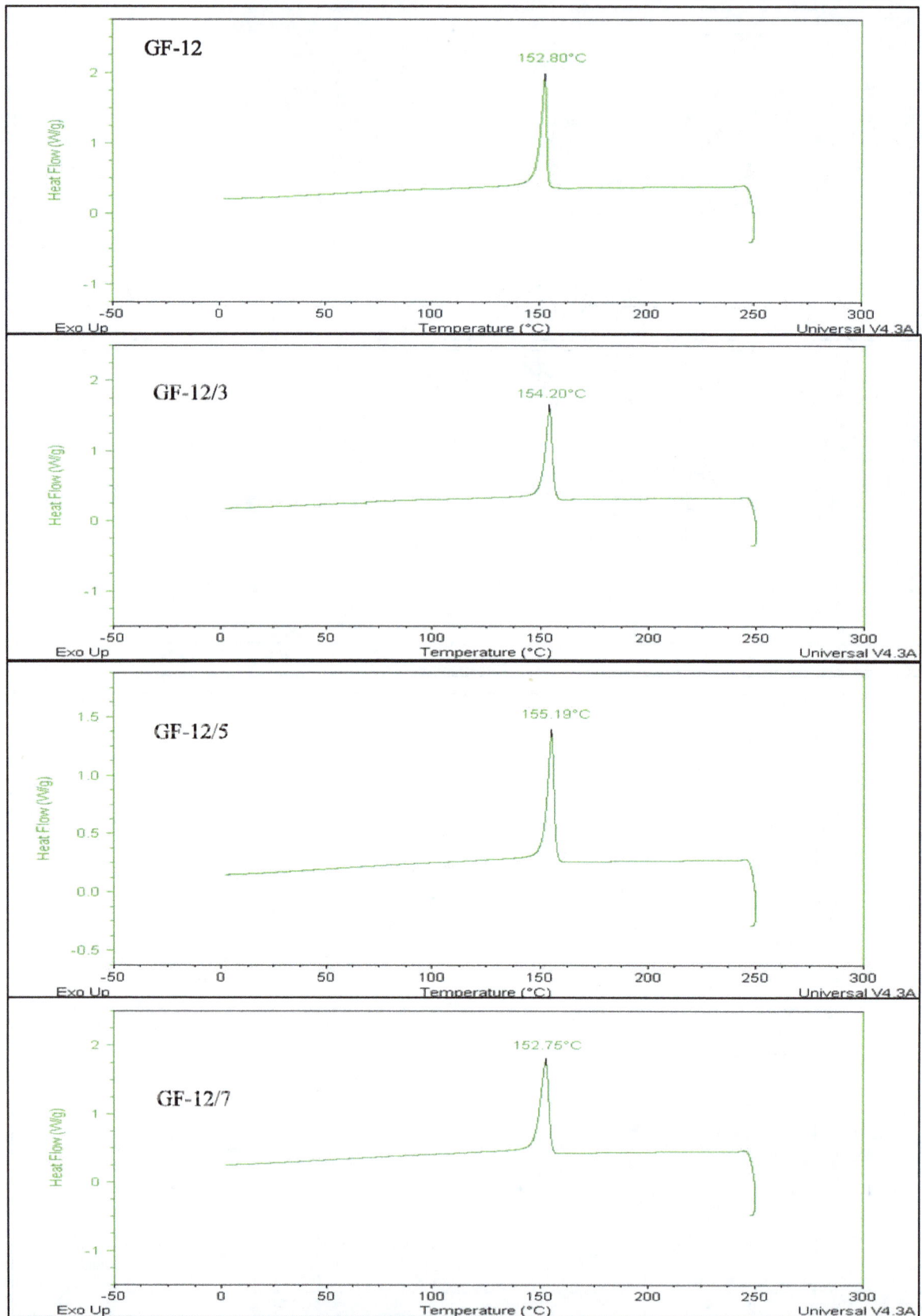

Figure 4. DSC cooling scans of the GF-12 samples

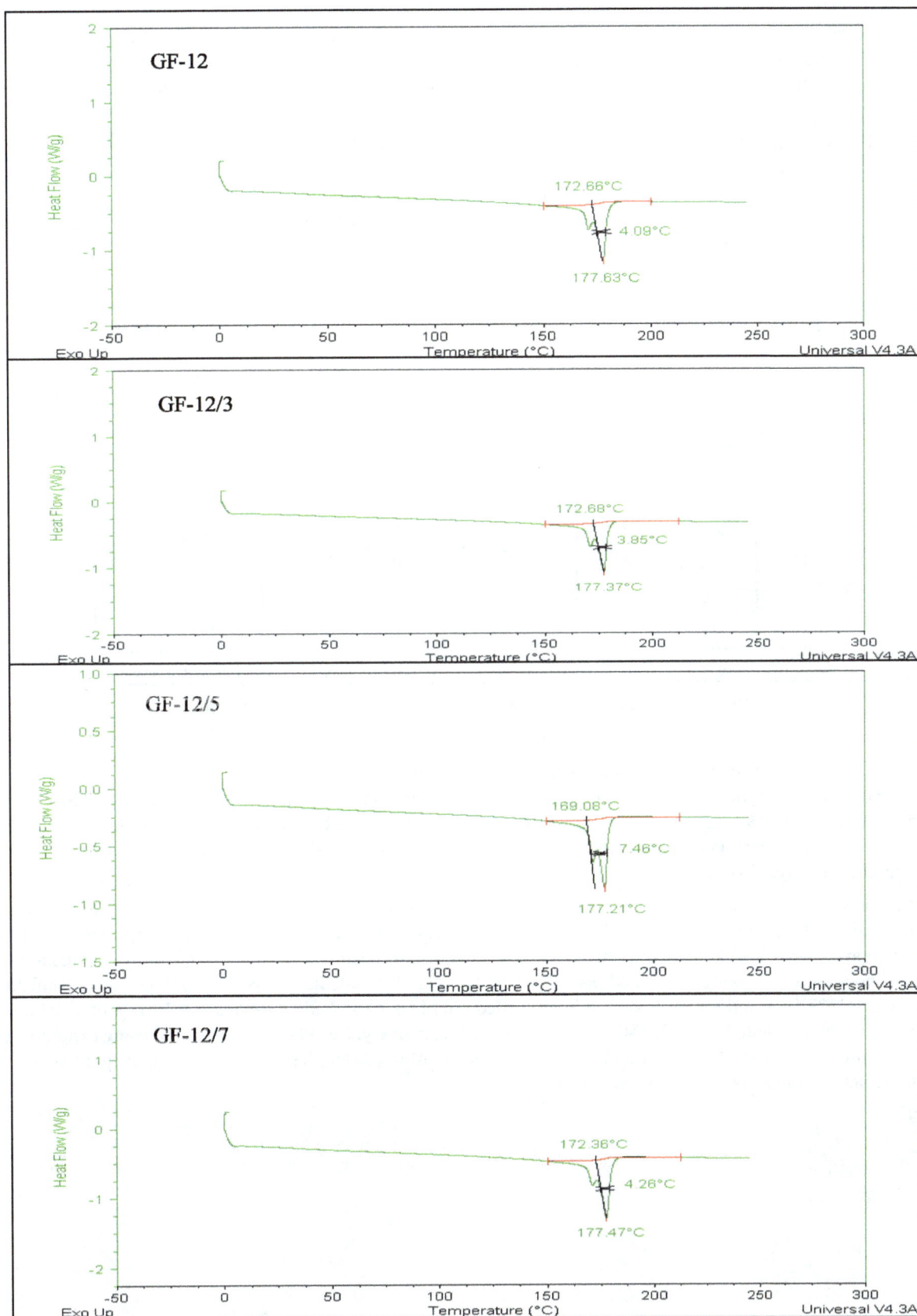

Figure 5. DSC heating curve of the GF-12 samples

3.3 Nanoindentation: Maximum Depth, Hardness and Elastic Modulus

The average values of the experimental data extracted from the loading/unloading curves from the Nano-indentation tests are shown and summarized in Figures 6 & 7 and Table 3. As can be seen, with an increase in the layer silicate reinforcement, the materials' resistance to nanoindentation was found to increase monotonously.

The neat recycle GF-12 had the highest indentation depth (1472 nm); hence, it had the lowest hardness (0.0922 MPa). The depth at maximum load for 3%, 5% and 7 wt.% nanocomposite samples were 1099, 1147 and 1007 nm respectively which were lower than the unreinforced recycle GF-12 sample, as can clearly be seen in Figure 6. At low/high levels of layered silicate, it can clearly be seen that the nanocomposite samples decreased in penetration depth compared to the unreinforced sample.

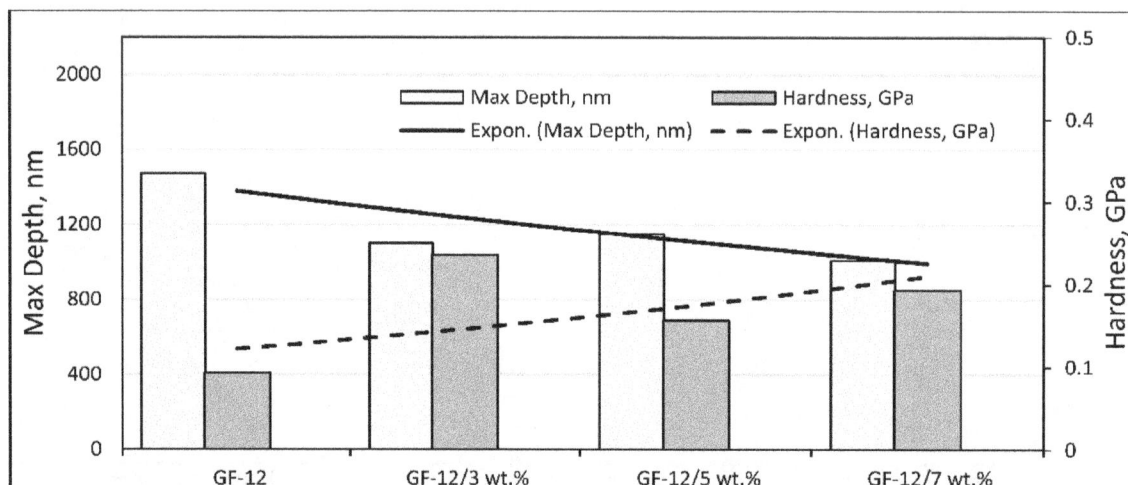

Figure 6. Maximum depth and Non-hardness of the GF-12 samples

As expected, the addition of layered silicates into the recycled GF-12 had a marked effect on the nanohardness. The nanohardness for the GF-12 nanocomposite samples increased with the increased content of the layered silicates (up to 7 wt.%). Nanohardness values showed the ability of the material to resist plastic deformation. Unlike the unreinforced GF-12 sample, all GF-12/nanocomposite systems showed more internal obstruction due to the presence of layered silicates.

The reduced modulus measured for recycled GF-12 was 1.28744 GPa. For 3%, 5% and 7 wt. % clay loaded samples, the reduced moduli were 2.95051 (an approximately 129% increase), 2.08470 (an approximately 62% increase) and 2.96331 GPa (an approximately 130% increase), respectively. By using Equation 3 the theoretical elastic modulus values were calculated, as shown in Table 3. The calculated elastic modulus for unreinforced GF-12 is 1.2373 GPa, and for the 3 wt. % of clay loaded sample is 2.8397 GPa. For 5% and 7 wt. % of clay loaded samples, the elastic moduli were 2.0049 and 2.8521 GPa, respectively. Results were calculated using experimental load information for recycled GF-12 and various nanoclay reinforced samples and the results compared favorably with the practical indentation results, as shown in Figure 7.

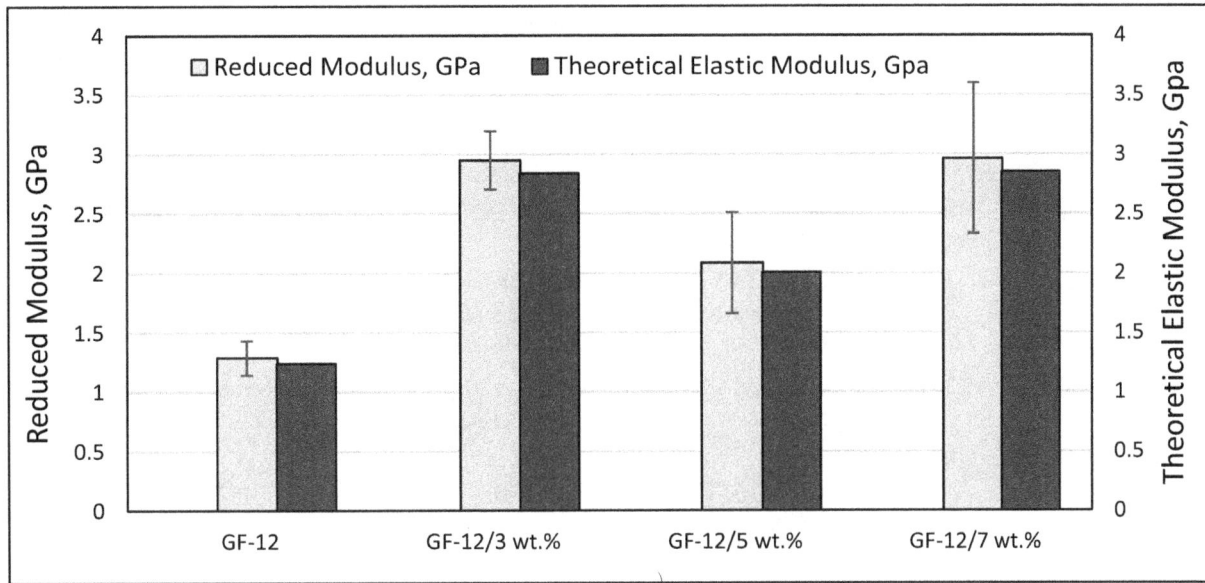

Figure 7. Experimental and theoretical reduced modulus of the GF-12 samples

The mechanisms of improvement in the nanohardness for the GF-12 matrix can be explained as the positive effect of high aspect ratio nanoclay reinforcement. The addition of the "high aspect ratio" layered silicate into GF-12 leads to an increase of hardness (decrease in indentation depth) in comparison with the neat GF-12. The nanohardness results were found to be in close agreement with the work carried out by other studies (Aldousiri et al., 2011; Dhakal, Aldousiri, Zhang, Alhubail, & Bennett, 2015).

Table 3. Summary of nanoindentation results for GF-12 nanocomposites

Sample	Max Depth (nm)	Hardness (GPa)	Reduced Modulus (GPa)	Theoretical Elastic Modulus (GPa)
GF-12	1472.26 (±129.64)	0.09223 (±0.018)	1.28744 (±0.146)	1.2373
GF-12/3 wt.%	1099.79 (±429.01)	0.23557 (±0.141)	2.95051 (±1.873)	2.8397
GF-12/5 wt.%	1147.81 (±132.91)	0.15664 (±0.041)	2.08470 (±0.426)	2.0049
GF-12/7 wt.%	1007.56 (±116.46)	0.19317 (±0.051)	2.96331 (±0.633)	2.8521

4. Conclusion

This study was conducted to examine the effect of various loading levels of layered silicates reinforcement on the thermal and nanoindentation properties of GF-12 and GF-12/nanocomposites. This investigation leads to the following conclusions:

- All the GF-12/nanocomposite samples showed d-spacing values that are modestly higher than that for the unreinforced GF-12 sample.

- The addition of layered silicate to GF-12 composite led to remarkable positive effects on the penetration resistance and hardness of the nanocomposite samples. Moreover, the nanoindentation properties thus created were more sensitive to the amount of clay than the nanostructure for GF-12 and GF-12/nanocomposites.

- The improvement in the glass transition temperature up to 5 wt. % was almost proportional to the clay loading. The further addition of layered silicate resulted in the reduction of the T_g. which was attributed to the intercalation level as demonstrated by XRD.

- The crystallization temperature showed an improvement from the incorporation of layered silicate compared to the base polymer. The enhancement of crystallization was attributed to the addition of clay loading, which functioned as a nucleating agent which can increase the rate of crystallization.

- The heat scanning cycle by the DSC showed that the addition of layered silicate resulted in a slight decrease in the melting temperature, which was traced to the reduced crystallite size which was influenced by the incorporation of layered silicate in the polymer matrix.

5. References

Alateyah, A. I., Dhakal, H. N., & Zhang, Z. Y. (2013). Processing, Properties, and Applications of Polymer Nanocomposites Based on Layer Silicates: A Review. *Advances in Polymer Technology, 32*(4), n/a-n/a. http://dx.doi.org/10.1002/adv.21368

Alateyah, A. I., Dhakal, H .N., & Zhang, Z. Y. (2014). Water Absorption Behavior, Mechanical and Thermal Properties of Vinyl Ester Matrix Nanocomposites Based on Layered Silicate. *Polymer-Plastics Technology and Engineering, 53*(4), 327-343. http://dx.doi.org/10.1080/03602559.2013.844246

Aldousiri, B., Dhakal, H. N., Onuh, S., Zhang, Z. Y., & Bennett, N. (2011). Nanoindentation behaviour of layered silicate filled spent polyamide-12 nanocomposites. *Polymer Testing, 30*(6), 688-692. http://dx.doi.org/10.1016/j.polymertesting.2011.05.008

Aldousiri, B., Shalwan, A., & Chin, C. W. (2013). A Review on Tribological Behaviour of Polymeric Composites and Future Reinforcements. *Advances in Materials Science and Engineering, 2013*, 8. http://dx.doi.org/10.1155/2013/645923

Alexandre, M., & Dubois, P. (2000). Polymer-layered silicate nanocomposites: preparation, properties and uses of a new class of materials. *Materials Science and Engineering: R: Reports, 28*(1–2), 1-63. http://dx.doi.org/10.1016/S0927-796X(00)00012-7

Bakar, M., Kostrzewa, M., Hausnerova, B., & Sar, K. (2010). Preparation and property evaluation of nanocomposites based on polyurethane - modified epoxy/montmorillonite systems. *Advances in Polymer Technology, 29*(4), 237-248. http://dx.doi.org/10.1002/adv.20192

Dhakal, H. N., Aldousiri, B., Zhang, Z. Y., Alhubail, M., & Bennett, N. (2015). Characterization and Properties of Layered Silicate Reinforced Spent DuraForm EX Nanocomposites. *Polymer-Plastics Technology and Engineering, 54*(5), 484-493. http://dx.doi.org/10.1080/03602559.2014.955198

Djebara, Y., El Moumen, A., Kanit ,T., Madani, S., & Imad, A. (2016). Modeling of the effect of particles size, particles distribution and particles number on mechanical properties of polymer-clay nano-composites: Numerical homogenization versus experimental results. *Composites Part B: Engineering, 86*, 135-142. http://dx.doi.org/10.1016/j.compositesb.2015.09.034

Fischer, H. (2003). Polymer nanocomposites: from fundamental research to specific applications. *Materials Science and Engineering: C, 23*(6–8), 763-772. http://dx.doi.org/10.1016/j.msec.2003.09.148

Giannelis, E. P. (1996). Polymer Layered Silicate Nanocomposites. *Advanced Materials, 8*(1), 29-35. http://dx.doi.org/10.1002/adma.19960080104

Ginzburg, V. V., Singh, C., & Balazs, A. C. (2000). Theoretical phase diagrams of polymer/clay composites: the role of grafted organic modifiers. *Macromolecules, 33*(3), 1089-1099 .

Hu, Y., Shen, L., Yang, H., Wang, M., Liu, T., Liang, T., & Zhang, J. (2006). Nanoindentation studies on Nylon 11/clay nanocomposites. *Polymer Testing, 25*(4), 492-497. http://dx.doi.org/10.1016/j.polymertesting.2006.02.005

Hussain, F., Hojjati, M., Okamoto, M., & Gorga, R. E. (2006). Review article: polymer-matrix nanocomposites, processing, manufacturing, and application: an overview. *Journal of composite materials, 40*(17), 1511-1575. http://dx.doi.org/10.1177/0021998306067321

Kiliaris, P., & Papaspyrides, C. D. (2010). Polymer/layered silicate (clay) nanocomposites: An overview of flame retardancy. *Progress in Polymer Science, 35*(7), 902-958. http://dx.doi.org/10.1016/j.progpolymsci.2010.03.001

Kim, J., & Creasy, T. S. (2004). Selective laser sintering characteristics of nylon 6/clay-reinforced nanocomposite. *Polymer Testing, 23*(6), 629-636. http://dx.doi.org/10.1016/j.polymertesting.2004.01.014

Kotal, M., & Bhowmick, A .K. (2015). Polymer nanocomposites from modified clays: Recent advances and challenges. *Progress in Polymer Science, 51*, 127-187. http://dx.doi.org/10.1016/j.progpolymsci.2015.10.001

Krishnamoorti, R., Vaia, R. A., & Giannelis, E. P. (1996). Structure and Dynamics of Polymer-Layered Silicate Nanocomposites. *Chemistry of Materials, 8*(8), 1728-1734. http://dx.doi.org/10.1021/cm960127g

Liu, T., Ping Lim, K., Chauhari Tjiu, W., Pramoda, K. P., & Chen, Z.-K. (2003). Preparation and characterization of nylon 11/organoclay nanocomposites. *Polymer, 44*(12), 3529-3535. http://dx.doi.org/10.1016/S0032-3861 (03)00252-0

Pinnavaia, T. J., & Beall, G. W. (Eds.). (2000). *Polymer-clay nanocomposites.* John Wiley.

Shalwan, A., & Yousif, B. F. (2013). In State of Art: Mechanical and tribological behaviour of polymeric composites based on natural fibres. *Materials & Design, 48*, 14-24. http://dx.doi.org/10.1016/j.matdes. 2012.07.014

Shalwan, A., & Yousif, B. F. (2014). Influence of date palm fibre and graphite filler on mechanical and wear characteristics of epoxy composites. *Materials & Design, 59*, 264-273. http://dx.doi.org/10.1016/j.matdes. 2014.02.066

Shen, L., Phang, I. Y., Liu, T., & Zeng, K. (2004). Nanoindentation and morphological studies on nylon 66/organoclay nanocomposites. II. Effect of strain rate. *Polymer, 45*(24), 8221-8229. http://dx.doi.org/10. 1016/j.polymer.2004.09.062

Shen, L., Tjiu, W. C., & Liu, T. (2005). Nanoindentation and morphological studies on injection-molded nylon-6 nanocomposites. *Polymer, 46*(25), 11969-11977.http://dx.doi.org/10.1016/j.polymer.2005.10.006

Sinha Ray, S., & Okamoto, M. (2003). Polymer/layered silicate nanocomposites: a review from preparation to processing. *Progress in Polymer Science, 28*(11), 1539-1641. http://dx.doi.org/10.1016/j.progpolymsci. 2003.08.002

Thomas, S., & Pothan, L. A. (2009). *Natural fibre reinforced polymer composites from macro to nanoscale.* Paris; Philadelphia (Pa): Éd. des Archives contemporaines; Old City publishing.

Yasmin, A., Luo, J. J., Abot, J. L., & Daniel, I. M. (2006). Mechanical and thermal behavior of clay/epoxy nanocomposites. *Composites Science and Technology, 66*(14), 2415-2422. http://dx.doi.org/10.1016/ j.compscitech.2006.03.011

Youssef, A. M. (2013). Polymer Nanocomposites as a New Trend for Packaging Applications. *Polymer-Plastics Technology and Engineering, 52*(7), 635-660. http://dx.doi.org/10.1080/03602559.2012.762673

Yu, S., Zhao, J., Chen, G., Juay, Y. K., & Yong, M. S. (2007). The characteristics of polyamide layered-silicate nanocomposites. *Journal of Materials Processing Technology, 192–193*, 410-414. http://dx.doi.org/10. 1016/j.jmatprotec.2007.04.006

Zeng, Q. H., Yu, A. B., Lu, G. Q., & Paul, D. R. (2005). Clay-based polymer nanocomposites: research and commercial development. *Journal of nanoscience and nanotechnology, 5*(10), 1574-1592.

16

Morphology of Excess Carbides Damascus Steel

D. A. Sukhanov[1], L. B. Arkhangelsky[2], N. V. Plotnikova & N. S. Belousova[3]

[1] ASK-MSC Company, Moscow, Russia

[2] President Union Smiths, Korolev, Russia

[3] Novosibirsk State Technical University, Novosibirsk, Russia

Correspondence: D. A. Sukhanov, ASK-MSC Company, 117246, Moscow, Nauchny proezd 8, Building.1, office 417, Russia. E-mail: sukhanov7@mail.ru

Abstract

Considered the nature of changes in the morphology of carbides of the unalloyed high-carbon alloys type Damascus steel depending on the degree of supercooling of the melt, heat treatment and plastic deformation. It is shown that iron-carbon alloy with carbon content as in white cast iron at high degrees of supercooling can crystallize as a high-carbon steel. Considered three hypotheses for the formation of the eutectic carbides in pure iron-carbon alloys. The first hypothesis is based on the thermal process of dividing plates of secondary cementite or of ledeburite on isolated single grain. The second hypothesis is based on the deformation process of crushing of secondary cementite or of ledeburite into separate fragments (the traditional view on the formation of eutectic carbides). The third hypothesis is based on the transformation of metastable ledeburite in a stable phase of eutectic carbide prismatic morphology. Found that some types of wootz, which carbon content as in of white cast irons not is contain its structure of ledeburite. It is shown that the structure of consists entirely of the eutectic carbides prismatic morphology.

Keywords: Bulat, Damascus steel, wootz, Indian steel

1. Introduction

Modern tool steels are produced by complications of chemical composition and heat treatment. The highest achievement in the field of cutting tools include are high-speed steels and die steels as ledeburitic steels. Heat treatment of these steels is a complex technological process, including multi-stage heating and cooling and cold treatment. High-speed steels and die steels because of their physico-chemical peculiarities of the structure have low plasticity, which constrains their application as a cutting tool subject to dynamic loads. The aim of this work is to develop an alternative steels ledeburite without expensive alloying elements and the creation of resource-saving technologies of producing cutting tools of high elasticity under dynamic loads. Such conditions are responsible unalloyed high-carbon alloys on the carbon content, in the field of white cast irons. After thermomechanical processing of these alloys acquire the structure of the ledeburitic steels, which are called Damascus steel (Bulat, Wootz). Cutting properties and elasticity of these steels will depend on morphological features of excess carbide phase and ability of the matrix pearlite to take the load.

Today, it is increasingly common belief that Damascus steel lost its practical value. From our point of view, this is because there is no single agreed upon theory about the chemical composition and the microstructure the Damascus steel, which would define its technological properties and expand the field of application in modern industry. Therefore, in order to understand why high cutting capacity the Damascus steel of the blade of combined with high elasticity, it is necessary to reveal its true structure and to establish the main signs of alloys.

The morphology of the excess carbide phase in the Damascus steel has always been of interest to researchers. After analyzing the works that have studied samples of ancient Damascus steels blades (Anosov, 1841; Gaev, 1965; Tavadze, Amaglobeli, Inanishvili, & Eterashvili, 1984; Basov, 1991; Gurevich, 2007; Schastlivtsev, Gerasimov, & Rodionov, 2008; Schastlivtsev, Urtsev, Shmakov, 2013; Arkhangelsky, 2007; Sherby & Wadsworth, 1985; Wadsworth & Sherby, 2001; Wadsworth, 2015; Verhoeven, Pendray, & Gibson, 1996; Barnett, Sullivan, & Balasubramaniam, 2009), we can make a preliminary conclusion that the excess carbides is a special morphological type of cementite, is fundamentally different from the excess phases of secondary cementite, the ledeburite and primary cementite in iron-carbon alloys. Virtually all authors have noted that the morphological feature of the cementite is in the size abnormality of the carbides having the shape of an irregular octahedral and prisms, which from our point of view, more similar in morphology to the carbides of the ledeburite steels (Sukhanov, 2014; Sukhanov & Arkhangelski, 2016). The size of carbides in the Damascus steel in different parts of the deformation are in the range of 0.005 - 0.1 mm. It is also noted that the heterogeneity of the distribution and morphology of abnormally large carbides remain after annealing at temperatures above 900 °C. The nature of these abnormally large carbide formations the authors to identify and failed. There were attempts, but none successful.

In the periodical literature, there is no consensus about how to call an abnormally large carbide formation. In the field of ledeburite steel, for all values of large carbides having the shape of an irregular octahedral and prisms, are such names as "angular carbides", "prismatic carbides", "faceted carbides" and "eutectic carbides" (Geller, 1968; Kremnev & Zabolotsky, 1969; Golikov, 1958; Taran, Nizhnekovskaya, & Mironova, 1981). Most often used, the term "eutectic carbide". It is formed from the eutectic the transformation of metastable complexes type M_6C and M_3C in stable carbide formation type MC, M_2C and M_7C_3 with hexagonal structure.

The mechanism of formation of faceted carbides in pure iron-carbon alloys type Damascus steel is still not clear. Question about the transformation of the cementite in the carbide prismatic of the morphology is one of the most interesting and important in the analyzed problem. It has not only scientific, but also practical value, because knowing the answer to this question and affecting the process of conversion of cementite, it is possible to control the whole complex of mechanical and physical properties of high carbon alloys of the Damascus steels (wootz).

2. Materials and Methods

The object of the study was chosen high-carbon alloy type BU22A (2.25 %C; 0.065 %Si; 0.024 %Mn; 0.002 %P; 0.004%S, all the rest elements in hundredths and thousandths fractions). On the initiative of L.B Arkhangelsky, experiments have been carried out for melting in a crucible high-carbon alloys a vacuum furnace without deoxidation in the Federal State Unitary Enterprise (FSUE) I.P. Bardin Central Research Institute for Ferrous Metallurgy. In the marking of alloy letters and numbers signify the following: BU is carbon Bulat (Damascus steel, wootz) containing not more than 0.1% of manganese and silicon (each individually); 22 is the average carbon weight fraction (2.25 wt.%); A is a high-quality alloy containing not more than 0.03% sulfur and phosphorus (each individually); all the rest elements in hundredths and thousandths fractions. Deformation of the alloy was carried out with the help of forging under an oblique angle of 45 degrees in the temperature range from 850 °C to 650 °C. The essence of forging under an oblique angle lies in the broach of the metal initially at a right angle, and then at an acute angle of 45 degrees to the front of the dies. Structural investigations were carried out using an optical microscope of a series METAM RV-21-2 in the zoom range from 50 to 1100 fold. Deeper structural investigations were carried out on scanning electron microscope CarlZeiss EV050 XVP using microanalyzer EDS X-Act. The chemical composition of the alloy controlled with the help of optical emission spectrometer ARL 3460 type on the Novosibirsk State Technical University (NSTU). The sizes of the analyzed samples was 15x15x30 mm. Phase analysis was performed by x-ray diffractometer ARL X'tra. The diffraction patterns of samples were recorded using copper x-ray tube as an x-ray source at a voltage of 40 kV and current of 40 mA. Analysis of samples was performed in the reflection geometry without monochromatization of the incident and reflected radiation. Average recorded energy dispersive Si(Li) detector wave length of the beam was $\lambda = 0,15406$ nm. Diffraction patterns were recorded repeatedly in the time mode (t = 4...9 seconds.) with step $\Delta 2\theta = 0.02$ and 0.05 degrees.

3. Results and Discussion

It is known that the elastic properties and the cutting ability of the tool depend on the morphology and volume fraction of excess carbide phase. The main technological parameters determining the morphology of excess cementite are heating temperature, cooling rate and degree of deformation. In this article we are interested in the question of how the degree of supercooling of the melt, temperature regimes of heat treatment and the degree of plastic deformation of the ingot affect the morphology of the excess carbides in pure iron-carbon alloys containing 2,25% carbon.

During melting in a vacuum, the degree of melt supercooling is quite high, since a crucible is cooled at room temperature (+20 °C). Due to the high degree of undercooling of the melt, there is a matrix of lamellar pearlite with of the excess phase of Widmanstatten cementite with a volume fraction of about 20%. The volume fraction of Widmanstatten cementite was determined by a random set of points marked on the microstructure of the sample. The number of points per area of Widmanstatten cementite divided by the total number of points was their relative volume fraction. The shape of crystals of the excess of cementite was determined by constructing the dependence of the area of their cross section from the degree of elongation. The character of distribution of values allows you to define elongated crystals of cementite, like the plates, because with increasing elongation of the observed cross sections for their area decreases.

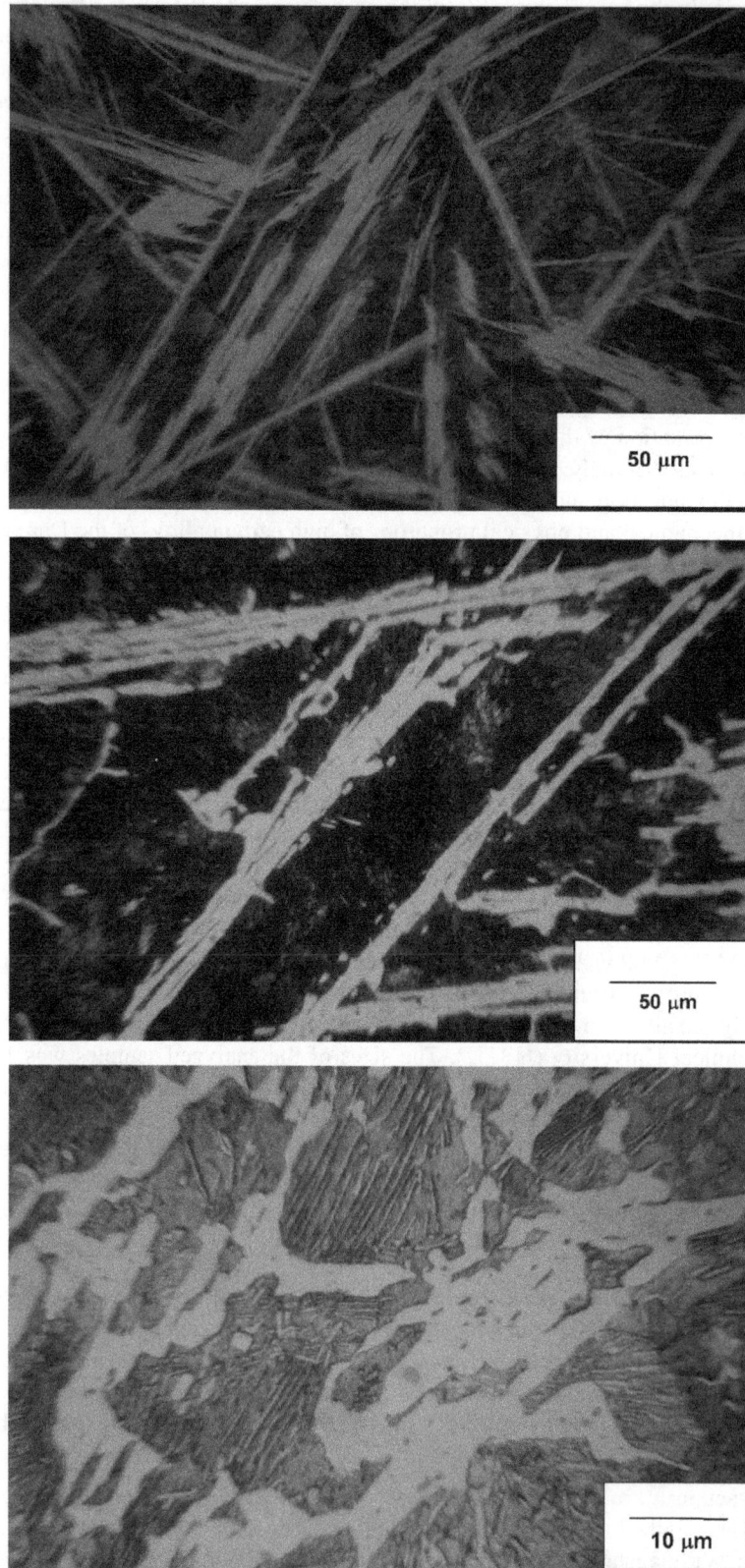

Figure 1. Morphology of excess cementite in the alloy BU22A
a – vacuum melting, is cooled at room temperature (+20 °C) (Widmanstatten cementite);
b – annealing at 700 °C, for 2 hours (Widmanstatten cementite);
c – annealing at 1150 °C, for 2 hours (metastable ledeburite)

In alloy BU22A, excess phase of Widmanstatten cementite of thickness is about 7-10 μm (Figure 1, a). Explicit plots with ledeburite eutectic in not observed. Which gives the basis of take of the maximum saturation of the primary crystals of austenite carbon during the crystallization of the melt. At the time of rapid cooling of the melt, the diffusion processes cannot develop fully. There is a situation in which critical size nuclei is extremely small, which leads to the formation of a huge number of crystallization centers. As a result, the alloy with carbon as white cast iron is crystallized as high carbon steel. Upon further cooling of the alloy, all the excess carbide supersaturated austenite is allocated mostly public in the form of secondary cementite Widmanstatten type as shown in Figure 1, a.

As there is a change in the morphology of the excess cementite Widmanstatten type at annealing? In particular, we were interested in changing the shape of the cementite at temperature at which not occurs the dissolution of excess phases (below 727 °C) and at which occurs the dissolution the excess phase (above 1147 °C). To answer this question, we have conducted studies of samples of alloy BU22A, which were subjected to isothermal annealing at 700 °C for 2 hours and at 1150 °C for 2 hours.

During annealing at 700 °C for 2 hours, the structure of the alloy does not undergo polymorphic transformations. In the pearlite matrix residual stresses are removed which appear during crystallization of the melt. In excessive carbide phase smoothed the sharp peaks Widmanstatten cementite. Microstructural studies showed that on the surface of the plates of Widmanstatten cementite appear protrusions in the form of spikes (Figure 1, b). In the working (Baranov, Bunin, Evsukov, & Pritomanova, 1969) it was noted that the opening angle of spines-protrusions is about 60 degrees and their presence is associated with the separation of the plates of Widmanstatten cementite to pieces, starting the process of spheroidization carbides. In these annealing conditions, morphology of the excess phases Widmanstatten cementite does not change drastically.

In the process of annealing at 1150 °C for 2 hours, the structure of the alloy experiences phase transformation. After slow cooling with the furnace, the matrix alloy BU22A becomes coarse structure of pearlite with interlamellar spacing of about 0.6-1.0 μm (Figure 1, c). Noticeable changes occur in the morphology of the excess carbide phase. In the plane of the thin section observed coarse conglomerates excess carbide formations (Figure 1, c). Which is identify as metastable ledeburite. Characteristic morphological feature of this metastable ledeburite is that it, compared to lamellar and unmodified (cell) ledeburite white cast iron (Sukhanov & Arkhangelski, 2016) contains in its structure a reduced number of micropores and not has a pronounced layering (Golikov, 1958), i.e. it is already not of ledeburite white cast iron , but still not quite eutectic carbide.

Widmanstatten cementite has low thermal stability and contributes to a drastic embrittlement of the steel. In order to preserve the elasticity and the cutting edge of the tool for a longer time, needed another morphology of the carbide contributing to the increase in thermal stability. Such a morphological feature of the carbides have a faceted prismatic shape, which are formed in the structure of the unalloyed high-carbon alloys during deformation by forging.

During the deformation of alloy BU22A with excess phase in the form of metastable ledeburite (Figure 1, c), which was performed using with the help of forging under an oblique angle of 45 degrees in the temperature range from 850 °C to 650 °C, with the beginning of the deformed austenitic matrix. Metastable ledeburite is under the influence of the normal compressive stress of austenite and the shear stress of deformation. As noted in article (Nizhnikovskaya, 1982), in the process the deformation around carbide accumulated defects such as dislocations. When the dislocation density reaches a critical value in metastable ledeburite changes occur, because of which, less stable carbides of ledeburite turning in the sustainable eutectic carbides prismatic shape (Figure 2).

Analysis of the diffraction lines showed that the major phase in the alloy BU22A are ferrite (α-Fe) and cementite (Fe3C). X-ray phase analysis did not specifically identify this excess carbide phase due to interference of - overlapping lines of secondary cementite. At this stage of research it is possible to speak only about what we are faced with the particular morphology of cementite.

From our point of view, we can distinguish three fundamental hypotheses that explain why, under certain thermomechanical conditions, the cementite is stabilized and acquires a trigonal-prismatic morphology in Damascus steel (wootz).

The first hypothesis is based on the thermal process of dividing plates of secondary cementite or of lamellar ledeburite on isolated single grain. The mechanism of the process heat division is described in detail in (Baranov, Bunin, Evsukov, & Pritomanova, 1969). The authors of this article it is shown that in the process of annealing on the surface of the cementite at the junction of the boundaries of austenite appear protrusions in the form of spikes. In the process of isothermal of annealing of high-carbon alloys, between the spikes (protrusions) in thin places of plates of cementite are divided into parts. As a result, between the individual particles formed of grains boundaries of austenite. Angular carbides have is faceted of the close to equilibrium and, as a rule, are located along the former cementite plates.

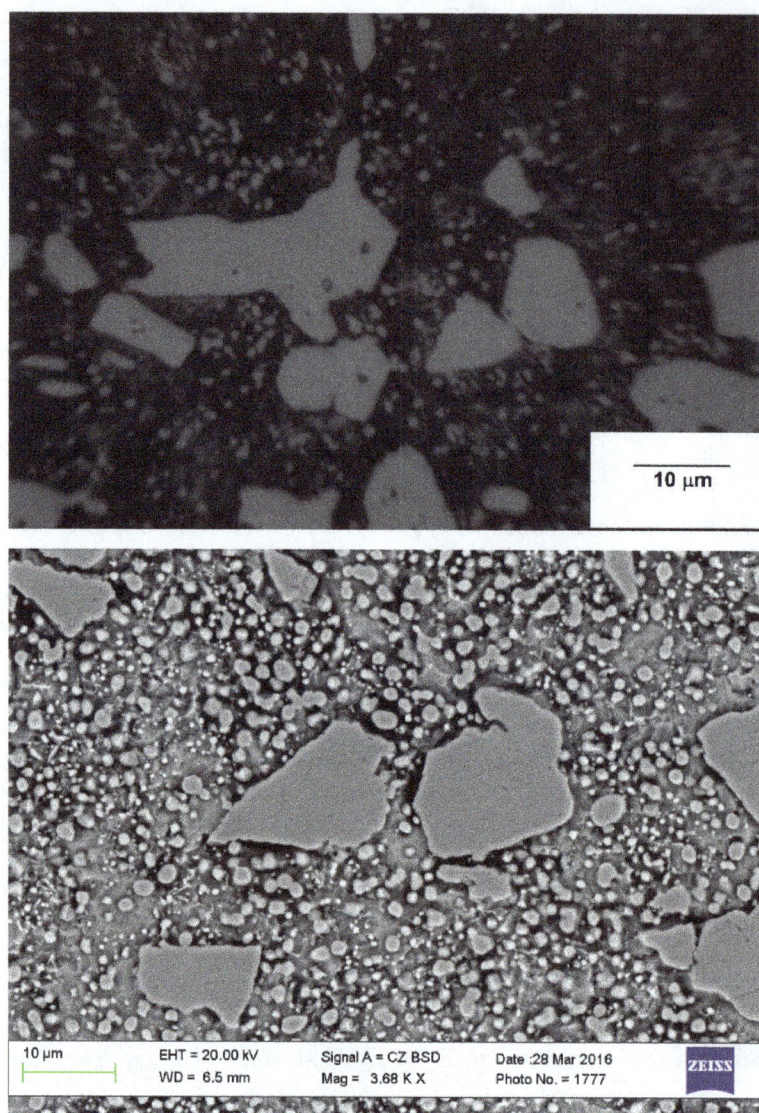

Figure 2. Morphology of excess carbides in the alloy BU22A
(forging was performed on samples with metastable ledeburite):
a – optical microscope (eutectic carbides); b - electron microscope (eutectic carbides).

The second hypothesis is based on the deformation process of crushing of secondary cementite or of lamellar ledeburite into separate fragments. This traditional view on the formation of angular carbides are described in detail in article (Geller, 1968). It is assumed that the anomalously large angular carbides are formed as a result of fragmentation of cementite of lamellar ledeburite in the process of plastic deformation into separate parts (fragments). The greater the degree of deformation during forging, the more are of crushing of the carbides on parts. Crushing carbide conglomerates occur in places of a congestion of dislocations. It can be assumed that the data of the angular carbide formation are of splinters of the ledeburite eutectic.

The basis of the third hypothesis are laid of the conclusion about what the plasticity of white cast irons, possible as a result to the carbide transformation (Nizhnikovskaya, 1982). The essence of the hypothesis about the formation of angular carbides consists in transformation of metastable ledeburite in the process of plastic deformation in a stable phase of eutectic carbide prismatic morphology. According to author of articles (Nizhnikovskaya, 1984), decrease resistance inside the lattice of cementite in the process of transformation happen due to the weakening of the barriers of Peierls-Nabarro and the formation depleted of carbon areas of cementite.

All three mechanisms of transformation of metastable ledeburite in eutectic carbides prismatic shape, in varying degrees, encountered in practice. The above explanations would have been enough had we not met some of the circumstances that are contrary to these explanations. We are talking about the fact that some grains of the eutectic carbides prismatic morphology exceed or commensurate with ledeburite colonies. This may indicate that the formation of eutectic carbides prismatic morphology happen is not crushing of metastable ledeburite at the deformation but is a transformation of metastable ledeburite in the of stable eutectic carbide. Otherwise, one should assume that there are additional energy factors associated with the growth mechanisms of cementite to abnormal size during plastic deformation. Deformation accelerates the process of complete of transformation of metastable ledeburite in the eutectic carbides prismatic shape.

Carbide phase at the number, at size and at morphology is distributed in volume of pearlite matrix is uneven (Figure 3). Professor N.I. Golikov in their work (Golikov, 1958) shows that almost all tool steel industrial production is chemically heterogeneous. In this work, we demonstrated that some types of Damascus steel, which carbon content as in of white cast irons not is contain its structure of ledeburite. The structure Damascus steel of consists entirely of the eutectic carbides prismatic morphology.

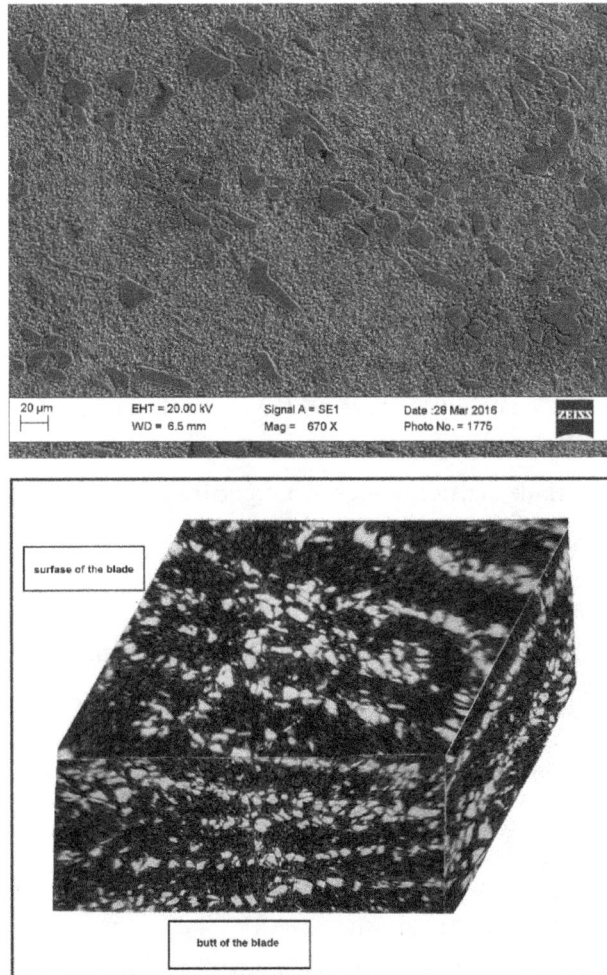

Figure 3. Carbide inhomogeneity of the alloy BU22A:
a - electron microscope (eutectic carbides); b - diagram of the structure of blade Damascus steel.

4. Conclusions

4.1 In some types of Damascus steel excess carbides are a special morphological type of cementite, is fundamentally different from the excess phases of secondary cementite, phases of ledeburite and primary cementite in iron-carbon alloys. The morphological features of excess cementite in these samples Damascus steel consists of anomalous particle size of the carbides having the shape of an irregular octahedra and prisms, which is similar to eutectic ledeburite carbides alloyed steels.

4.2 The major phases in the alloy structure BU22A in which excess carbides are in the form of faceted crystals of trigonal-prismatic morphology are ferrite (α-Fe) and cementite (Fe3C).

4.3 There are two types of process of formation of excess carbides in the form of separate particles: first, the spheroidization of carbides - carbides have the correct round or oval, without distinct angles; secondly, the faceted of carbides – carbides have an irregular trigonal-prismatic morphology.

4.4 It is revealed that the main mechanism of formation of the faceted crystals of cementite trigonal-prismatic morphology is the recrystallization of metastable ledeburite complexes in the process of plastic deformation in a stable phase of the eutectic carbide.

4.5 It is shown that some types of Damascus steel, which carbon content as in of white cast irons not is contain its structure of ledeburite. Structure Damascus steel (wootz) of consists entirely of the eutectic carbides prismatic morphology.

References

Anosov, P. P. (1841). Bulats. In D. M. Nakhimov, & A. G. Rakhstadt (Eds.) (1951), *Russian Scientists - Metallurgists* (pp. 38-112). GNTI Mash. Lit., Moscow.

Arkhangelsky, A. V. (2007). *Secrets of Bulat*, Metallurgy, Moscow.

Baranov, A. A., Bunin, K. P., Evsukov, M. F., & Pritomanova, M. I. (1969). The mechanism of division of crystals of cementite during spheroidizing annealing. *MiTOM, 6*, 1-3.

Barnett, M. R., Sullivan, A., & Balasubramaniam, R. (2009). Electron backscattering diffraction analysis of an ancient wootz steel blade from central India. *Materials Characterization, 60*, 252-260.

Basov, V. I. (1991). Bulat: lifeline. *Metallurg, 7*, 16-23.

Gaev, I. S. (1965). Bulat and contemporary iron-carbon alloys. *MiTOM, 9*, 17-24.

Geller, Y. A. (1968). *Tool steels.* Metallurgy, Moscow.

Golikov, I. N. (1958). *Dendritic liquation in steel.* Metallurgizdat, Moscow.

Gurevich, Y. G. (2007). Classification of bulat with respect to macro- and microstructure. *MiTOM, 2*(620), 3-7.

Kremnev, L. S., & Zabolotsky, V. K. (1969). Large carbides in the structure of high-speed steels. *MiTOM, 1*, 54-56

Nizhnikovskaya, P. F. (1982). Carbide transformations in eutectic iron-carbon alloys. *Metals, 6*, 105-110.

Nizhnikovskaya, P. F. (1984). Structure and ductility of eutectic type iron-carbon alloys. *MiTOM, 9*, 5-9.

Schastlivtsev, V. M., Gerasimov, V. Y., & Rodionov, D. P. (2008). Structure of three Zlatoust bulats (Davascus-steel blades). *FMM, 106*(2), 179-185.

Schastlivtsev, V. M., Urtsev, V. M., Shmakov, A. V. (2013). Structure of bulat. *FMM, 114*(7), 650-657.

Sherby, O. D., & Wadsworth, J. (1985). Damascus steel. *Scientific American, 252*(2), 112-120.

Sukhanov, D. A. (2014). Bulat - carbide class unalloyed steel. *Metallurg, 2*, 93-96.

Sukhanov, D. A., & Arkhangelski, L. B. (2016). Damascus steel microstructure. *Metallurgist, 59*, 818-822.

Taran, Y. N., Nizhnekovskaya, P. F., & Mironova, T. M. (1981). Structural changes in eutectic steels R6M5 with hot plastic deformation. *Izv. Vyssh. Uchebn. Zaved., Cher. Met., 5*, 109-113.

Tavadze, F. N., Amaglobeli, V. G., Inanishvili, G. V., & Eterashvili, T. V. (1984). Electron microscopic studies of bulat steel. *Soobsh. AN GSSR, 113*(3), 601-604.

Verhoeven, J. D., Pendray, A. H., & Gibson, E. D. (1996). Wootz Damascus Steel Blades. *Materials Characterization, 37*, 9-22.

Wadsworth, J. (2015). Archeometallurgy related to swords. *Materials Characterization, 99*, 1-7.

Wadsworth, J., & Sherby, O. D. (2001). Response to Verhoeven comments on Damascus steel. *Materials Characterization, 47*, 163-165.

17

Novel Process Approach for in-situ Insertion of Functional Elements in RTM-Applications

Mathias Bobbert[1], Florian Augethaler[1], Zheng Wang[2], Thomas Tröster[2] & Gerson Meschut[1]

[1] Laboratory for Material and Joining Technology (LWF), Paderborn University, Paderborn, Germany

[2] Chair for Automative Lightweight Design (LiA), Paderborn University, Paderborn, Germany

Correspondence: Mathias Bobbert, Laboratory for material and joining technology (LWF), Paderborn University, Pohlweg 47-49, 33098 Paderborn, Germany. E-mail: mathias.bobbert@lwf.upb.de

Abstract

Lightweight design in vehicles leads to an increasing use of fibre-reinforced-plastics (FRP). To ensure the possibility of detachable joints in those constructions specific joining technologies have to be developed. Available technologies are not suitable for mass production or are leading to damage in the FRP. The aim of this paper is to show the possibilities and the general feasibility of a new process approach for the application of functional elements in FRP or hybrid materials during resin transfer moulding (RTM) processes. Therefore the movement of the RTM-tool-punch is utilised to produce an interlock between the splay stud and the fibres of the non-consolidated preform. This enables the application of the elements during the RTM-process. The resin infusion leads to the consolidation of the CFRP and a further fixation of the splay stud. Pull-out-tests as well as torsion tests are showing the mechanical performance of the joints.

Keywords: hybrid structures, inserts, functional elements, RTM-process, CFRP

1. Introduction

In the course of progressive developments to reduce the mass of vehicles, fibre-reinforced-plastics and hybrid materials getting increasingly important as a lightweight construction material for the body in modern cars. For use in large scale productions an efficient joining method is needed to realise detachable screw connections for attachments in the assembly line. In the mass production of conventional metal bodies weld studs / nuts and mechanically set functional elements have been established for realizing releasable connections. Such support threads can be found in all areas with thin sheets where a screw connection with additional components is needed (Grandt, 1998). For example 959 functional elements are located in the Mercedes-Benz E-Class W212 series for applications as cable-clamps, pipe supports, mounting cover and liquid containers (Hahn & Westhoff, 2015). In contrast to the thermally inserted functional elements like welding studs or welding nuts, mechanical functional elements provide a wide range of applications inclusive non-metal materials contingent on a universal function principle (Müller, 1992; Heimlich, 2008). They are therefore also suitable for use in CFRP components. Nevertheless, the related piercing process is problematic for composite materials. Machined introduced holes, e.g. by means of circular milling, constitute a relatively gentle processing method thereby, but always mean an additional process step. A direct joint is achieved by self-piercing functional elements, but lead through the shear cutting process to local material failures like delaminations of the layer structure. Despite of the efficient procedure the components are weakened and the residual compressive strength in particular of the CFRP is reduced (Meschut et al., 2015). Moreover the deformation of the CFRP during riveting causes similar damage due to the low ductility. Based on the viscoelastic properties of the matrix system, it may further come to reduction of the clamping forces of the functional element by creeping, so that a stable anchorage in the CFRP cannot be guaranteed in the long term.

Furthermore adhesive based functional elements are available for composite applications and in use in BMW series i3 and i8 for example, see Figure 1. The use of these adhesive studs requires an elaborate surface pre-treatment, a complicated introducing process and is restricted to lightly loaded points as the forces cannot be transmitted directly into the fibres.

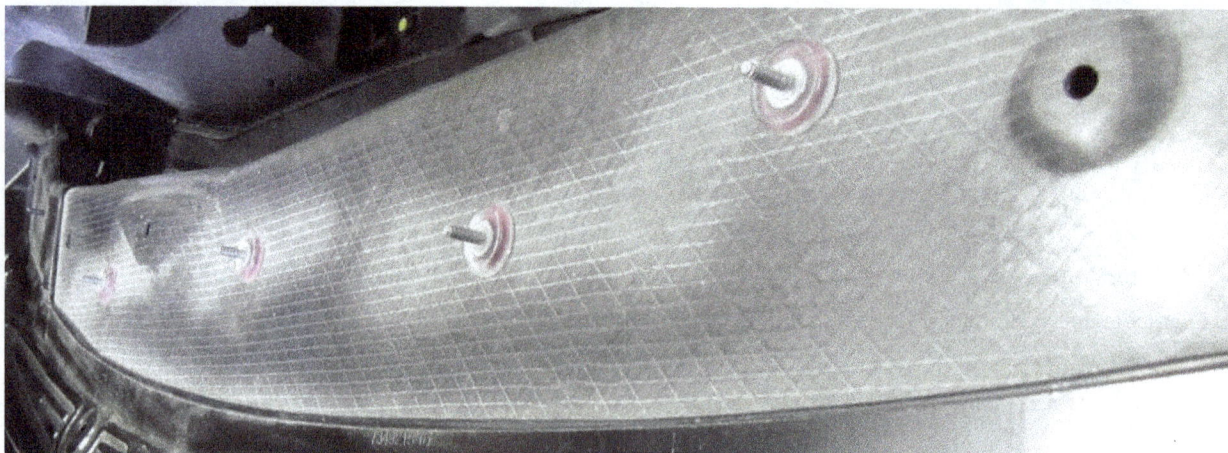

Figure 1. Dashboard carrier of a BMW i3 body in white with adhesive studs

While an element insertion in metal sheets is favoured in the press shop for reasons of efficiency, this method is not possible with CFRP components, as these do not exist in a consolidated state at the closing of the pressing tools. Admittedly various systems exist for this purpose. Currently they are laid between the fibre layers stitched with them or placed on the laminate and bonded by the matrix system (Schürmann, 2007). However, these variants are only suitable for the time-consuming manual lamination process and thus for small batch series.

This results in the development of a novel functional element for applications in FRP components that can be set in the non-consolidated part during the RTM process. Additionally process steps and damage of the composite can be avoided. Currently, there is no possibility to integrate functional elements in the series production of CFRP components in the RTM process. Especially the anchoring of the functional element to the reinforcing fibres at the press stroke and suitable sealing against resin still remains a challenge. Furthermore the closing of the RTM tools, flexibility of the element positioning regardless of the location in the top and bottom mould push the development of this innovative technology. This paper presents a suitable functional element for the described application, the associated receiving tools and the insertion process. Furthermore the mechanical performance is evaluated by pull-out and torsion tests.

2. Method

2.1 Splay-Stud-Process

To attach studs or other functional elements to a CFRP-component, the closing step of the RTM press can be utilised. Therefore, modifications to established studs, which are applied during the stacking or preforming step or after the CFRP-consolidation, have to be made. Studs which are applied after the curing of the matrix by drilling holes or cutting processes damage the composite, e.g. by delamination. An application before consolidation avoids damage. This leads to the requirement of a process with small manipulations to the fibres architecture. Due to the resin infusion gaps or other geometrical inhomogeneities, which grow during the shift, can be closed. To fulfil these requirements:

- Insertion during the closing of the RTM-tool and
- Minimization of damage in the CFRP fibres

a novel process (Figure 2) with an innovative type of functional element (Figure 4) has been developed. This consists of a threaded stud, a force transmission support and a shell. The process can also be used for CFRP-steel-hybrids.

Within the first step the attachment bushing is loaded with the splay stud. Furthermore the release agent has to be applied and the preform can be positioned on the RTM-tool. In the second step the RTM-tool starts to be closed and the splay stud touches the preform. At this point the splay element braces on the residual fibre layers and the fibres are allocated locally. Through the compression, the stud shapes four shanks which are bended and splayed sideways and dip between the fibre layers. Therefore, the splay element is pushed inside the semitubular shell and separates it in four shanks. No die is needed for this procedure. Because of this, elaborate tool adjustments are unnecessary. When the end position of the RTM-tool is reached, the forming process of the splay stud ends and an interlock between the overhead 2-4 fibre layers and the splayed shanks is generated. To protect the thread against contamination of the low-viscous resin, the studs are pushed into sealings in the bushing during the complete process. After that, in the 4th step, the resin is infused and cured. The infusion and curing of the resin results in additional adhesive bonding between the stud and the CFRP component. During the opening step of the RTM-tool the stud is demoulded from the attachment bushing. The attachment of the stud is completed. This new process depicts a significant improvement of the in (Koch et al., 2016) described process.

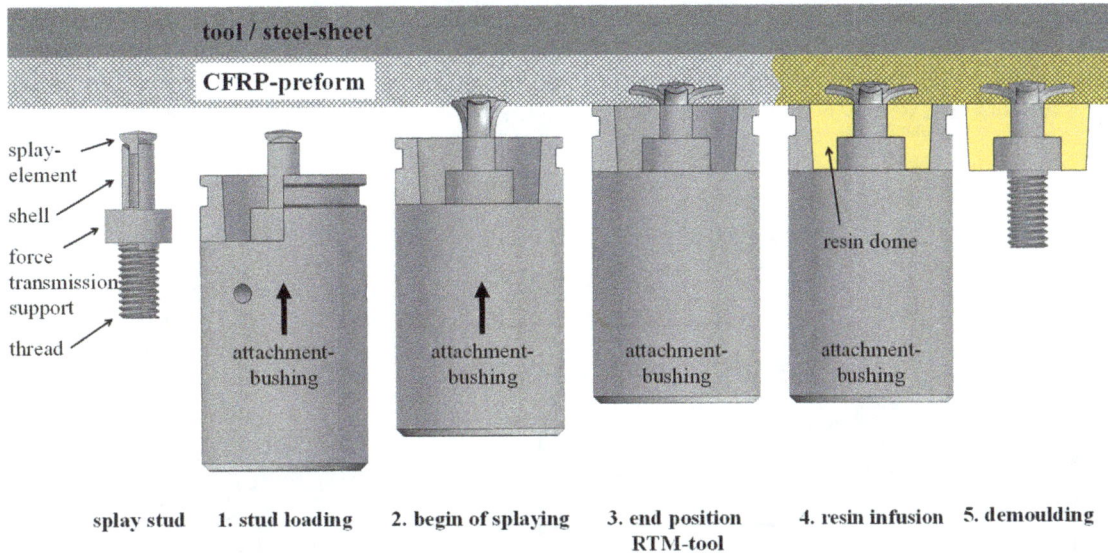

Figure 2. Principle process of the in-situ insertion of splay-studs

2.2 Tooling Concept

The current experimental setup consists of a planar RTM-tool with four implemented attachment bushings (Figure 3). The attachment bushings are designed for adaption to every kind of RTM-tool. Requirement is a vertical closing step and an area with a planar tool surface. Therefore a blind hole is positioned in the RTM-tool between the channels for the heating cartridges. A chamfer supports the insertion of the bushing into the hole. The bushing is fastened to the tool via screwing. To seal the gap between the RTM-tool and the bushing a silicone ring is used. The sealing material is chosen due to the low adhesion to the used resin.

The inner shape of the attachment bushing forms a defined cavity with a draft angle which will be filled with resin in the impregnation process and defines the shape of the resin dome. In the centre of this cavity the seating for the stud is located. The stud is placed in a slot and locked due to a lug. Therefore small steel balls are inserted in radial holes followed by prestressed springs and setscrews. They fix and centre the functional elements. To ensure the sealing of the thread against contamination of the low viscous resin the contact between the sealing ring and the stud has to be uphold through the whole process. The process force is transmitted through the ring until a defined compression of the sealing is reached. The force transmission support passes the residual process force directly from the bushing to the element, to avoid damage of the sealing. The sealing rings are located close to the outer diameter of the stud, to minimize the contaminated areas by resin. The depicted bushing is an advanced version of the in (Wang et al., 2016) described attachment bushing.

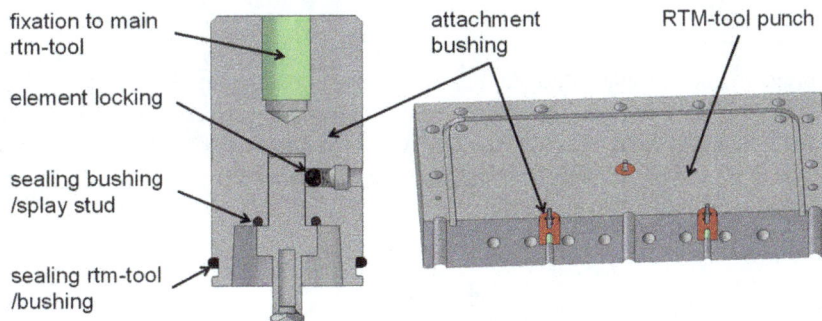

Figure 3. Attachment bushing and integration in RTM-tool

3. Experimental Setup

3.1 Materials

The substrate materials of the test plates are consisting of 2 mm steel sheet and 2 mm CFRP with the sizes of 400 mm length, 400 mm width and a total thickness of 4 mm. For the CFRP a 5-minutes epoxy matrix system (Momentive, 2012) and six layers carbon fibre scrim, which has a surface weight of 320 g/m² (SGL, 2016), are used. The layer structure of the bidirectional carbon fibre scrim is (90/0/90/0/90/0). For the steel component a HC340LA alloy is used. The splay stud consists of a turn part of AlMg5 and a splay element of steel. The geometry is shown in Figure 4a. To get a higher stability of the resin-dome after injection of the resin, this area can be additionally reinforced with winded fibres according to Figure 4b.

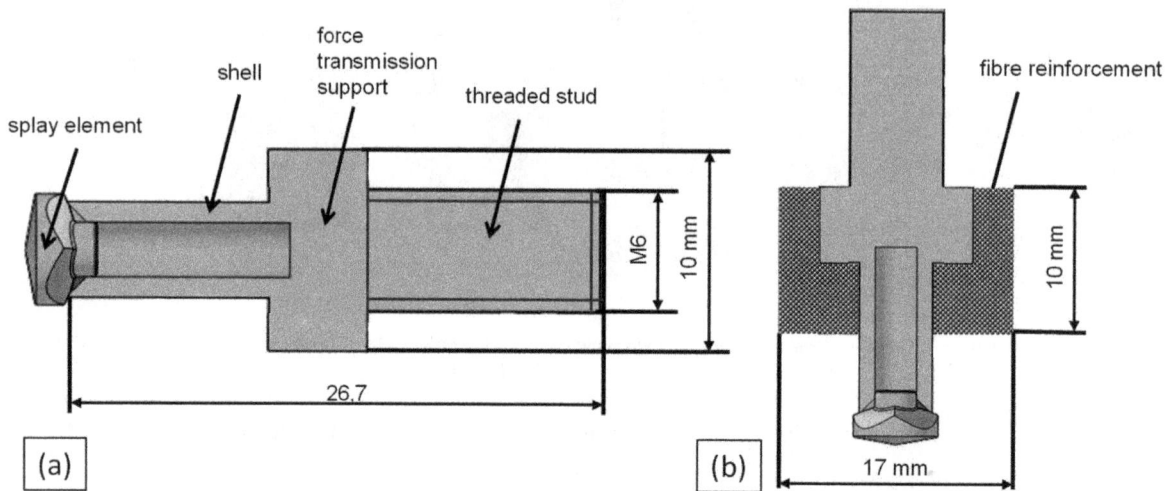

Figure 4. Geometry of the used splay stud (a) and additionally attached reinforcements (b)

3.2 Experimental Setup for RTM-Process and Process Parameters

The process parameters during manufacturing of the test plates are summarised in Table 1. By the RTM process the epoxy resin is injected into the mould with the temperature of 25 °C under a constant injection pressure by 0.5 MPa for 300 sec. The mould is heated at 80 °C. The subsequent curing process takes 3600 sec. in order to achieve a good demouldability of the parts.

Table 1. Process parameters

Nomenclature		Value	Unit
T_m	Temperature of matrix	25	[°C]
T_t	Temperature of tool	80	[°C]
p_i	Injection pressure	0.5	[MPa]
p_v	Vacuum pressure	-0.095	[MPa]
t_i	Injection time	300	[s]
t_c	Curing time	3600	[s]

The position of the preform and the splay studs in their attachment bushings in the lower tool is shown in Figure 5. The tool is heated with heating cartridges. The thickness of the plates is defined with a 4 mm distance frame. The epoxy matrix is injected through four sprues into the cavity, which are located at the corners of the upper tool. The matrix impregnates the dry fibres by an annular duct und the rest flows from the centrally positioned riser into a resin trap, which is connected with a vacuum pump.

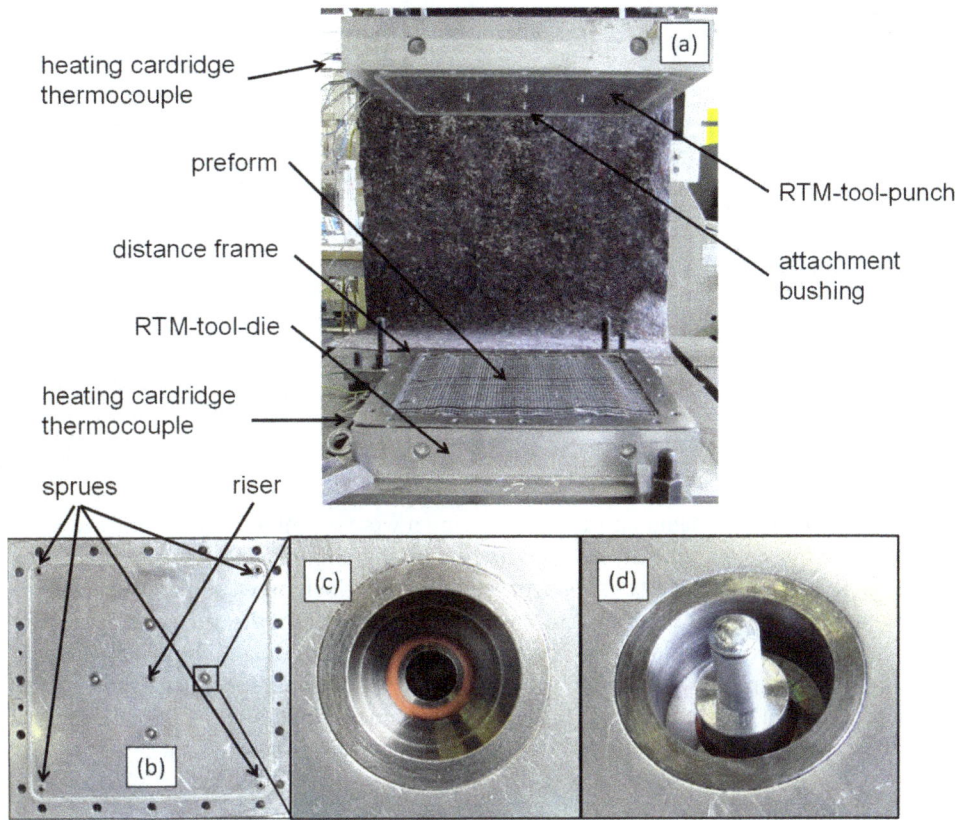

Figure 5. Experimental setup for the manufacturing of the specimen: (a) RTM-tool, (b)RTM-tool-punch with attachment bushings, (c) attachment bushing without stud, (d) attachment bushing with stud

3.3 Experimental Setup for Testing

In order to test the mechanical performance of the manufactured studs, specimens have to be cut out of the test plates. This allows to test each stud on his own and avoids prior damage which might occur when each of the four studs are tested when connected to one hole plate. In previous projects a specimen size of 50 mm x 50 mm was suitable. To avoid damage through the cutting process a water-cooled precision cut-off machine with a low feed rate was used. The specimen is shown in Fig 6.

Figure 6. Specimen cut from the plate

To obtain a good overview over the mechanical performance of the studs, two different critical test scenarios were carried out. The first scenario is the pull out test. Therefore, a special test device, described in (Meschut and Süllentrop, 2014), was used which is shown in Figure 7 (a). The tests are done at the universal test machine (Zwick Z 100). Furthermore a torsion test was performed. To this a screw test bench (Schatz) was used. The test arrangement is shown in Figure 7 (b).

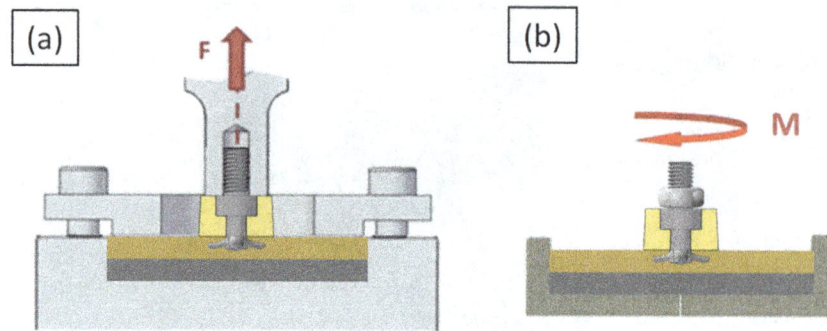

Figure 7. Experimental setup for the testing of the specimen: (a) pull out and (b) torsion

4. Results and Discussion

In the first step after the component manufacturing the specimen were examined concerning the achieved quality. Therefore, the joints were inspected optically and by micro sections. Figure 8 depicts a good quality of the joint (Figure 8 (a)). The micro section in Fig 8 (c) shows the required interlock between the studs and the CFRP. The four shanks are lying under the second fibre layer of the substrate material. In this micro section a small groove in the resin dome can be recognized as well. This appears due to the design of the sealing with a silicone ring, which is partially pressed in the cavity of the attachment bushing. Furthermore, some specimens show shrinkage cavity's in the top section of the joint (Figure 8 (b)). In some cases cracks in the resin dome appeared during the demoulding step. Those specimens were excluded from the testing of the joint strength.

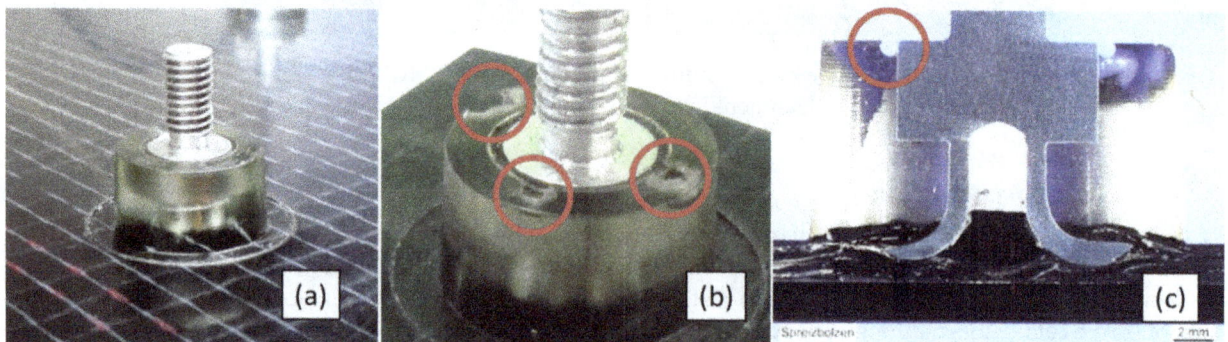

Figure 8. Photos (a, b) and microsection (c) of the joints without reinforcement

The modification of the stud with a fibre-reinforced resin dome leads to an improved quality of the joint. Demoulding cracks and shrinkage cavities haven't appeared anymore.

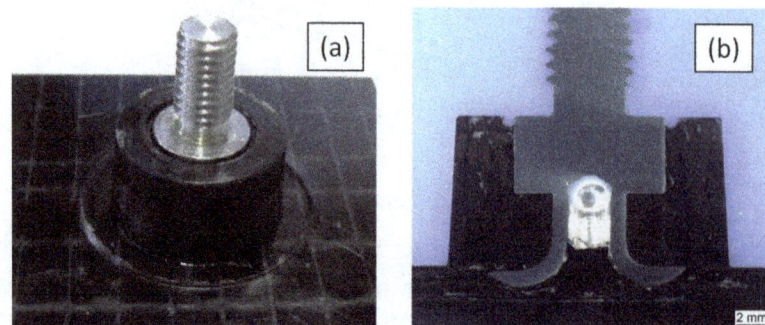

Figure 9. Photo (a) and microsection (b) of the joints with reinforcement

According to chapter 3.3 the tests for the joint strength were carried out. The pull out behaviour of the stud without reinforced resin dome can be seen in Fig 10. The graph shows at first a linear elastic behaviour up to the maximum. The reached maximum forces averages at 2.2 kN with a standard deviation of 0.1 kN. After the maximum a failure area with a decreasing force can be observed. The fracture behaviour was ruled by the rupture of the resin dome in the area of the connection to the CFRP-plate. Furthermore a delamination of the CFRP can be recognized which is caused by the interlock between CFRP and splay stud. On the basis of this behaviour the reinforcement of the resin dome was invented. The aim was to avoid the failure of the dome and get a higher utilisation of the stud and the interlock.

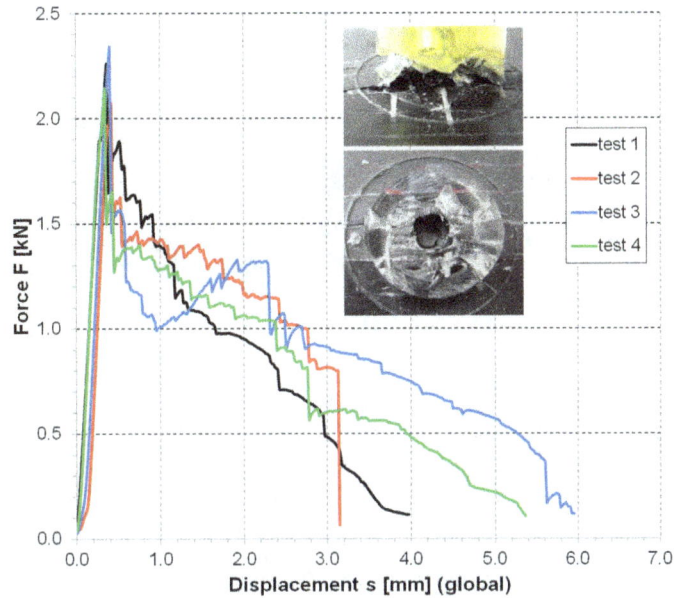

Figure 10. Experimental results of pull out tests without reinforced dome

The corresponding results to this optimisation step are shown in Figure 11. Here it can be seen that the resin dome with reinforcement stays intact during the tests. The failure appears exclusively due to delamination of the CFRP. This leads to higher maximum forces which average at 2.8 kN with a standard deviation of 0.2 kN. The failure area which is caused by the delamination is on the same level as without reinforced resin dome.

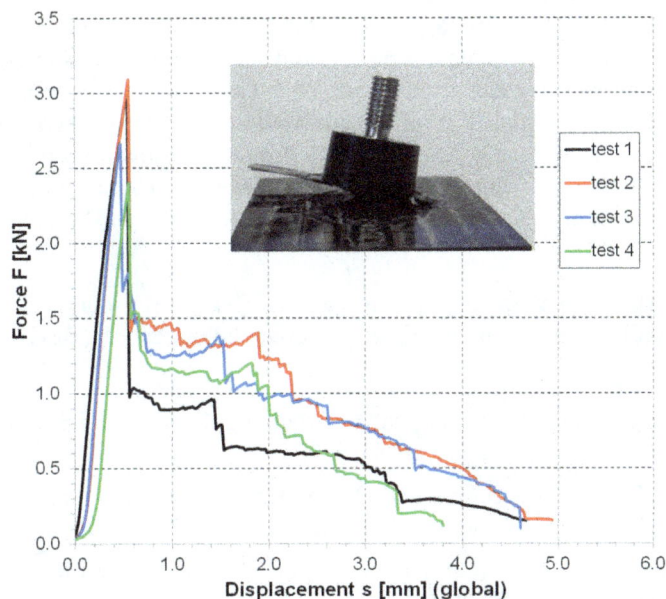

Figure 11. Experimental results of pull out tests with reinforced dome

Furthermore, torsion tests were done. These tests showed that in this load case the strength of the aluminium splay stud was utilized completely. The failure appears only on the thread of the splay stud. The rest of the joints remains undamaged. Because of this the tests where only performed with the unreinforced stud. The maximum reached torque moments average 4.9 Nm with a standard deviation of 0.4 Nm.

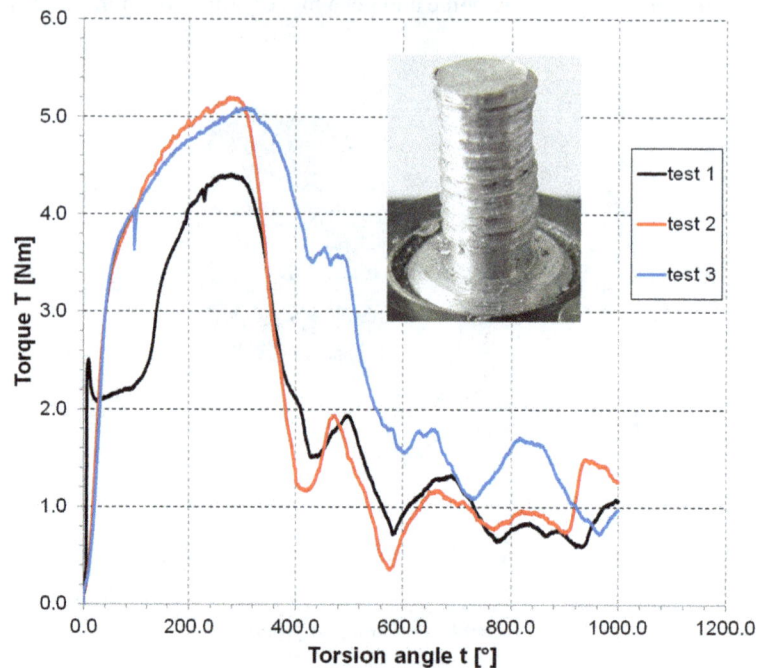

Figure 12. Experimental results of torsion tests without reinforced dome

5. Conclusion

A new kind of functional element for the use in CFRP and hybrid components has been described. This element can be introduced in a non-consolidated textile during the RTM process efficiently. Thus further process steps and material damages like delaminations can be avoided. With special bushings located in the RTM-tools the studs splays in the scrim under axial compression of the press stroke and forms an interlock in the CFRP. Additionally a resin dome that can be reinforced with extra winded fibres causes adhesively bonding of the stud and the component. The thread of the functional element is protected from the low-viscous resin by sealings located in the bushing. Manufactured specimens showed a good quality of the interlock and impregnation so that loading tests (pull out and torsion) have been carried out. The splay studs reached high load capabilities with partly failure of the aluminum thread which mirror the high material utilization. Further investigations will look at a better anchoring of the stud to the fibre scrim through an improved splay process or stud materials with higher strength. Further on the geometry of the resin dome will be optimised to minimize the required construction height.

Acknowledgments

This paper is based on investigations of the priority programme 1712 „Intrinsic hybrid composites for lightweight load-bearings", which is kindly supported by the German Research Foundation (DFG).

References

Grandt, J. (1998). Blindnietgewindesysteme: Typen, Verarbeitung, Einsatzbereiche, Die Bibliothek der Technik Bd. 159; Verlag Moderne Industrie, Landsberg/Lech.

Hahn, O., & Westhoff, D. (2015). Entwicklung und Qualifizierung eines Funktionselementes für das einseitige, vorlochfreie Fügen im Karosserierohbau, Dissertation Universität Paderborn, Shaker Verlag Aachen.

Heimlich, F. (2008). Beitrag zum Einsatz von Aluminium-Funktionselementen in Leichtbauwerkstoffen; D466 Dissertation Universität Paderborn; Shaker Verlag Aachen.

Koch, S. F., Barfuss, D., Bobbert, M., Gross, L., Gruetzner, R., Riemer, M., ... Wang, Z. (2016). Intrinsic hybrid composites for lightweight structures: New Process Chain Approaches. Advanced Materials Research, 1140, 239-246. Trans Tech Publications. http://dx.doi.org/10.4028/www.scientific.net/AMR.1140.239

Meschut, G., & Süllentrop, S. (2014). The use of radiation-cured adhesives on adhesive stud assembly process, *Weld World, 58*, 755. http://dx.doi.org/10.1007/s40194-014-0176-1

Meschut, G.; Gude, M., Geske, V., & Augenthaler, F. (2015). Bewertung der Schädigung beim Stanznieten von Faser-Verbund-Kunststoffen, AiF-Forschungsvorhaben 17667 BG/1, EFB-Abschlussbericht.

Momentive Specialty Chemicals Inc. (2012). EPIKOTE Resin, 05475 EPIKURE Curing Agent 05443, Technical DATA Sheet, Columbus, OH, USA.

Müller, R. (1992). Schraubenverbindungen, Selbststanzende Hilfsfügeteile, Dynamisch hoch belastbar; Industrieanzeiger Heft 22.

Schürmann, H. (2007). Konstruieren mit Faser-Kunststoff-Verbunden, page 554, Springer, Berlin.

SGL. (2016). HIGH-PERFORMANCE TEXTILES. Textile Produkte aus Carbon-, Glas- oder Aramidfasern. SGL GROUP.

Wang, Z., Riemer, M., Koch, S. F., Barfuss, D., Grützner, R., Augenthaler, F., & Schwennen, J. (2016). Intrinsic Hybrid Composites for Lightweight Structures: Tooling Technologies. *Advanced Materials Research, 1140*, 247-254. Trans Tech Publications. http://dx.doi.org/10.4028/www.scientific.net/AMR.1140.247

18

Modeling of the Weld Strength in Spot Weld Using Regression Analysis of the Stress Parameters based on the Simulation Study

Sachin Arun Patil[1], Farzad Baratzadeh[2], & Hamid Lankarani[1]

[1] Wichita State Univ, KS, United States

[2]Advanced Joining and Processing Laboratory, National Institute for Aviation Research, Wichita, KS, USA

Correspondence: Sachin Arun Patil, Wichita State Univ, KS, United States. E-mail: sapatil1@wichita.edu

Abstract

To enhance the performance of spot weld joints, various improvement methods are used to strengthen the properties of welded joints. Spot welding process is very well suited for welding of various steels grades. Regression analysis is the statistical modeling technique, and it is suitable for predicting strength of welded joints. It is valuable for quantifying the impact of various loading types upon a spot weld rupture.

In the present study quantification of impact strength of spot welded EHSS steel, Mild Steel (DC05) and AHSS (DP780) were carried out, by developing the regression models. The analysis includes Material, Thickness, Test type, Test Speed as process parameters. The complete analysis will be helpful in deciding the best combinations for desired performance characteristics. Taguchi technique revealed that the impact speed is the most significant factor in weld strength followed by thickness and material grade.

Keywords: Spot Welding, EHSS Steel, mild steel, DP780, DOE Method, Regression Analysis and ANOVA

1. Introduction

The joining method has high share of costs in product manufacture and has Impact that a link failure for the functionality of the product in various industries. The Automobile industry is a driving force in the development of existing and new joint techniques. A literature survey of the behavior and modeling of weld connections has shown a limited number of relevant articles. Some information is available from Donders, Brughmans, Hermans, and Tzannetakis (2005) and Schweizerhof, Schmid, and Klamser (2000) gives recent modeling information on spot weld process as well as the advantages of using spot weld compared with the most common joining techniques.

Several investigations are reported with respect to the behavior and modeling of spot weld in steel. Lin et al. (1998) have studied the ultimate tensile strength of resistance spot welds in mild steel subjected to combined loading tension and shear loads. Sebastian et al. (Sebastian & Silke, 2012; Silke & Frederick, n.d.; Sommer, 2010) have presented a study of a spot weld for numerical analysis of automotive applications under crash loading conditions.

Although there is currently no universally accepted measure of weld performance, there are several metrics available that provide a single numerical characterization of weld performance. The weld efficiency, the weld failure time, weld separations are such metrics. A welded structure usually represents a weakness in various strength and surface hardness at weld. These low properties are limiting the use welded joints in the design of structural components. Hence to upgrade the performance of welded joints the best combination of parameters need to be used to enhance mechanical properties. Generally, the most important characteristic of quality at the welds is their strength. In its own the experiment was taken into account the strength of welds in normal, bending and shear.

2. Objective

In the present study the mathematical models are developed for impact strength in terms of weld parameters. The investigation is helpful to the safety evaluation of structure, performance. In this investigation, parameters on spot weld strength were performed by means of Taguchi method (Montgomery, 1984). Four factors, each at three levels, were selected and the experiment was designed using L_{27} orthogonal array (Chen et al., 1996). Taguchi methodology was also successfully applied for parameter optimization.

3. Method

The general approach for this study was to select weld coupon tests, compute the weld strength metrics, estimate the weld failure based on the response of the weld internal forces, and then compare the ability of the metrics to predict failure risk. Experimental validation is the important step in the modelling process to investigate the accuracy and robustness of the established model. Coupon tests were selected from a database maintained by the NAIR WSU based on the previous studies for Mild Steel (DC05), and EHSS steel (ASTM international, 2003). Each material has 3 deformation modes. A total of 6 tests were randomly selected after applying the above constraints. An effort was made to choose tests with varied weld loading types. AHSS (DP780) spot weld evaluated using simulation study only. For each test, the weld performance metrics were computed using forces measured in the test. All data traces used were checked against redundant sensor traces to ensure data accuracy; corrections for sensor bias were made as necessary. The failure was computed using the following relations (Hallquist & Manual, 1998):

$$\left(\frac{\sigma_n}{S_N}\right)^2 + \left(\frac{\sigma_s}{S_S}\right)^2 + \left(\frac{\sigma_b}{S_B}\right)^2 \leq 1 \tag{1}$$

This failure function used to define weld behavior in Ls-Dyna as a power law combination of three stress components.

S_N, S_S and S_B = Denominator, namely law parameters, is fed into the card material in the input file to LS -DYNA (usually taken from test)

σ_n, σ_s and σ_b = Numerators in the equation are resultants as calculated in the local coordinate system in cross section planes during simulation time. Denominator shows again three stress components which are critical to rupture or failure welds. σ_n, σ_s and σ_b are the axial load and the shear load acting on the spot weld, respectively. This failure model has become widely adopted in commercial software such as ABAQUS/Explicit and LS-DYNA3D (Hallquist & Manual, 1998; SIMULIA, 2011). Usually Failure exponents of 2 give results reasonable accuracy. Failure exponents of 3 used for shear dominant cases.

In Ls-dyna, opt=8 select failure model based on stress. Resultant of stress considered are bending torsion , shear default values are denominator are 1e+20 with exponent 2.As numerator value are stress created in weld due to loading are lower than denominator. This maintain overall value less than 1 to ensure good weld conditions. However, weld failure should include the effects of weld preload which is currently not included in the analysis.Yield criteria and failure surface are treated as an elliptical boundary contour. This form elliptical shape for weld yield surface in the system of fs, fn and fb. Numerator in this equation are bounded by an elliptical boundary contour depending on the exponent

MAT_100_DA define effective plastic strain or criteria incorporating a combination axial, shear, bending stresses (Hallquist & Manual, 1998). With option OPT=8 it shows bilinear elasto-plastic behavior enhanced by state of art failure concept. Damage type=4 consider fading energy based damage. Damage will be initiated once f > 0 will occur when damage growth w which is function of plastic strain is greater than 1. Damage type 4 considers internal work done by spot weld after its failure and is supposed to be more realistic than other damage type. An inverse parameter identification used to plot failure curve. Its mathematically formulated Procedure for determining the appropriate values for the failure model (Sachin, 2014). The failure model is based on equation of an ellipse whose exponents a and b are the intersection of failure and failure by not setting a trend. As the spot weld strength limit is exceeded, numerator this equation becomes higher than denominator critical values parameter for stresses. Hence left side of equation is larger than unity and the joint is damaged and after certain time spot weld ruptured.

4. Simulation Set up and Analysis

4.1 Materials of Interest in this Study

In this paper, the mechanical properties and spot weld-ability of newly developed steels are discussed. High-strength steels have been developed for the improvement of weight reduction, crash-worthiness, and anti-corrosion properties of an auto body. All of the specimens are made of high-strength steel viz, *EHSS steel, Mild Steel (DC05) and AHSS (DP780)* of the varying thickness.The quasi-static material properties of three steel , including yield strength, ultimate tensile strength, total elongation (TE), and strain hardening exponent/index (n) are indicated in Table 1. Table 1 is a list of the steel grades that were evaluated

Table 1. Base Material properties of steel Investigated

Grade	YS (MPa)	UTS MPa)	TE (%)	n Strain hardening coefficient
EHSS steel	368	445	34	0.19
DC05 Cold rolled steel	180	350	38	0.21
DP780	430	810	22	0.16

Yield stress is undefined as 0.2 % proof stress and n value calculated as a point of 2% and 5 % total strain. The base metal used in this study is a EHSS steel initially due to its economical aspect when compared to high grade steel such as AHSS OR UHSS.. The microstructural feature of the base exhibits a yield strength of 368 Mpa. Later this steel compare with AHSS steel DP 780 and lower grade steel DC05. DP-780 has 430 MPa yield Strength whereas for grades DC05 assumed to be 180 MPa .Applications of all three steel finds in difficult interior and exterior automotive parts. A high level of consistency in the mechanical properties of this steel grade guarantees by database maintained by the NAIR WSU.The validity of the test results is discussed in detail (Sachin, 2014), here we mentioned in brief on the basis of comparison with the standard specimens, the finite element analysis and regression analysis.

The LS-DYNA material model MAT_PIECEWISE_LINEAR_PLASTICITY model (Type 24), which utilizes the VP=0.Rate effects are accounted for with the use of scaling by the Cowper and Symonds model which scales the yield stress with a strain rate dependent factor. The accurate prediction of failure under different loading conditions is desirable. This corresponds to a plasticity with rate dependency. Inclusion of strain rate in crash analysis reduces the dynamic crush. For dynamic loadings typical of an impact, strain rate effects tend to increase the loading capacity of the material and decrease the elongation. The Cowper-Symonds formula is the standard method in LS-DYNA for taking account of this strain rate effect. It scales yield stress by the factor:

$$\frac{\sigma_d}{\sigma_s} = 1 + \left(\frac{\dot{\varepsilon}}{C}\right)^{\frac{1}{P}}$$

(2)

here: σ d – dynamic yield stress, σ s – static yield stress, . ε – strain rate, C,P – constants of Cowper-Symonds relation

C and P are material factors determined empirically, and off-set the post-0.2% proof stress strain curve in the stress axis by the same amount.

To reproduce the spot weld behavior, a 2-D spot weld simulation is carried out. A solid element spot weld is used with shell elements for the sheet metal part, as illustrated in Figure 1. Mesh elements are selected as an optimal size for accurate results and a reduction in computational cost. The numerical analysis of this shell component with new spot weld provides a direct check of accuracy of model. Failure was defined for shell elements using the MAT 24 material model, which is equivalent to MAT 105 (PAMCRASH) (ESI Group, 2014). This will allow the material around the weld nugget to fail according to the strain and strain rate that it experienced.

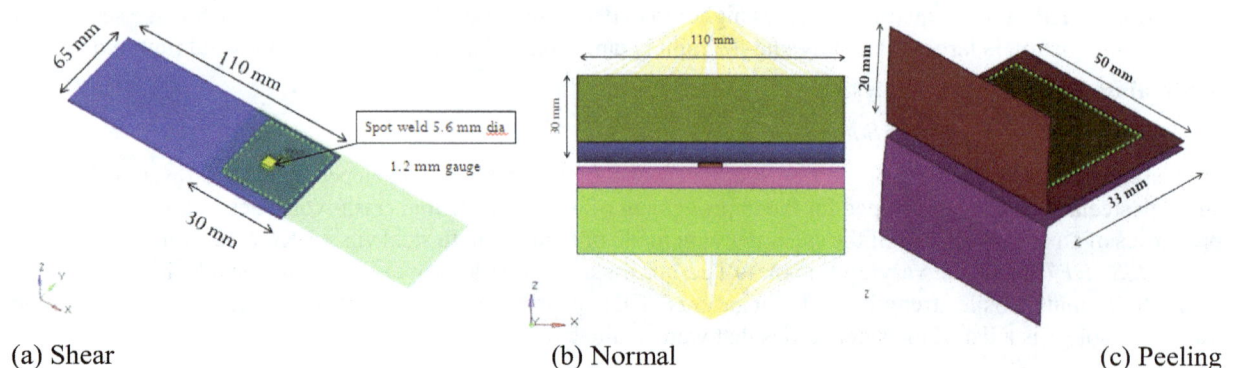

(a) Shear   (b) Normal   (c) Peeling

Figure 1. Geometry of model specimen for 2-D weld simulation (sheet thickness 1.2 mm)

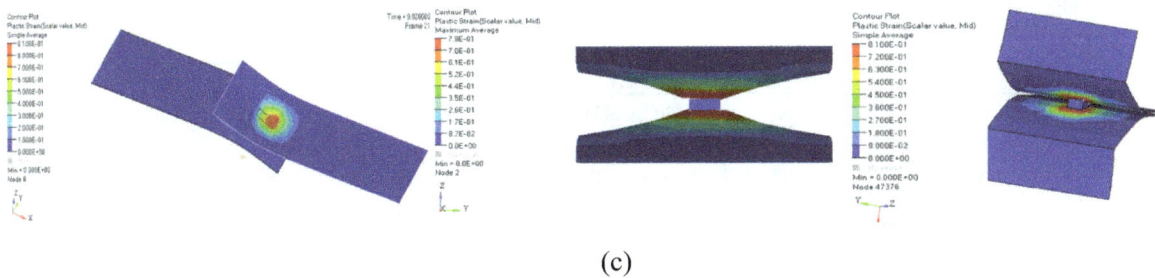

(c)

Figure 2. D3plot results of LS-DYNA analyses, and comparison of equivalent plastic strain [ε]

The stress components for normal, bending and shear loading were then collected at the observed peak loads. In substance for one point there is a plane where the shear stress is zero. The 3 principal stresses define the stress in this point respect the plane and his 3 direction. All tests on the single spot-weld specimens have indicated that failure occurs in the plate around the weld and the behavior of the weld was apparent elastic (ESI Group, 2014). Failure strain option in material card can be used to model fracture of the material. The behavior of the connection changed as a function of the loading angle. As the load angle increases, from pure shear to pure pull-out loads, the ductility of the connection increases and load carrying capacity is reduced. An interaction curve was found to adequately represent the behavior of specimens under combined pull-out and shear loading. The weld model includes failure criteria based on a critical plastic-failure strain, as well as on a force envelope. The weld model calibrated from the tests results for better correlation. Figures 3 report the fracture initiation force for each weld region which experienced failure prior to the achievement of the first peak force.

5. Quantification of various Strength using Regression Analysis

The Taguchi method is a unique statistical experimental design approach that greatly improves the engineering productivity. Taguchi suggests the production process to be applied at optimum levels with minimum variation in its functional characteristics. In general, the signal-to-noise (S/N) ratio (η, dB) represents quality characteristics for the observed data in the Taguchi method (He, Li, & Chen, 2012). S/N ratio is an index to evaluate the quality of manufacturing process. Here, the 'signal' represents the desirable value and the 'noise' represents the undesirable value, where signal to noise ratio expresses the scatter around the desired value.

5.1 Design of Experiments (DoE) by Taguchi

Number of possible attempts of full factorial experiment (what a try a unique combination of factors in setting certain levels) can be calculated by the formula. Regression analysis is well established approach to develop complex non-linear model to predict the performance characteristics. The researchers (Tang et al., 2001; He, Li, & Chen, 2012) developed a mathematical model and the adequacy of the model was verified using ANOVA. To establish the prediction model, a software package MiniTab has used to determine the regression coefficients (Minitab software, n.d.). The present work is a three factor three level problem, the available Taguchi orthogonal array is L9 & L27.In order to determine which one is suitable, and degrees of freedom (D.O.F) in both cases have to be determined. D.O.F tells about the minimum number of test runs required for a particular problem. The following formula is used to determined D.O.F

$$D.O.F = m (L-1) + n (L-1) (L-1) + 1 \qquad (3)$$

Where m= number of variables, L= number of levels, n=number of interaction terms. From equation (3) it can be seen that number of interaction terms n = 6 (MT, MC, MV, TC, TV, CV). So m=4, L=3, n=6. After putting these values in equation (3) D.O.F becomes=2. So the number of test runs required for this problem is 33. The DoE was based on full factorial design considering five factors each at three levels. While full factorial evaluates all combinations of factors at all specified levels, it requires many runs. Therefore orthogonal array used partial full factorials set to reduce number of experiments required while still providing effective information. The appropriate orthogonal array is L27 which provides 27 test runs. Standard L 27 table selected from taguchi set of orthogonal array for DOE. The selection of process parameters is most important step in Design of Experiments (DoE). These spot welding parameters along with their levels are shown in Table 2.

Table 2. Process Parameter and their Levels

Process Parameter	Parameter Designation	Levels		
		L1	L2	L3
Material	M	Mild Steel DC05	EHSS 340	DP780
Thickness	T	0.8	1	2
Test type	C	normal (pull test)	shear (lap test)	bending (peel test)
Test Speed	V	0.001	0.1	100

The DOE variables and the values (sheet gage, test type, weld material, and strain rate) for statistical analysis are listed in Tables 2. A summary of all types of stress for the weld is also reported in DOE Table 3.The DOE statistical study was done using MINITAB statistical analysis software and major effects formulations have been developed and will be discussed further.

The purpose of the investigations described in this paper is a simulation approach for weld stresses generated in different types of coupon test. The simulation results show a weld deformation that is quite similar to the coupon test. Also the calculation of the dynamic strains correlates well with the coupon test. Therefore simulation weld model proves to be test specific robust and can be used for future analyses.

Table 3. Simulated stress output Results for Different spot weld Parameters/ DOE table and Simulated Results for Different Shot welding Parameters

Exp. No.	M	T	C	V	Normal (pull test)	Shear (lap test)	Bending (peel)
1	1	1	1	1	0.295	0.022	0.597
2	1	1	1	1	0.110	0.874	0.002
3	1	1	1	1	0.194	0.685	0.132
4	1	2	2	2	0.454	0.224	0.028
5	1	2	2	2	0.570	0.012	0.000
6	1	2	2	2	0.152	0.841	0.064
7	1	3	3	3	0.295	0.345	0.437
8	1	3	3	3	0.105	1.048	0.000
9	1	3	3	3	0.194	0.898	0.053
10	2	1	2	3	0.454	0.578	0.009
11	2	1	2	3	0.570	0.278	0.000
12	2	1	2	3	0.152	1.022	0.024
13	2	2	3	1	0.295	0.043	0.887
14	2	2	3	1	0.070	0.909	0.005
15	2	2	3	1	0.155	0.729	0.272
16	2	3	1	2	0.409	0.303	0.106
17	2	3	1	2	0.539	0.023	0.001
18	2	3	1	2	0.096	0.878	0.105
19	3	1	3	2	0.571	0.880	0.656
20	3	1	3	2	0.697	1.061	0.570
21	3	1	3	2	0.545	0.876	0.976
22	3	2	1	3	0.669	1.049	1.133
23	3	2	1	3	0.516	0.855	1.092
24	3	2	1	3	0.633	1.035	1.457
25	3	3	2	1	0.483	0.839	1.014
26	3	3	2	1	0.562	1.083	1.609
27	3	3	2	1	0.562	1.083	1.609

5.2 Regression Analysis for Quantification of Impact Strength

The ANOVA is a common statistical technique to determine the percent contribution of each factor for the experimental results (Montgomery, 1984). It is used to calculate the parameters known as sum of squares (SS), corrected sum of squares (SS'), degree of freedom (D), variance (V), and percentage of the contribution of each factor (P). The less-significant coefficients were determined from further analysis using the t-test. Also, to check the adequacy of each model, the analyses of variance were carried out by using the F-ratio test.

The ANOVA results are presented in Table 4 to 6. The Table 4 shows the following correlation between the impact strength and the process parameters as:

$$\sigma NS = 0.85 + 0.00958M + 0.00199T + 0.00133C + 0.000972V - 0.895M^2 - 0.482T^2 - 0.395C^2 - 0.293V^2 + 0.188MT$$
$$+ 0.022MC + 0.032MV - 0.065TC - 0.087TV - 0.085CV \tag{4}$$

Similarly, the regression analysis results for other strength component are tabulated in Table 5. It shows the following correlation between the bending strength and the process parameters as:

$$\sigma BB = 0.42 + 0.00006 (M) + 0.0007 (T) + 0.358 (C) + 0.0004 (V) - 0.507M^2 - 0.582T^2 - 0.288C^2 - 0.593V^2 +$$
$$0.688MT + 0.011MC + 0.016MV - 0.027TC - 0.08TV - 0.04CV \tag{5}$$

The shear strength component of spot weld is as follows:

$$\sigma SS = 2.15 + 0.00286 (M) + 0.0369 (T) + 0.358 (C) + 0.0781 (V) - 0.321M^2 - 0.342T^2 - 0.698C^2 - 0.223V^2 +$$
$$0.212MT + 0.015MC + 0.01MV - 0.023TC - 0.03TV - 0.035CV \tag{6}$$

The resulting regression analysis equations (4), (5) and (6) determine the approximate values of impact strength of welded steel before failure.

These all equations provide a useful guideline for setting proper values of process parameters so as to obtain desired performance characteristics of three material components viz, Mild Steel (DC05), EHSS steel, and AHSS (DP780). ANOVA, R-sq value and Adjusted R Square value are used for the validation of the models obtained by regression analysis. The ANOVA is the statistical treatment applied to determine the significance of the regression model. The R-sq is used in the context of statistical models whose main purpose is the prediction of future outcomes on the basis of other related information. It gives the information about goodness of fit for a model. In regression, the R-sq is a statistical measure of how well the regression line approximates the real data points. An R-sq of 1.0 indicates that the regression line perfectly fits in the data. Unlike R-sq, an Adjusted R Square allows for the degrees of freedom associated with the sums of the squares. Therefore, even though the residual sum of squares decreases or remains the same as new independent variables are added, the residual variance does not. For this reason, Adjusted R Square is generally considered to be a more accurate goodness-of-fit measure than R-sq. Adjusted R Square ,is a modification of R-sq that adjusts for the number of explanatory terms in the model.

Table 4. ANOVA for Normal Strength of spot weld

Coefficients and Intercepts for Tensile Normal Strength				
Predictor	**Coef**	**SE Coef**	**T**	**P**
Constant	0.79	0.003879	256.89	0
Material	0.00858	0.0008801	10.78	0
Thickness	0.00199	0.000304	9.38	0
Test	0.00133	0.0001598	9.47	0.006
Test Speed	0.000972	0.000114	3.88	0
S = 0.016 R-Sq = 91.0% R-Sq(adj) = 90.0%				

SUMMARY OUTPUT	
Regression Statistics	
Multiple R	0.86
R Square	0.91
Adjusted R Square	0.90
Standard Error	0.016

ANOVA for Normal Strength				
Source	**DF**	**SS**	**MS**	**F**	**P**
Regression	4	0.15291	0.03421	83.69	0
Residual Error	39	0.0149	0.00029		
Total	43	0.1812			

Table 5. ANOVA for Shear Strength of spot weld

Coefficients and Intercepts for Lap shear Strength				
Predictor	**Coef**	**SE Coef**	**T**	**P**
Constant	2.15	0.08291	0.12	0.909
Material	0.00286	0.01094	9.28	0
Thickness	-0.0369	0.02263	-3.82	0.001
Test	0.3238	0.05069	10.73	0
Test Speed	0.0781	0.01802	4.06	0
S = 0.0182 R-Sq = 94.0% R-Sq(adj) = 91.0%				

SUMMARY OUTPUT	
Regression Statistics	
Multiple R	0.88
R Square	0.94
Adjusted R Square	0.91
Standard Error	0.018

ANOVA for Shear Strength					
Source	**DF**	**SS**	**MS**	**F**	**P**
Regression	4	65.34	3.95	51.17	0
Residual	42	4.123	0.0032		
Total	46	66.11			

Table 6. ANOVA for Bending Strength of spot weld

Coefficients and Intercepts for Bending Strength				
Predictor	**Coef**	**SE Coef**	**T**	**P**
Constant	0.51	0.003879	256.89	0
Material	0.00049	0.0008801	10.78	0
Thickness	0.0042	0.000304	9.38	0
Test	0.0031	0.0001598	9.47	0.006
Test Speed	0.000013	0.000114	3.88	0
S = 0.0216 R-Sq = 93.0% R-Sq(adj) = 91.3%				

SUMMARY OUTPUT	
Regression Statistics	
Multiple R	0.82
R Square	0.93
Adjusted R Square	0.91
Standard Error	0.021

ANOVA for Bending Strength					
Source	**DF**	**SS**	**MS**	**F**	**P**
Regression	4	0.072	0.0121	56.64	0
Residual	57	0.0043	0.00026		
Total	61	0.087			

5.3 Analyzing the adequacy of the developed model

The results of ANOVA, R-square and Adjusted R Square are obtained by regression analysis using MINITAB 14 and are shown in the previous sections (Minitab software, n.d.). The results show the significance of the analysis. It is observed from Tables 4 that p-values for the response impact strength is less than 0.05, which shows that it is at 95% confidence level. R-square is the statistical measure of the exactness at which the total variation of dependent variables is explained by regression analysis. The obtained values of R-sq and R-sq (adj) are more than 0.90 and quite near to 1.0 for the performance characteristics, it indicate a good fit. This confirms that the model adequately describes the observed data. Statistical parameters for each of the models are summarized in Table 4 to 6. As another measure of model fit, the adjusted R^2 value is reported in the rightmost column of Table 4. The p-value for each term tests the null hypothesis that the coefficient is equal to zero (no effect). A low p-value (< 0.05)

indicates that you can reject the null hypothesis. In other words, a predictor that has a low p-value is likely to be a meaningful addition to your model because changes in the predictor's value are related to changes in the response variable. Conversely, a larger (insignificant) p-value suggests that changes in the predictor are not associated with changes in the response. As per the technique, if the calculated F-ratio values of the model exceeds the standard tabulated value of the F-ratio for a desired level of confidence (say 95%), then the model may be considered adequate within the confidence limit (Stat-Ease Inc, 2000).

6. Confirmation study

A simplified model of the weld joint that enables the user to describe the global response for a component-level study is proposed. Component level structures verified for this regression results using Finite element studies.

6.1 Model Setup

Square beam parts are very common in automotive systems for absorbing energy during impact events like front and rear rails, cross members in the B-pillar structure, bumpers and B and C pillar reinforcements. Failure analysis of a T-joint is useful to improve vehicle safety since it mimics the B-pillar and sill cross-member welding region. The T-joint specimens are used for the stress in the transverse direction and also under load speeds simulating 1 m/s corresponds to strain rate 100 m/s to match with real accident scenario. For this purpose, a slide mass is identified in the amount of 192 kg to realize the failure of spot welds (Oeter, Kenan Özdem, & Hahn, n.d.). To examine spot weld failure, six side spot welds are connected, as shown in Figure 3. Validation of the simulation model is done as described in the following section. The simulations are carried out only with spot weld parameters. Furthermore, contact problems are carried out by scaling the contact thickness, as in simulations of the lap shear coupon test.

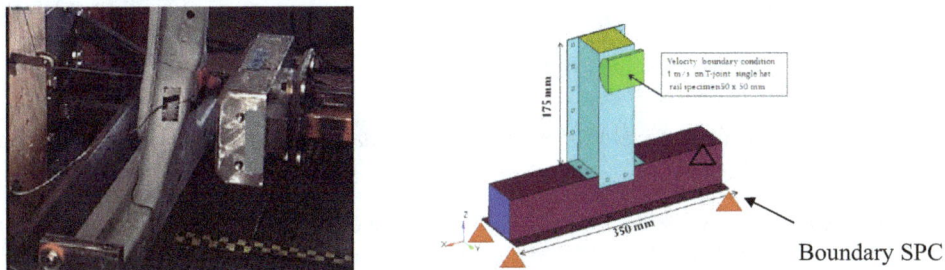

(a) Experiment loading for B-pillar weld test   (b) Transverse loading of joint

Figure 3. Simulation setup illustration of T-joint

Instead of selection of the spot weld tests data, regression model output was chosen for calibrating the material input for the spot weld model. Failure strains were obtained by benchmarking directly against the regression model.

FEA models in Ls-Dyna format were generated to simulate the crash responses of the straight rail tests. The shell elements were used to model the STEEL sheet metal. The SPOT weld was modeled by the hex type according the developed regression method. The results obtained for optimum process parameters by earlier three regression equations are used to define spot weld failure parameters on the card Define_Connections_Properties. Force resultants for Mat_Spotweld_Da are written to the spot weld force file, SWFORC, and the ELOUT file for element stresses and resultants for designated elements. In this database the resultant moments are not available, but they are in the binary time history database.

MAT_100_DA define effective plastic strain or criteria incorporating a combination axial, shear, bending stresses. With option OPT=8 it shows bilinear elasto-plastic behavior enhanced by state of art failure concept. Damage type=4 consider fading energy based damage. Damage will be initiated once f > 0 (equation (1)) will occur when damage growth w which is function of plastic strain is greater than 1. Damage type 4 considers internal work done by spot weld after its failure and is supposed to be more realistic than other damage type. Also structural integrity of hat-type welded structures are generally controlled by the strength of the spot welds which commonly fail under combined loading. This universal formulation applies for all of the deformation modes, i.e. U-tension, lap shear and coach peel. For each deformation mode, there is a unique set of formulation for the coefficient factors as seen in equations 5, 6,7 specific conditions of the joint design (materials, gages, loading and strain rate).

6.2 T-Shock Simulation Results

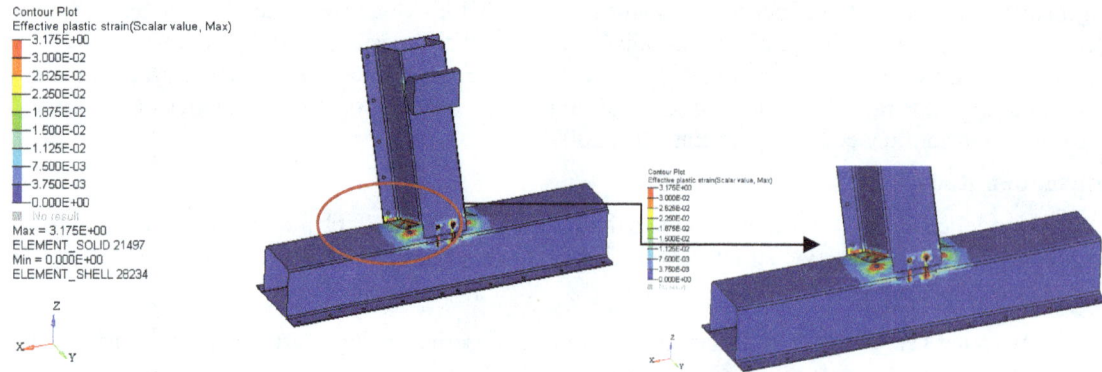

fracture happened after the peak force has reached.

Figure 4. Deformation contour of referenced model

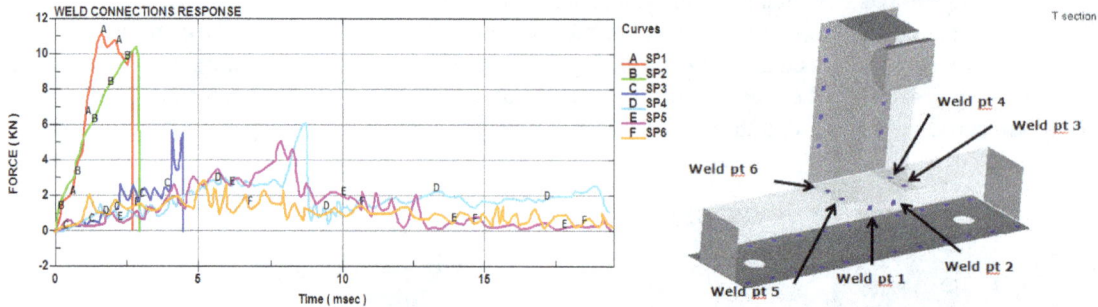

Figure 5. (a) Local spot weld forces from the simulation, (b) location of spot weld on specimen

From Figure 5, it is clear that the behavior of the force-time curves from the simulation approaches smaller peaks after the first force peaks. The SP1 and SP2 specimens fail on the same high level of shear force and significantly low in normal force. This is due to the type of loading, because the shear component influences results of the welding spots. The force levels vary little from one another. This suggests a good set of close failure criteria as per references (AVIF, 2005).

6.3 Extraction of weld strength Characteristics in This Study

Dimensionless Number was used to see failure of spot weld (Refer Table 7).This number is nothing but ratio of applied load to spot weld strength. Baseline model showing higher values for force component of shear, normal and peel as given below. Regression weld data shows lower value for theses dimensionless Number. To fully capture the deformation behavior, adjacent figure show Spot-weld Rupture/ failure is very important.

Finally graph of weld failure criteria vs time plotted to compare performance of weld seen in Figure 6. Baseline weld model to correlate with test failed at 88 ms while regression weld model failed at 81 ms. Test specimen failed at 86 ms (Sachin, P.,2014).Hence all three weld fails at approximately same time which confirm that this force based criteria can be used with accuracy for automotive steel spot weld analysis. New regression spot weld Model shows weld deformation close to test.

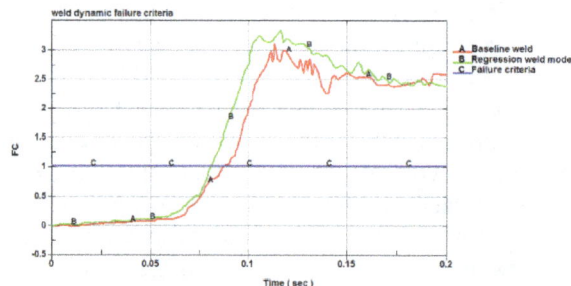

Figure 6. Comparison of weld failure time in Baseline model and Regressed weld model

Table 7. Force component history for T-joint specimen

Normalized components of failed spot weld (baseline test spec)			
sw element id	shear	normal	peel
1	0.466	1.66	0.477
2	0.477	1.044	0.530
3	0.862	0.669	0.862
4	0.466	-1.72	0.542
5	0.342	0.378	0.319
6	0.387	0.322	0.309

Normalized components of failed spot weld (Improved Design)			
sw element id	shear	normal	peel
1	0.242	0.205	0.265
2	0.419	0.046	0.500
3	0.544	0.212	0.546
4	0.455	-0.94	0.511
5	0.286	0.296	0.211
6	0.267	0.254	0.206

7. Conclusion

In the present work, a methodology of arriving at suitable combination of spot weld parameters to achieve better weld response in impact cases study is highlighted. Using regression analysis, the weld contact forces were found to offer an advantage of using previous test database for the prediction of weld failure. Although limited to an analysis of 9 coupon tests for 3 material Mild Steel (DC05), EHSS steel, and AHSS (DP780), we believe this analysis provides an important first look into how the weld contact forces compare in terms of ability to predict failure. This methodology can be extended to optimize the structures at system or subsystem level in different impact conditions.

Some tools of Taguchi method such as, Orthogonal array (OA), experimental design, analysis of variance (ANOVA) are implemented. These models are based on simulations carried out on shell spot welded specimen as well as previous experiment data. All analysis results, including, best parameter level combinations, 95% confidence intervals, R-sq and R-sq (adj) of the regression models are estimated using MINITAB 14.These regression models are correlating impact strength with process parameters. They have obtained their R-sq and R-sq (adj) values more than 0.90.

The CAE analysis based on the new formulation consistently predicts accurate crush responses of T -section type rails section specimen. The weld strength results showing here is a statistical representation of these tests. The results obtained for optimum process parameters by these equations are near to the experimental values as observed in this component level simulation study. Hence regression equations provide a useful guide for setting proper values of process parameters so as to obtain desired impact strength of steel.

A sensitivity study may be conducted on the CAE model to forecast the range of predicted results and a consideration of where the tested results lie relative to this range made. In many cases this will entail the use of additional parameters in the simulation. Sensitivity of the predicted results to idealized parameters such as friction, joint models etc., should be studied.

References

ASTM international. (2003). *Annual book of ASTM standards, Metals test methods and analytical procedures.* ASTM international.

AVIF. (2005). Characterization and Modeling of the Viability of Spot Welded Sheet Steel Connections under Crash Loading using Advanced Damage Models. *AVIF Proceedings Report.* AVIF, Frankfurt, Germany, 09/03-08/05.

Chen, Y., Tam, S. C., Chen, W. L., & Zheng, H. Y. (1996). Application of Taguchi method in the optimization of laser micro-engraving of photomasks, Intern. *J. Materials and Product Tech, 11*(3-4).

Donders, S., Brughmans, M., Hermans, L., & Tzannetakis, N. (2005). The effect of spot weld failure on dynamic vehicle performance. *Sound and Vibration, 39*(4), 16-25.

ESI Group. (2014). *PAM-CRASH, ESI Group, Paris, France.* Retrieved from http://www.esi-group.com

Hallquist, J. O., & Manual, L. D. T. (1998). Livermore Software Technology Corporation, LS-DYNA Keyword Manual, version 971, 1998. Retrieved from http://www.lstc.com

He, K., Li, Q., & Chen, J. (2012). Regression Analysis of the Process Parameters Effect on Weld shape in Twin-Arc SAW. *Journal of Convergence Information Technology, 7*(13).

Lin, S. H., Pan, J., Wu, S. R., Tyan, T., & Wung, P. (2002). Failure loads of spot welds under combined opening and shear static loading conditions. *International Journal of Solids and Structures, 39*(1), 19-39.

Malcolm, S., & Nutwell, E. (2007, May). Spotweld failure prediction using solid element assemblies. In *6th European LS-Dyna users' conference. Gothenburg, Sweden.*

Minitab software. (n.d.). Retrieved from https://www.minitab.com/support/documentation/ Minitab's Help Section

Montgomery, D. C. (1984). *Design and Analysis of Experiments* (2nd Ed.). New York: John Wiley & Sons.

Oeter, M., Kenan Özdem, K., & Hahn, O. (n.d.). Experimental determination of Structural behavior of spot welded Components made of sheet steel connections. *AVIF A172, 17*(05), 63-60079.

Pakalnins, E., Lu, M. W., Morrissett, T. W., Wehner, T. J., & Lee, Y. L. (1998). Ultimate strength of resistance spot welds subjected to combined tension and shear. *Journal of Testing and Evaluation, 26*(3), 213-219.

Sachin, P. (2014). *Modeling and Characterization of Spot Weld for Crash Analysis.* Phd Dissertation, Wichita State Univ.

Schweizerhof, K., Schmid, W., & Klamser, H. (2000). Improved simulation of spotwelds in comparison to experiments using LS-DYNA. In *18th CAD-FEM Users' Meeting–International Congress on FEM Technology, Friedrichshafen (Lake Constance), Germany.*

Sebastian, B., & Silke, S. (2012). Characterization and modeling of fracture behavior of spot welded joints. *11 th Ls-dyna Forum, Ulm 2012.*

Silke, S., & Frederick, K. (n.d.). Modelling of the deformation and fracture behaviour of laser welds. *7th Ls-dyna European Conference.*

SIMULIA. (2011). *ABAQUS 6.11: Analysis User's Manual.*

Sommer, S. (2010). Modeling of the fracture behavior of spot welds using advanced micro-mechanical damage models. In *IOP Conference Series: Materials Science and Engineering* (Vol. 10, No. 1, p. 012057). IOP Publishing.

Stat-Ease Inc. (2000). *Design-Expert software, v6, user's guide, Technical manual.* Stat-Ease Inc., Minneapolis, MN, 2000.

Tang, D., Branson, D., Thomas, B., Perez, C., Liu, Y., & Loosle, D. (2001). Evaluation of Aluminum Laser and MIG Connections in Crash Environment. *Report No. GCET503, 2001, Aluminum Body Structures Dept., Ford Motor Company.*

19

Electrochromic Properties of Sputtered Iridium Oxide Thin Films with Various Film Thicknesses

Yoshio Abe[1], Satoshi Ito[1], Kyung Ho Kim[1], Midori Kawamura[1] & Takayuki Kiba[1]

[1] Department of Materials Science and Engineering, Kitami Institute of Technology, Kitami, Japan

Correspondence: Yoshio Abe, Department of Materials Science and Engineering, Kitami Institute of Technology, Kitami, Japan. E-mail: abeys@mail.kitami-it.ac.jp

Abstract

Iridium oxide is an anodic electrochromic material, which takes on a blue-black color through electrochemical oxidation and turns to transparent via reduction. Hydrated amorphous Ir oxide thin films with various thicknesses from 20 to 400 nm were prepared by reactive sputtering in a H_2O atmosphere, and their transmittance spectra in both the bleached and colored states as well as their response times were examined in this study. The bleached and colored transmittances decreased with increasing film thickness according to Lambert's law, and the absorption coefficients in the bleached and colored states were estimated to be 3.2×10^3 and 1.1×10^5 cm^{-1}, respectively, at a wavelength of 600 nm. The results point to almost all the Ir atoms being electrochemically active and contributing to the color change. A maximum transmittance change of 81% was obtained for the 400 nm-thick film. Further, there was a trade-off between the response speed and the transmittance change. The response speed slowed down with increasing the film thickness, while the coloring and bleaching response time for the thick films was several tens of seconds.

Keywords: electrochromic, iridium oxide, optical transmittance

1. Introduction

Electrochromic (EC) materials change their optical transmittance reversibly via electrochemical oxidation and reduction (Granqvist, 1995). A typical EC device consists of an EC layer, an ion-conducting layer (electrolyte), and an ion-storage layer, which are all sandwiched between two transparent conducting electrodes. WO_3 is the most widely studied inorganic EC material; it turns dark blue upon reduction and becomes transparent upon oxidation. Two other well-known EC materials, nickel oxide and iridium (Ir) oxide, turn brown and blue-black upon oxidation and become transparent upon reduction; they are used as ion-storage layers to enhance the transmittance variation of EC devices (Granqvist, 1994; Granqvist, 2012). Such EC devices are used in displays, antiglare automobile rearview mirrors, and smart energy efficient windows (Granqvist, 2014; Motimer, Rosseinsky, & Monk, 2015).

Ir oxide has been reported to possess a superior chemical stability in both acidic and basic electrolyte solutions, a long cycle durability, and a fast optical response (Dautremont-Smith, 1982). The coloration mechanism of Ir oxide has been proposed to be governed by the following equations (Mortimer, 1997).

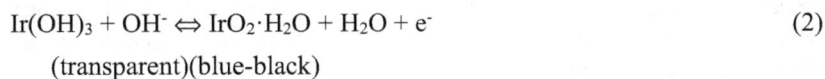

$$Ir(OH)_3 \Leftrightarrow IrO_2 \cdot H_2O + H^+ + e^- \qquad (1)$$
$$\text{(transparent)} \quad \text{(blue-black)}$$

$$Ir(OH)_3 + OH^- \Leftrightarrow IrO_2 \cdot H_2O + H_2O + e^- \qquad (2)$$
$$\text{(transparent)} \text{(blue-black)}$$

IrO_2 has a rutile-like structure and is a conducting oxide with a bulk resistivity of 35 $\mu\Omega \cdot$cm (Ryden, Lawson, & Sartani, 1970). Its electronic structure has been studied using X-ray photoelectron spectroscopy and theoretical band calculations (Mattheiss, 1976; Kahk et al., 2014). Ir atoms are surrounded by a nearly octahedral array of six oxygen atoms, with the metal d level split into triply degenerate t_{2g} and doubly degenerate e_g components due to the effect of the crystal field. Five $5d$ electrons in Ir^{4+} partly fill the t_{2g} bands while the e_g bands are empty. The optical properties of IrO_2 can be explained via free-carrier and intraband transitions within the t_{2g} bands below ~2.5 eV as well as interband transition from the filled O $2p$ to the Ir d bands above ~3 eV (Goel, Skorinko, & Pollak, 1981). Transitions between t_{2g} and e_g states are partly-forbidden. When Ir^{4+} is reduced to Ir^{3+}, the t_{2g} bands are filled with six d electrons, which results in optical transparency.

Ir oxide thin films have been prepared by sputtering (Schavone, Dautremont-Smith, Beni, & Shay 1979; Schavone, Dautremont-Smith, Beni, & Shay, 1981; Backholm & Niklasson, 2008; Wen, Niklasson, & Granqvist, 2014), anodic oxidation (Shay, Beni & Schiavone, 1978), electrodeposition (Yoshino, Baba, & Arai, 1987; Yamanaka, 1991; Jung, Lee, & Tak, 2004), thermal oxidation (Sato, Ono, Kobayashi, Wakabayashi, & Yamanaka, 1987), sol-gel (Nishio, Watanabe, & Tsuchiya, 1999), and spray pyrolysis methods (Patil, Mujawar, Sadala, Deshmukh, & Inamdar, 2006). The EC properties of the Ir oxide thin films were found to be dependent on the preparation methods and conditions, which determine the structure and composition of the Ir oxide thin films. Their EC performances also depend on the cell structures, such as film thickness and cell area. The properties of other components, such as the ion conductivities of the electrolytes and the electrical resistances of transparent electrodes, also affect the EC properties, which makes it difficult to quantitatively compare the characteristics of the EC films.

In our laboratory, a reactive sputtering technique using water vapor (H_2O) as a reactive gas was developed (Ueta, et al., 2009; Li et al., 2012, Lee et al., 2012), and the technique was applied to Ir oxide thin films (Ito et al., 2015). The hydrated Ir oxide thin films deposited in an H_2O atmosphere on cooled substrates at −30 °C were found to have good EC properties, namely a large transmittance variation and high bleached transmittance in the visible wavelength region. In the present study, we prepared Ir oxide thin films with various thicknesses and examined the effects of film thickness on the bleached and colored transmittances as well as the response times. We believe that these data are useful in order to quantitatively compare the EC performances of Ir oxide films prepared with different methods and under different conditions and to design EC devices with appropriate optical properties.

2. Method

2.1 Sample Preparation

Hydrated Ir oxide thin films were prepared using a radio-frequency (RF) magnetron sputtering system with an Ir metal target (99.9% purity, 2 inches diameter) in an H_2O atmosphere. A glass coated with fluorine-doped tin oxide (FTO) with a sheet resistance of 200 Ω/sq was used as the substrate, and the substrate was cooled to −30 °C with a Peltier device during sputtering. The sputtering gas pressure and RF power were 6.7 Pa and 50 W, respectively. These were the optimum conditions for our sputtering system to deposit Ir oxide thin films with good EC properties. The films deposited under these conditions were found to consist of highly hydrated amorphous Ir oxide. The film density was estimated to be approximately 6 g/cm³, which is nearly half the density of bulk IrO_2, 11.666 g/cm³ (ICDD, 2009). The low density amorphous structure of the hydrated Ir oxide thin films is considered to be the reason for their high electrochemical activity. The details of the structure as well as the electrochemical and electrochromic properties of the films have been reported in our previous paper (Ito et al., 2015). In this study, Ir oxide thin films with various thicknesses from 20 to 400 nm were prepared by changing the deposition time assuming a constant deposition rate of 5.5 nm/min. The size of the FTO-coated glass substrate was 3×1 cm², and the active area of the Ir oxide films was 1 cm².

2.2 Characterization Methods

The EC properties of the Ir oxide thin films were studied in an electrolyte (0.5 M H_2SO_4 aqueous solution), while Ag/AgCl and Pt were used as the reference and counter electrodes, respectively. A potentiostat/galvanostat (Hokuto Denko, HSV-100) was also used, and a constant voltage of either −0.15 or +1.25 V (vs Ag/AgCl) was applied to bleach and color the Ir oxide working electrodes, respectively. The optical transmittance of the films was measured in situ using a tungsten halogen light source and a multi-channel charge coupled device (CCD) detector (Ocean Optics, USB2000+). The total transmittance of a sample with an Ir oxide/FTO/glass in a quartz-glass cell filled with the aqueous electrolyte was measured using a quartz-glass cell filled with the aqueous electrolyte as a reference.

3. Results and Discussion

3.1 Bleached and Colored Transmittance

The bleached and colored transmittance spectra of the Ir oxide thin films with different thicknesses are shown in Figure 1. The spectra were measured after bleaching at −0.15 V and coloring at +1.25 V for 10 min. The colored transmittance spectra of all the samples are almost flat in a visible wavelength region, which means that the films' color was gray or black. The thinnest film has a very high transmittance and a flat transmittance spectrum in the bleached state; the film looks fairly transparent. However, the bleached transmittance in the short wavelength region decreased and the color of the films became yellow with increasing film thickness. Slight oscillations in the bleached transmittance spectra are thought to be caused by optical interference originating from the multi-layered structure of the samples (Ir oxide/FTO/glass).

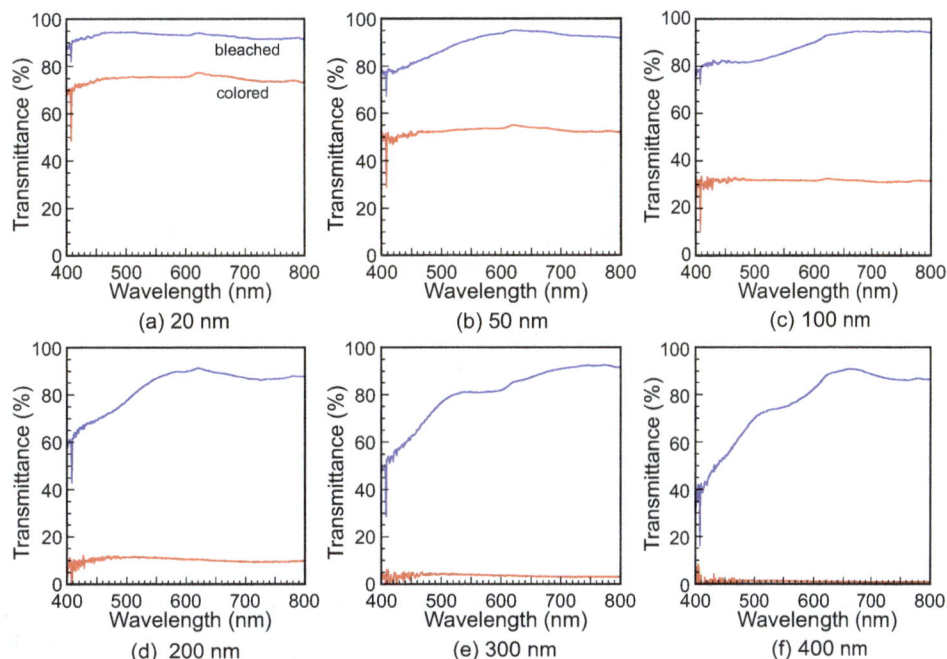

Figure 1. The bleached and colored transmittance spectra of Ir oxide thin films with various film thicknesses. The spectra were measured after bleaching at −0.15 V and coloring at +1.25 V for 10 min

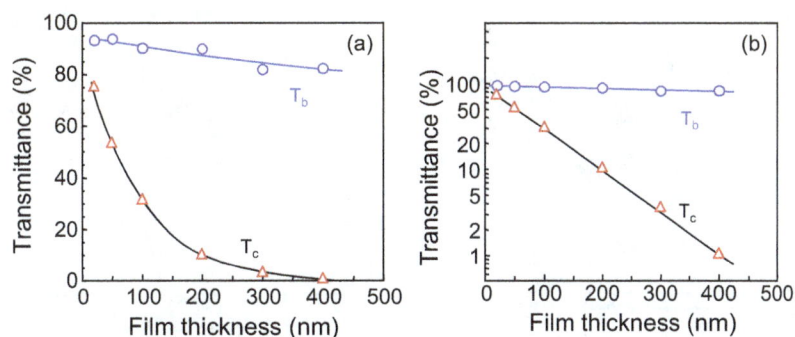

Figure 2. Bleached and colored transmittances (T_b and T_c) at a wavelength of 600 nm as a function of film thickness. Transmittances are plotted on a linear scale (a) and a logarithmic scale (b)

Figure 2 shows the bleached and colored transmittances of the films as a function of film thickness at a wavelength of 600 nm, at which the bleached transmittances almost reached their maximum values. Figure 2(a) shows that the colored transmittance, T_c, decreases steeply with increasing film thickness; in contrast the bleached transmittance, T_b, remains high, and the maximum transmittance change, $T_b - T_c$, and the maximum contrast ratio, T_b/T_c, were obtained (81% and 79, respectively) for the 400 nm-thick film. A logarithmic plot of the bleached and colored transmittances are shown in Figure 2(b), and a linear relationship between the transmittance and the film thickness was clearly obtained. This result indicates that the transmittance of the films, T, can be described by Lambert's law.

$$T = \exp(-\alpha \cdot d), \tag{3}$$

where d is the film thickness and α is the absorption coefficient. The absorption coefficients of the films in the bleached and colored states were estimated to be 3.2×10^3 and 1.1×10^5 cm^{-1}, respectively, from the slope of the straight lines shown in Figure 2(b). The extrapolated transmittances at zero film thickness (approximately 94%) correspond to the loss of the transmitted light by the absorption of the glass substrates and the reflection at the interfaces between the aqueous electrolyte and the samples.

The absorption coefficient can be expressed by the following equation:

$$\alpha = 4\pi k / \lambda, \tag{4}$$

where k is the extinction coefficient (imaginary part of the refractive index) and λ is the wavelength. The extinction coefficients of the Ir oxide films in their bleached and colored states were calculated to be 0.015 and 0.5, respectively. The extinction coefficient of single-crystal IrO$_2$ for the electric field vector of the light polarized parallel and perpendicular to the c-axis were reported to be approximately 1.1 and 0.8, respectively, at a wavelength of 600 nm (2.07 eV) (Goel et al., 1981). These values are comparable to those of our Ir oxide thin films in the colored states. Gottesfeld and Srinivasan reported extinction coefficients of 0.01 and 0.10 for the Ir oxide films formed on the Ir electrode via electrochemical reduction and oxidation, respectively, in 0.5 M H$_2$SO$_4$ (Gottesfeld & Srinivasan, 1978). Backholm & Niklasson reported extinction coefficients of 0.12 and 0.19 for the bleached and colored states, respectively, of the IrO$_x$ films prepared by reactive sputtering in an Ar + O$_2$ atmosphere at a wavelength of 660 nm (Backholm & Niklasson, 2008). The larger difference of the extinction coefficients in the bleached and colored states in our study indicates a much higher proportion of Ir atoms that are electrochemically active. These results demonstrate that almost all the Ir atoms in our Ir oxide thin films can be considered to be contributing to the coloration change. These optical constants are useful to estimate the EC activity of devices using Ir oxide films.

3.2 Response Time

The transmittance change of the Ir oxide films after applying constant voltages of -0.15 V for bleaching and $+1.25$ V for coloring is shown in Figure 3. The transmittance was measured at a wavelength of 600 nm. The bleaching and coloring response times were defined as the time required to obtain a 90% transmittance change from the fully colored state to the fully bleached state, and vise versa. Figure 4 shows the response times as a function of film thickness. From these figures, we found that the response time increased from 2–3 s to several tens of seconds with increasing film thickness.

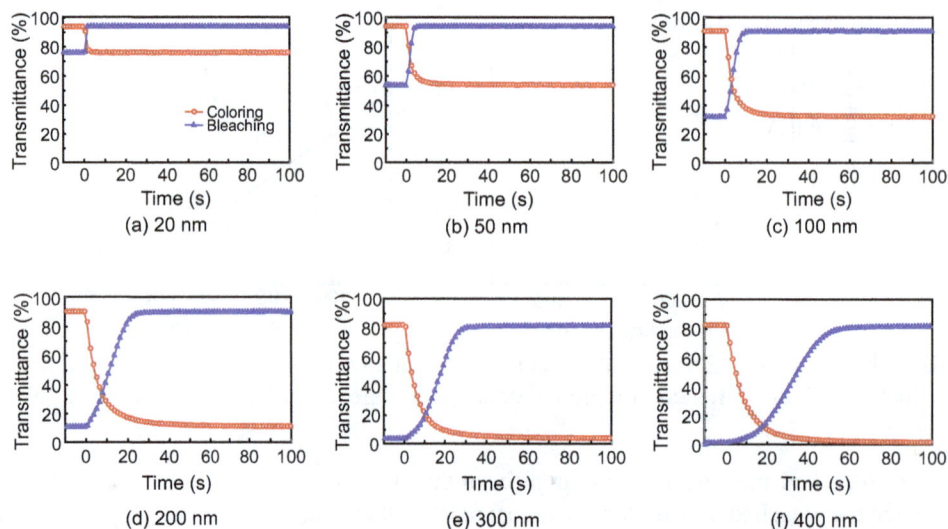

Figure 3. Transmittance change of the Ir oxide films with various film thicknesses by applying a constant voltage of -0.15 V for bleaching and $+1.25$ V for coloring. The measurement wavelength was 600 nm

Compared with the very fast response reported in previous papers, such as ~0.2 s (Shay et al, 1978) and less than 50 ms (Schiavone et al, 1979), the response speed of our samples is much slower. This discrepancy can be explained by the high sheet resistance of our FTO electrode. The electric charge required to color and bleach the Ir oxide films increases with increasing film thickness; however, the maximum current is restricted by Ohm's law, which results in longer response times for EC cells using transparent electrodes with high sheet resistances.

The response times of EC cells depend on the properties of the EC materials, such as the porosity and surface area, which affect the diffusion coefficients of ions in the films. The ion conductivity of the electrolyte, the electrical resistance of the transparent electrodes, and the device structure, such as the thickness of the EC layer and the device area, also affect the response times of EC devices (Viennet, Randin, & Raistrick, 1982; Kamimori, Nagai, & Mizuhashi, 1987). Although quantitative discussions of the response times are difficult, Figures 2 and 4 clearly show that there is a trade-off between the response time and the optical contrast ratio. These data are considered to be useful for the design of EC devices with appropriate performances.

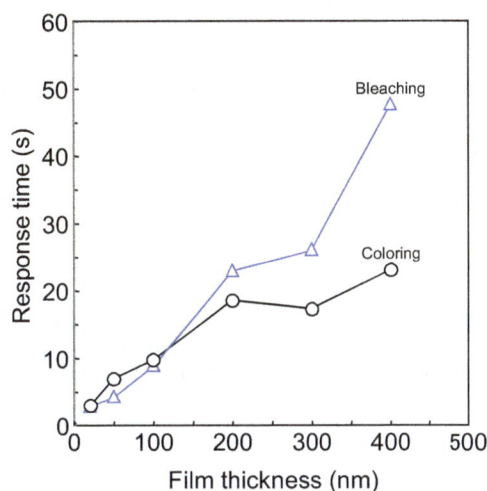

Figure 4. Bleaching and coloring response times as a function of film thickness

4. Conclusions

Hydrated Ir oxide thin films with thicknesses from 20 to 400 nm were prepared by reactive sputtering in a H_2O atmosphere, and their transmittance spectra in the bleached and colored states as well as their response times were studied. The bleached and colored transmittances decreased with increasing film thickness according to Lambert's law, and the absorption coefficients in the bleached and colored states were estimated to be 3.2×10^3 and 1.1×10^5 cm^{-1}, respectively, at a wavelength of 600 nm. These results demonstrate that almost all the Ir atoms in the oxide thin films can be considered to be electrochemically active and to be contributing to the color change. The transmittance change and the contrast ratio of the bleached and colored transmittances increased with increasing film thickness with maximum values of 81% and 79, respectively, obtained for the 400 nm-thick Ir oxide film. We found that there is a trade-off between the response speed and the transmittance change, i.e., the response is slower with increasing film thickness.

References

Backholm, J., & Niklasson, G. A. (2008). Optical properties of electrochromic iridium oxide and iridium-tantalum oxide thin films in different colouration states. *Solar Energy Materials and Solar Cells, 92*(11), 1388–1392. http://dx.doi.org/10.1016/j.solmat.2008.05.015

Dautremont-Smith, W. C. (1982). Transition metal oxide electrochromic materials and displays: a review, Part2: oxide with anodic coloration. *Displays, 3*(2), 67–80. https://dx.doi.org/10.1016/0141-9382(82)90100-7

Goel, A. K., Skorinko, G., & Pollak, F. H. (1981). Optical properties of single-crystal rutile RuO_2 and IrO_2 in the range of 0.5 to 9.5 eV. *Physical Review B, 24*, 7342–7350. https://doi.org/10.1103/PhysRevB.24.7342

Granqvsit, C. G. (1994), Electrochromic oxides: a bandstructure approach, *Solar Energy Materials & Solar Cells, 32*(4), 369–382. http://dx.doi.org/10.1016/0927-0248(94)90100-7

Granqvist, C. G. (1995). *Handbook of inorganic electrochromic materials.* Amsterdam, The Netherlands: Elsevier.

Granqvist, C. G. (2012). Oxide electrochromics: An introduction to devices and materials, *Solar Energy Materials & Solar Cells, 99*, 1–13. http://dx.doi.org/10.1016/j.solmat.2011.08.021

Granqvist, C. G. (2014). Electrochromics for smart windows: Oxide-based thin films and devices. *Thin Solid Films, 564*, 1–38. http://dx.doi.org/10.1016/j.tsf.2014.02.002

International Center for Diffraction Data (ICDD). (2009). *Powder Diffraction File (PFD)-2*, Release 2009, Reference Code 00-015-0870.

Ito, S., Abe, Y., Kawamura, M., & Kim, K. H. (2015). Electrochromic properties of iridium oxide thin films prepared by reactive sputtering in O_2 or H_2O atmosphere, *Journal of Vacuum Science and Technology B, 33*, 041204-1–5. http://dx.doi.org/10.1116/1.4923227

Jung, Y., Lee, J., & Tak, Y. (2004). Electrochromic mechanism of IrO_2 prepared by pulsed anodic electrodeposition. *Electrochemical and Solid-State Letters, 7*(2), H5–H8. http://dx.doi.org/10.1149/1.1634083

Kahk, J. M., Poll C. G., Oropeza, F. E., Ablett, J. M., Céolin, D., Rueff, J.-P., ... Payne, D. J. (2014). Understanding the electronic structure of IrO_2 using hard-X-ray photoelectron spectroscopy and density-function theory. *Physical Review Letters, 112*, 117601-1–6. http://dx.doi.org/10.1103/PhysRevLett.112.117601

Kamimori, T., Nagai, J., & Mizuhashi, M. (1987). Electrochromic devices for transmissive and reflective light control, *Solar Energy Materials, 16*(1–3), 27–38. http://dx.doi.org//10.1016/0165-1633(87)90005-0

Lee, K.M., Abe, Y., Kawamura, M., & Kim, K. H. (2012). Effects of substrate temperature on electrochromic properties of cobalt oxide and oxyhydroxide thin films prepared by reactive sputtering using O_2 and H_2O gases. *Japanese Journal of Applied Physics, 51*(4R), 045501-1–4. http://dx.doi.org/10.1143/JJAP.51.045501

Li, N., Abe, Y., Kawamura, M., Kim, K. H., & Suzuki, T. (2012). Evaluation of ion conductivity of ZrO_2 thin films prepared by reactive sputtering in O_2, H_2O, and $H_2O+H_2O_2$ mixed gas. *Thin Solid Films, 520*(16), 5137–5140. http://dx.doi.org/10.1016/j.tsf.2012.04.024

Mattheiss, L. F. (1976). Electronic structure of RuO_2, OsO_2, and IrO_2, *Physical Review B, 13*, 2433–2450. http://dx.doi.org/ 10.1103/PhysRevB.13.2433

Mortimer, R. J. (1997). Electrochromic materials, *Chemical Society Reviews, 26*, 147–156. http://dx.doi.org/10.1039/CS9972600147

Motimer, R. J., Rosseinsky, D. R., & Monk P. M. S. (2015). *Electrochromic Materials and Devices*. Weinheim, Germany: Wiley-VCH.

Nishio, K., Watanabe, Y., & Tsuchiya, T. (1999). Preparation and properties of electrochromic iridium oxide thin film by sol-gel process. *Thin Solid Films, 350*(1–2), 96–100. http://dx.doi.org/10.1016/S0040-6090(99)00290-4

Patil, P. S., Mujawar, S. H., Sadale, S. B., Deshmukh, H. P., & Inamdar, A. I. (2006). Effect of film thickness on electrochromic activity of spray deposited iridium oxide thin films, *Materials Chemistry and Physics, 99*(2–3), 309–313. http://dx.doi.org/10.1016/j.matchemphys.2005.10.029

Ryden, W. D., Lawson, A. W., & Sartain, C. C. (1970). Electrical transport properties of IrO_2 and RuO_2, *Physical Review B, 1*, 1494–1500. http://dx.doi.org/10.1103/PhysRevB.1.1494

Sato, Y., Ono, K., Kobayashi, T., Wakabayashi, H., & Yamanaka, H. (1987), Electrochromism in iridium oxide films prepared by thermal oxidation of iridium-carbon composite films. *Journal of the Electrochemical Society, 134*(3), 570–575. http://dx.doi.org/10.1149/1.2100510

Schiavone, L. M., Dautremont-Smith, W. C., Beni, G., & Shay, J. L. (1979). Electrochromic iridium oxide films prepared by reactive sputtering. *Applied Physics Letters, 35*, 823–825. http://dx.doi.org/10.1063/1.90950

Schiavone, L. M., Dautremont-Smith, W. C., Beni, G., & Shay, J. L. (1981). Improved electrochromic behavior of reactively sputtered iridium oxide films. *Journal of the Electrochemical Society, 128*(6), 1339–1342. http://dx.doi.org/10.1149/1.2127632

Shay, J. L., Beni, G., & Schiavone, L. M. (1978). Electrochromism of anodic oxide films on transparent substrates. *Applied Physics Letters, 33*, 942–944. http://dx.doi.org/10.1063/1.90227

Ueta, H., Abe, Y., Kato, K., Kawamura, M., Sasaki, K., & Itoh, H. (2009). Ni oxyhydroxide thin films prepared by reactive sputtering using O_2 + H_2O mixed gas. *Japanese Journal of Applied Physics, 48*(1R), 015501-1–4. http://dx.doi.org/10.1143/JJAP.48.015501

Viennet, R., Randin, J.-P., & Raisrrick, D. (1982). Effect of active surface area on the response time of electrochromic and electrolytic displays. *Journal of the Electrochemical Society, 129*(11), 2451–2453. http://dx.doi.org/10.1149/1.2123565

Wen, R.-T., Niklasson, G. A., Granqvist, C. G. (2014). Electrochromic iridium oxide films: Compatibility with propionic acid, potassium hydroxide, and lithium perchlorate in propylene carbonate, *Solar Energy Materials & Solar Cells, 120*, 151–156. http://dx.doi.org/10.1016/j.solmat.2013.08.035

Yamanaka, K. (1991). The electrochemical behavior of anodically electrodeposited iridium oxide films and the reliability of transmittance variable cells, *Japanese Journal of Applied Physics, 30*(6), 1285–1289. http://dx.doi.org/10.1143/JJAP.30.1285

Yoshino, T., Baba, N., & Arai, K. (1987). Electrochromic IrO_x thin films formed in sulfatoiridate (III, IV) complex solution by periodic reverse current electrolysis (PRIROF), *Japanese Journal of Applied Physics, 26*(9), 1547–1549. http://dx.doi.org/10.1143/JJAP.26.1547

Permissions

List of Contributors

Hiroyuki Matsukizono and Takeshi Endo
Molecular Engineering Institute, Kinki University, Japan

Kalpana Singh, Ashok Kumar Baral and Venkataraman Thangadurai
Department of Chemistry, University of Calgary, 2500 University Drive NW, Calgary, Alberta, T2N 1N4, Canada

XiaoMing Qu, YuFeng Guo, FuQiang Zheng, Tao Jiang and GuanZhou Qiu
School of Minerals Processing and Bioengineering, Central South University, Changsha 410083, Hunan, China

Reese E. Jones and Jonathan A. Zimmerman
Sandia National Laboratories, Livermore, CA 94550, USA

Giacomo Po
University of California, Los Angeles, CA 90095, USA

Arunima Sarkar, Duygu Kocaefe, Yasar Kocaefe, Dipankar Bhattacharyay and Brigitte Morais
Department of Applied Sciences, University of Quebec at Chicoutimi, Québec, Canada

Patrick Coulombe
Aluminerie Alouette Inc., 400, Chemin de la Pointe-Noire, C.P. 1650, Sept-Îles, Québec, Canada

Jose J. Chavez, Sergio F. Almeida, Rodolfo Aguirre and David Zubia
Department of Electrical and Computer Engineering, the University of Texas at El Paso, Texas 79968, USA

Xiao W. Zhou
Mechanics of Materials Department, Sandia National Laboratories, Livermore, California 94550, USA

Shahril Anuar Bahari
Department of Wood Biology and Wood Products, Faculty of Forest Sciences and Forest Ecology, Georg-August-Universität Göttingen, Büsgenweg 4, 37077 Göttingen, Germany

Andreas Krause
Centre for Wood Science, Hamburg University, Leuschnerstraße 91, 21031 Hamburg, Germany

Mansour A. Al-Shafei, Ahmed K. Al-Asseel, Hasan A. Al-Jama, Amer A. Al-Tuwailib and Shouwen X. Shen
Research & Development Center, Saudi Aramco, Dhahran, Saudi Arabia

Abdulhadi M. Adab
Shedgum Gas Plant Department, Saudi Aramco, Shedgum, Saudi Arabia

Mohamed Atef Mohamed Gebril
Department of material engineering, Faculty of petroleum and minning engineering, Suez University, Egypt

Mahmudun Nabi Chowdhury and Dinh Thi Kieu Anh
School of Mechanical Engineering Department, University of Ulsan, Ulsan, South Korea

Yoshimi Seida
Dept. Policy Study, Natural Science Laboratory, Toyo University, Japan

Mitsuteru Ogawa
Fuji Silysia Chemical Ltd., Japan

Denise Loder, Susanne K. Michelic and Christian Bernhard
Chair of Ferrous Metallurgy, Montanuniversitaet Leoben, 8700 Leoben, Austria

Haidar F. AL-Qrimli and Ahmed M. Abdelrhman
Department of Mechanical Engineering, Curtin University of technology, Miri, Sarawak, Malaysia

Karam S. Khalid2,
Department of Mechanical Engineering, Ibra College of Technology, Ibra, Oman

Roaad K. Mohammed A
Oil Products Distribution Company, Ministry of Oil, Iraq

Husam M. Hadi
SIG Combibloc Obeikan, Iraq

Salokhiddin Nurmurodov, Alisher Rasulov and Kudratkhon Bakhadirov
Materials Science and Materials Technology department, Tashkent State Technical University, Tashkent, Uzbekistan

Lazizkhan Yakubov, Khusniddin Abdurakhmanov, Tokhir Tursunov and Nodir Turakhodjaev
Machine Building Department, Tashkent State Technical University, Tashkent, Uzbekistan

A. Shalwan and M. Alajmi
Manufacturing Engineering Technology Department, College of Technological Studies, Public Authority for Applied Education and Training, Kuwait City 13092, Kuwait

A. I. Alateyah
Mechanical Engineering Department, Qassim University, P.O.B.6677, Buraydah, Saudi Arabia

B. Aldousiri
Advanced Polymer and composites (APC) Research Group, Department of Mechanical & Design Engineering, University of Portsmouth, PO1 3DJ, UK

D. A. Sukhanov
ASK-MSC Company, Moscow, Russia

L. B. Arkhangelsky
President Union Smiths, Korolev, Russia

N. V. Plotnikova and N. S. Belousova
Novosibirsk State Technical University, Novosibirsk, Russia

Mathias Bobbert, Florian Augethaler and Gerson Meschut
Laboratory for Material and Joining Technology (LWF), Paderborn University, Paderborn, Germany

Zheng Wang and Thomas Tröster
Chair for Automative Lightweight Design (LiA), Paderborn University, Paderborn, Germany

Sachin Arun Patil and Hamid Lankarani
Wichita State Univ, KS, United States

Farzad Baratzadeh
Advanced Joining and Processing Laboratory, National Institute for Aviation Research, Wichita, KS, USA

Yoshio Abe, Satoshi Ito, Kyung Ho Kim, Midori Kawamura and Takayuki Kiba
Department of Materials Science and Engineering, Kitami Institute of Technology, Kitami, Japan

Index

R
Reaction Temperature, 29-36
Recycled Anode Butt, 58, 60-63, 67
Recycled Material, 57
Reduction Process, 19, 175
Rtm-process, 197, 200

S
Scanning Electron Microscopy, 20, 102, 112
Sedimentary Rocks, 98
Spinodal Reactions, 110-111, 116
Spot Welding, 206, 209
Stacking Faults, 52, 73, 77-79

Sulfuric Acid Leaching, 29-33, 35-36
Symmetrical Cell, 19-20, 25

T
Thermal Stability, 1-2, 9, 11, 102-103, 193
Thermoplastic Composite, 84, 96
Ti Extraction, 29, 31-36
Titanium-bearing Electric Furnace Slag, 29, 37
Titanomagnetite, 29-30, 36
Trimethylolpropane, 1-2, 11-12
Tungsten Oxide, 171, 174-175

W
Water Leaching, 29-36
Water Resistance, 83, 86-87, 93

www.ingramcontent.com/pod-product-compliance
Lightning Source LLC
Chambersburg PA
CBHW080530200326
41458CB00012B/4395

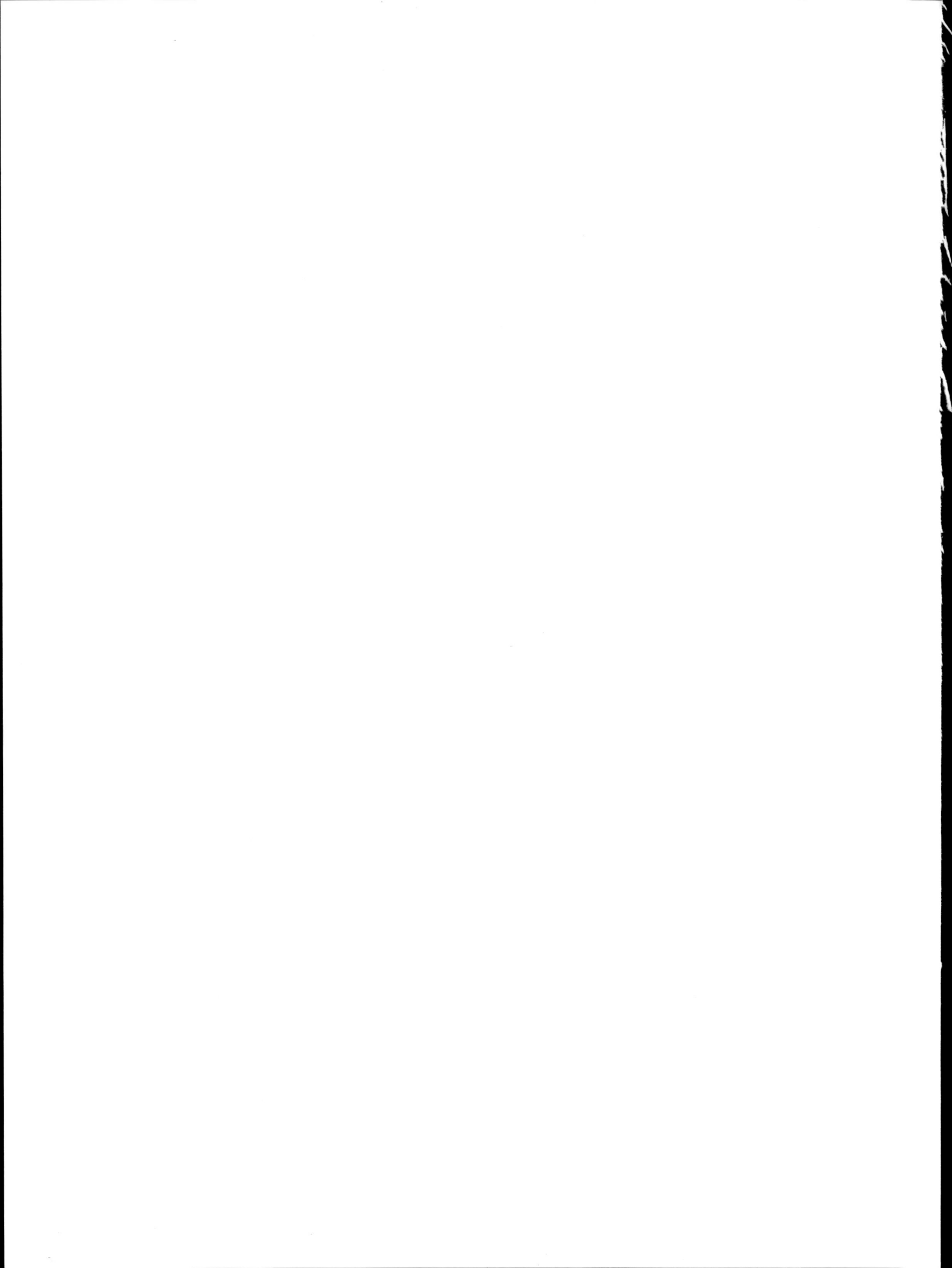